The Regions
and Global Warming

HARC Global Change Studies

VOLUME I

The Regions and Global Warming:
Impacts and Response Strategies

The Houston Advanced Research Center (HARC) is an independent, nonprofit corporation established to promote closer links between the producers and users of scientific knowledge.

The Center for Global Studies (CGS) is the policy research division of HARC. Its mission is to increase awareness of the social and policy implications of science and technology, focusing on global environmental issues and sustainable development. Issues of concern include population growth, environmental quality, climate change, water management, alternative fuels policy and U.S./Mexico environmental issues.

THE REGIONS
AND GLOBAL WARMING
Impacts and
Response Strategies

Edited by

Jurgen Schmandt
Director, Center for Global Studies
Houston Advanced Research Center
Professor of Public Affairs
University of Texas at Austin

Judith Clarkson
Consultant, Center for Global Studies
Houston Advanced Research Center

New York   Oxford
OXFORD UNIVERSITY PRESS
1992

Oxford University Press

Oxford   New York   Toronto
Dehli   Bombay   Calcutta   Madras   Karachi
Kuala Lumpur   Singapore   Hong Kong   Tokyo
Nairobi   Dar es Salaam   Cape Town
Melbourne   Auckland   Madrid

and associated companies in
Berlin   Ibadan

Published by Oxford University Press, Inc.
200 Madison Avenue, New York, New York 10016

Oxford is a registered trademark of Oxford University Press

Library of Congress Cataloging-in-Publication Data
The regions and global warming: impacts and response strategies/
edited by Jurgen Schmandt, Judith Clarkson.
p. cm.   Includes bibliographical references and index.
ISBN 0-19-507586-2
1. Global warming—Environmental aspects.
2. Climatic changes—Environmental aspects.
I. Schmandt, Jurgen.   II. Clarkson, Judith.
QC981.8.G56R44   1992
363.73'87—dc20   92-2826

1 3 5 7 9 8 6 4 2

Printed in the United States of America
on acid-free paper

PREFACE

This volume contains the essays prepared by the winners and finalists of the 1991 competition for The George and Cynthia Mitchell International Prize for Sustainable Development. The prize was established in 1975, long before sustainable development became a household word, to recognize individuals demonstrating exceptional creativity in clarifying the concept of sustainable development and designing workable strategies to achieve sustainable societies. The next prize competition will be held in 1994.

In 1991 the prize was awarded for outstanding papers on regional implications of global climate change. After a lengthy review process, a distinguished panel of judges selected eight authors to receive cash awards totaling $100,000. In addition to the main prize, the Young Scholars Competition was inaugurated to recognize young scientists who demonstrated creativity through research proposals related to the 1991 prize competition.

PANEL OF JUDGES

James Bruce
Chairman, Canadian Climate
  Program Board
Ottawa, Ontario, Canada

Paul Crutzen
Director, Division of Atmospheric
  Chemistry
Max-Planck-Institut for Chemistry
Mainz, Germany

Georgii Golitsyn
Director, Institute for Atmospheric
  Physics
Academy of Sciences
Moscow, Russia

Norman Hackerman
Professor Emeritus, Department of
  Chemistry
The University of Texas at Austin
Austin, Tex.

Sheila Jasanoff
Professor, Yale School of Forestry
  and Environmental Sciences
New Haven, Conn.

William Kellogg (retired)
National Center for Atmospheric
  Research
Boulder, Colo.

Roger Revelle (deceased)
Scripps Institution of Oceanography
University of California at San
  Diego
La Jolla, Calif.

Jurgen Schmandt
Director, Center for Global Studies
Houston Advanced Research Center
The Woodlands, Tex.

Jack Sommer
Deputy Director
Division of Science Resource Studies
National Science Foundation
Washington, D.C.

M.S. Swaminathan
Chairman
M.S. Swaminathan Research
  Foundation
Madras, India

1991 MITCHELL PRIZE WINNERS

First Prize Winner

Daniel Botkin
Professor of Biology and Environmental Studies
University of California, Santa Barbara, Calif.

Second Prize Winners

José Goldemberg
Minister of Education
Brasilia, DF, Brazil

Diana Liverman
Department of Geography
Pennsylvania State University
University Park, Pa.

Third Prize Winners

Ian Burton
Director, Department of Natural and
    Human Sciences Integration
Ottawa, Ontario, Canada

Michael Glantz
Head, Environment and Societal
    Impacts Group
National Center for Atmospheric
    Research
Boulder, Colo.

Antonio R. Magalhaes
Esquel Foundation-Brazil
Brasilia, DF, Brazil

William Moomaw
Director
Center for Environmental
    Management
Tufts University
Medford, Mass.

Vijaya Saha
Ministry of Environment and
    Land Use
Port Louis, Mauritius

CONTENTS

CONTRIBUTORS

ROBERT U. AYRES
Department of Engineering and
  Public Policy
Carnegie-Mellon Unviersity
Pittsburgh, Pa.

DANIEL B. BOTKIN
Department of Biology and
  Department of Environmental
  Studies
University of California
Santa Barbara, Calif.

MICHAEL BOWES
Resources for the Future
Washington, D.C.

IAN BURTON, Director
Department of Natural and Human
  Sciences Integration
Ottawa, Ontario, Canada

JUDITH CLARKSON, Consultant
Center for Global Studies
Houston Advanced Research Center
The Woodlands, Tex.

PIERRE R. CROSSON
Resources for the Future
Washington, D.C.

WILLIAM E. EASTERLING III
Department of Agricultural
  Meteorology
University of Nebraska
Lincoln, Nebr.

AMOS EDDY
Oklahoma State University
Oklahoma City, Okla.

KENNETH D. FREDERICK
Resources for the Future
Washington, D.C.

MICHAEL GLANTZ, Head
Environmental and Societal Impacts
  Group
National Center for Atmospheric
  Research
Boulder, Colo.

PROFESSOR JOSÉ GOLDEMBERG,
  Minister
Ministry of Education
Brasilia, DF, Brazil

N.S. JODHA, Head
Mountain Farming Systems Division
International Centre for Integrated
  Mountain Development (ICIMOD)
Kathmandu, Nepal

STEPHEN P. LEATHERMAN, Director
Laboratory for Coastal Research
University of Maryland
College Park, Md.

DIANA LIVERMAN
Department of Geography
Pennsylvania State University
University Park, Pa.

FASIH UDDIN MAHTAB
Planning and Development Services,
  Ltd.
Dhaka, Bangladesh

MARY S. McKENNEY
Resources for the Future
Washington, D.C.

ANTONIO R. MAGALHAES
Esquel Foundation-Brazil
Brasilia, DF, Brazil

WILLIAM MOOMAW, *Director*
Center for Environmental
   Management
Tufts University
Medford, Mass.

STELLA C. OGBUAGU
Department of Sociology
University of Calabar
Calabar, Nigeria

R.K. PACHAURI, Director
Tata Energy Research Institute
New Delhi, India

SUWANNA PANTURAT
Srinakharinwirot University
Bangkok, Thailand

NORMAN ROSENBERG, *Director*
Climate Resources Program
Resources for the Future
Washington, D.C.

VIJAYA SAHA
Ministry of Environment and
   Land Use
Port Louis, Mauritius

JURGEN SCHMANDT, Director
Center for Global Studies
Houston Advanced Research Center
The Woodlands, Tex.

FERENC TOTH
International Institute for Applied
   Systems Analysis
Laxenburg, Austria

KONSTANTIN YA. VINNIKOV
Princeton University, GFDL
Princeton, N.J.

The Regions
and Global Warming

1. INTRODUCTION: GLOBAL WARMING AS A REGIONAL ISSUE

Jurgen Schmandt and Judith Clarkson

Global climate change is an environmental problem of international proportions and vast economic and social implications. Never before has the world had to face a problem with these characteristics: The issue is global in scale; it results from mankind's very success in using science and technology as the foundation for economic development; and solutions to redress the situation will be costly and require far-reaching social and economic change. Consequently, few models exist for dealing with the issue. In 1988 a treaty was agreed upon for phasing out a group of chemicals—chlorofluorocarbons (CFCs)—that destroy the ozone layer. This was hailed as a major accomplishment in environmental diplomacy. Yet, in comparison, the agreement imposes restrictions only on a very small and specific part of each nation's economy. Addressing global climate change, on the other hand, goes to the very heart of each nation's economy, its energy production and use, the standard of living of its people and the size of its population. For the developing nations, in particular, these issues are critical as they attempt to raise the standard of living of their impoverished populations. They view restrictions on energy utilization as interfering with growth and progress.

While the theory underlying global climate change is widely accepted in the scientific community, there is equally widespread uncertainty about the extent, timing and regional implications associated with the predictions. Several global climate models have been developed. While these models are the principal tools for understanding global warming, they have many shortcomings. For one, there are significant differences between them. In general, with increasing concentration of greenhouse gases, significant warming is expected in polar regions, moderate warming in mid-latitudinal regions, and considerably

less toward equatorial zones. Changes in precipitation are harder to predict and some models even come to contradictory results. At the regional level it is difficult to make any sort of reasonable predictions based on global climate models, partly because of the coarseness of the grids in the models, and partly because geographic features in the landscape cannot be adequately represented in the models.

Two policies can be pursued to address global warming. One involves slowing and eventually reversing the effects by reducing the emission of gases, particularly carbon dioxide, that are responsible. The second involves adapting to the changes that can be anticipated, notably effects on vegetation and therefore crop production, effects on coastal areas resulting from sea level rise, and, in some places, reduced availability of water resources. In this volume we focus our attention on mitigative and adaptive strategies that can be implemented regionally. This approach appears to offer the most promise, at least for the short run, because it does not depend on lengthy international negotiations that may take a long time before yielding results. Action can be taken by individual political entities (or groups of such entities) who can take steps responsive to their own particular needs. In addition, adaptive strategies, if carefully designed, can help to reduce emissions and address other environmental and resource issues simultaneously.

Reversing the Global Warming Trend

"Solving" the greenhouse problem requires slowing and reversing the buildup of greenhouse gases in the atmosphere. The concentration of carbon dioxide has increased by almost one third compared to preindustrial

3

times, and its doubling is predicted to occur sometime around the middle of the next century. To halt and eventually reverse this trend will be a monumental task.

Quite possibly this will be beyond mankind's reach because painful limitations to growth must be agreed upon, implemented and enforced on a global scale. It is likely that almost any policy proposed will be unacceptable to certain nations; a consensus position will be hard to find. Improving energy efficiency would seem like an obvious goal for developed nations, and considerable progress has been made in this area in response to increasing energy costs. However, these steps alone will not be sufficient, and it will be necessary to develop an energy system increasingly less focused on fossil fuels. Controlling population growth might take first priority for developing countries. The experience of the past shows that population rates decline after a certain level of economic development has been reached. In the current debate, the position taken by many developing countries is quite different. From their perspective, industrialized nations should reduce consumption and welcome development, including increased use of fossil fuels, in the poor countries of the world. Thus, global warming may sharpen the conflict between north and south that has been building over the course of several decades. Because global warming is just one manifestation of overpopulation, overuse of limited resources and unsustainable growth, a new paradigm of development at national and international scales will be needed. The ultimate policy response must be sustainable development, which will be extremely difficult to implement.

Much has been written on sustainable development in recent years. The international Brundtland Commission concluded in its report that sustainable development was the most urgent task facing mankind (World Commission on Environment and Development, 1987). The commission defined sustainable development as "meeting the needs of the present without compromising the ability of future generations to meet their own needs." It is doubtful that we know how to translate this broad goal into workable policies. The knowledge, institutions and political resolve simply do not exist for transforming the world's economies from quantitative to qualitative or sustainable growth. But we certainly must try because failure to do so could mean extreme hardship to many of the world's populations. Global warming has added urgency to the call for sustainable development.

Although the Montreal Protocol of 1988 and the subsequent London agreement reached in the summer of 1990 for the protection of the ozone layer are limited as models for addressing the problem of global warming, they do provide some insights into the process of developing international treaties for protecting the world environment. For instance, the developing countries felt that the phaseout of CFCs would create an enormous financial burden for them as they work to enhance their industrial base and increase the standard of living of their people. The issue was resolved when the developing countries, led by China and India, agreed not to develop their nascent CFC industry on condition that the rich nations provide financial and technical help for the introduction of substitute technology (Roan, 1989). This suggests that industrialized nations— recognizing their own interest in reducing or controlling greenhouse gas emissions worldwide—may have to provide economic assistance to those nations ill-equipped to bear the financial burdens involved.

Negotiations toward an international convention are underway, and a timetable for completion by the time of the 1992 World Environment Conference has been established. Whether the resulting convention will be more than a declaration of good intentions remains to be seen. The ministerial declaration, issued at the conclusion of the Second World Climate Conference in November 1990, identified five principles around which the convention should be drafted:

- The goal of international action should be to hold greenhouse gases in the atmosphere to a safe level.
- Achieving such a goal will require a concerted international response initiated

without delay, despite scientific and other uncertainties.

- The developed countries should lead the way by reducing their emission of climate-altering greenhouse gases.
- The developing countries will require financial and technological cooperation to participate meaningfully in meeting international climate objectives.
- A global framework convention on climate change should be negotiated without delay.[1]

Regional Strategies for Mitigating and Adapting to the Effects of Global Warming

It is impossible to predict the outcome of current and future negotiations aimed at reaching a global agreement on reducing greenhouse gas emissions. In this volume we develop the thesis that regional response strategies to global warming must be developed in parallel, both as insurance (in case international agreements fail or remain too little to make a difference) *and* as part of prudent regional development policy. A regional approach designed to mitigate and adapt to global warming is useful for a number of reasons. As a rule, policies involving one or a small number of political entities are more easily implemented than are broad international treaties. Second, regional approaches are needed to take into account the specific impacts of global climate change. Not only are global models constructed on too large a scale to provide reliable information at the regional level, but they also do not have the capabilities necessary to take into account the topography of specific regions. Third, only at the regional level is it possible to link the effects of global warming to specific economic and social environments and to design comprehensive strategies for sustainable development.

---

[1]This summary of the less precisely worded official statement is from the foreword of the World Resources Institute, 1991.

Regional Impacts

Research on global warming has led to a broad consensus among experts that warming and sea-level rise will occur globally, and that human activities—industrialization, population growth, use of fossil fuels, deforestation and certain agricultural practices—are intensifying the natural greenhouse effect. The recent report by the Intergovernmental Panel on Climate Change (IPCC), compiled by a team of scientists from around the world under the sponsorship of the United Nations Environment Programme (UNEP) and World Meteorological Organization (WMO), represents the most authoritative summary of what is known and what remains unclear about manmade climate change. Using a business-as-usual emissions scenario (i.e., no major policy change compared to today), the report predicts warming of 0.3°C per decade and sea-level rise of 6 cm per decade (IPCC, 1990). These predictions, in the words of the IPCC report, are made "with confidence."

This statement does not mean that the effects of climate change can already be observed. Recent studies report that the earth was warmer in 1990 than in any year since 1880, and that the 10 warmest years on record have occurred since 1973. These findings from leading research teams in the United Kingdom and the United States are consistent with the theory of climate change. However, they cannot be used as proof that manmade climate change is now occurring. It will be several decades before the effects of climate change can be measured. For the time being, the statements about warming and sea level rise given in the IPCC report represent the best scientific information about the likely effects of climate change. But it should not be forgotten that they are predictions derived from theory about greenhouse warming.

The IPCC predictions are in the form of global averages. We know less about the regional effects of climate change. A number of generalizations, based on General Circulation Model (GCM) results, have been formulated at the Villach 2 and Bellagio conferences (WMO-UNEP, 1988). They are summarized in Tables 1.1 to 1.3. While these generalizations are the best ones available, they should be seen merely as a starting

TABLE 1.1 *Broad GCM Generalizations on Regional Effects of Global Warming*

Largest temperature changes will occur in high latitudes.
Mid-latitude changes will be more pronounced in the Northern Hemisphere.
Temperature will change more in winter than in summer.
Frostline will move 150 to 250 miles north or south.
Breadbaskets will move 200 miles north or south.
Canada and Russia will experience warmer winters.

TABLE 1.2. *GCM Generalizations on Temperature Change*

| | Seasonal temperature change[a] | |
	Summer	Winter
High latitudes (60°-90°)	0.5x - 0.7x	2.0x - 2.4x
Mid latitudes (30°-60°)	0.8x - 1.0x	1.2x - 1.4x
Low latitudes (0°-30°)	0.9x - 0.7x	0.9x - 0.7x

a. As a multitude of global average.

TABLE 1.3. *GCM Generalizations on Precipitation and Storms[a]*

High latitudes (60°-90°)	Enhanced in winter
Mid latitudes (30°-60°)	Reduced in summer
Low latitudes (0°-30°)	Enhanced in places with heavy rainfall today
More rain likely	Low latitudes, N. Europe, U.K. (?)
Less rain likely	Mediterranean, N. Africa, American Midwest, S. California (?)
Earlier snowmelt	Rocky Mountains, Himalayas
Summer desiccation	Interior of large continents in northern hemisphere
More extreme events	Droughts, floods, storms, hurricanes

a. Due to disagreements amoung models, these predictions are less reliable than those on temperature change.

point in building regional climate change scenarios. At present, predictions about global climate change cannot be translated directly into accurate statements about regional climate.

The statements in Tables 1.1 and 1.2 are supported by findings from several global models and are widely cited in the climate change literature. The statements on precipitation and storms, summarized in Table 1.3, are more tentative. In this case, several models report different results, even to the point that one model predicts more and the other less precipitation for the same region.

Overall, model results suggest temperature increases above global averages in regions closer to the poles and smaller changes closer to the equator. Because land masses are more concentrated in the Northern Hemisphere, temperature changes in the middle latitudes will be higher there than in the Southern Hemisphere. In addition, temperature increases in mid and high latitudes will be higher in winter than in summer.

Two important implications for agriculture are the expected retreat of the frostline toward higher latitudes by about 200 miles (320 km) and a similar movement of the most productive agricultural regions. The last statement, however, is a good example of how preliminary this kind of information is. The MINK study of the American Midwest (see Rosenberg and others, this volume) found compensating effects from a shorter growing season and increased carbon dioxide concentration acting as a fertilizer. Vinnikov, using a unique Soviet data set of soil moisture measurements, has questioned the association between higher temperatures and summer desiccation of Northern Hemisphere soils (Vinnikov, this volume).

The most serious impacts of reduced precipitation will be felt in regions with insufficient rainfall under current conditions where tomorrow's climate may bring a further reduction of usable moisture, either as a result of reduced runoff due to increased evaporation or to more severe or more frequent droughts (see Liverman and Magalhaes, this volume; Waggoner, 1990). Similarly, earlier snowmelt in mountain regions may reduce available moisture during the spring growing season.

The authors of the 1990 IPCC report summarize the uncertain state of knowledge about regional impacts of global warming as follows: "Our confidence in the prediction of the detail of regional changes is low." This conclusion is supported by a detailed review of global climate models (Grotch, 1988) and creates a serious dilemma for climate impact research and the development of regional response strategies. Without region-specific climate data, it is difficult to assess the likely economic and social impacts of future climate, such as changes in water supply, soil moisture, energy demand, agricultural productivity or infrastructure needs.

One approach used to overcome some of this uncertainty is to superimpose historical information on temperature change data obtained from the models. Climatic variation is a well-established phenomenon, with warmer periods known to have occurred between A.D. 900-1200, between 4,000 and 8,000 years ago, and during the late Tertiary (2.5 million B.P.), prior to the last ice age. Thus, any information that is available on climatic characteristics or vegetation patterns during these warmer periods may give insights into possible future trends. More recent weather patterns can also be useful. For instance, most of the models predict an increase in frequency and severity of extreme events, such as floods and droughts. Thus, those areas that are susceptible to droughts, for instance, the central United States, can look to their recent history, in this case the Dust Bowl days of the 1930s, for clues to likely future scenarios.

The assumption underlying this method is that past, current and future climates will differ in some regards, but remain similar in other respects. MacCracken (1989) suggests that future temperature and precipitation values can be superimposed on the known climatic variability of the region. Glantz (this volume) has used historical analogies in a number of case studies that assess social response to climatic change. Schlesinger (1990) uses nested simulations to link observed climate data with GCM predictions. Kasperson and his team at Clark University employ historical data in their studies of nine potentially critical environmental zones in different parts of the world. In our work, we use the ten-year period during which the drought of record occurred as the reference period for altered streamflows in Texas water basins (Schmandt and Ward, 1991). Rosenberg and his team used the warmer and drier climate of the 1930s in their study of climate change in four midwestern states (Missouri-Iowa-Nebraska-Kansas), described later in this volume.

Although the implications throughout most discussions of global warming are that all of the effects will be adverse, modelling specific regions could give insight into opportunities for some areas. For instance, model results suggest that there will be a retreat of the frostline toward higher latitudes by about 200 miles (320 km) and a similar movement of the most productive agricultural regions. This could provide additional opportunities for Canada and Russia, provided enough water and good soils are available to take advantage of the improved environment. It will also be of benefit to areas that produce citrus fruit, despite current susceptibility to frost. In this volume we describe two studies that address specific situations that lend themselves to adaptive responses. The first involves a study by Botkin of forest management strategies in the Great Lakes region. He identifies tree species that will be the most vulnerable and management techniques that can ensure continued forest productivity under changing environmental conditions. The second, by Panturat, focuses on identifying the strains of rice that will be best suited to the changed climate in specific areas of Thailand.

Identifying these opportunities and developing the necessary knowledge base in preparation for climate change is important. However, it is likely to be the harmful, or even devastating effects of climate change that attract the most attention. Regional impact studies and policies need to focus on the most serious risks. Bangladesh and the Netherlands, for example, are most vulnerable to sea-level rise and the resulting flooding, higher groundwater table, and intrusion of salt water into rivers. The Sahel, northeast Brazil and northwest Mexico are most vulnerable to drought. In developing a regional response strategy a first task for each region is to identify existing vulnerabilities, ask whether climate change is likely to increase or decrease these risks, and determine how new climatic conditions will change existing economic and social conditions.

A typology of vulnerable regions and key impact categories is presented in Table 1.4. The table emphasizes connections between climate change risks and existing environmental, economic and social problems. The table also makes the point that current and future vulnerabilities differ widely between regional categories, and therefore require specific response strategies. The key meassage conveyed by the table is this: Climate change will be most painful in regions that are poor, that suffer from conventional pollution, that are equipped with outdated infrastructure and that are least able to deal with additional stress. Many of the most vulnerable areas, in addition, suffer from rapid population growth. The combination of these factors in extensive areas of the world will make climate change difficult to deal with, even at the regional level, notwithstanding our earlier point that global solutions will be hardest to accomplish.

This is a recurring theme throughout this volume. Where flooding is a problem now, it is likely to get worse in the future; where drought conditions prevail, water availability will most likely become an increasing problem; and, most important, where poverty exists it will increase the vulnerability of an area by reducing its ability to cope with natural disasters.

Development of Regional Strategies

Although policies aimed at slowing, or eventually stopping, global warming will require concerted international action, most adaptive policies will be regional in nature because they respond to particular regional vulnerabilities and opportunities. However, some preventive measures can benefit the region while at the same time reducing global warming. If enough regions practice energy conservation or adopt reforestation programs they can help, however modestly, in slowing global warming. And some regions are large players in the greenhouse, making their action or inaction count. If Texas were an independent country, for example, it would rank seventh—ahead of the United Kingdom—as a producer of greenhouse gases (Lashof-Washburn, 1990). This reflects both the heavy concentration of petrochemical industry in the state and a higher than average use of motor vehicles. Texas is rich in natural gas, which has more benign greenhouse characteristics than coal and oil. Thus, increased reliance on natural gas as an energy source presents a desirable policy option, for at least several decades, combining economic with environmental benefits (Schmandt, Hadden and Ward, 1992). From this perspective, regional strategies will be particularly important should effective international agreements fail to materialize. They can serve double duty as an insurance policy for the world, and a sound development strategy for the region.

The most common regional entities are nation states. However, national borders do not follow geographical boundaries, except in the case of island nations, which may be too small to form effective political entities for international negotiations. Thus, we may have to be more flexible in our thinking. For large nations, such as the United States, Canada, Brazil, Russia and China, many areas with distinct geographical identity are evident and political authority is dispersed. For these countries, the more appropriate regional entity may be the state/province or groups of such entities working cooperatively. For the smaller nations, it may be necessary for them to pool their resources and develop regional strategies that are of mutual benefit. In some cases, political institutions are already in place for addressing common economic and social goals, as in the case of the European Common Market. In the case of the Southwest Indian Ocean Islands, the Indian Ocean Commission represents a less formal body for coordinating regional policies (see Saha, this volume).

TABLE 1.4. *Regional Impact Categories*

Category	Description	Examples
Coastal regions	Vulnerable to sea level rise. High-risk regions include low-lying and densely populated areas. May also be affected by land subsidence.	U.S. and Mexican Gulf coast, Netherlands, Bangladesh, Guyana, Baltic Sea, Mediterranean Sea, Caribbean Sea.
River deltas	Threatened by sea level rise, salt water intrusion, reduced fresh water streamflow and increased subsidence.	Ganges/Brahmaputra, Nile Mississippi.
Ocean islands	Vulnerable to sea level rise. Generally small islands with low elevation.	Maldivers, Indonesia, some Pacific islands.
Major food-producing	Under current climatic conditions these regions are well-suited to food production. May be vulnerable to more frequent and severe droughts, changes in monsoon patterns or timing of snow-melt, increased desertification or other climatic stresses.	U.S. Midwest, Soviet Ukraine, Argentina, parts of Australia, India, China. Under changed climatic conditions, other regions may become better suited to agricultural production and become the new breadbaskets or rice bowls of the world.
Marginal agricultural regions	Stressed agricultural systems are highly vulnerable to climatic variations. Typical practices include subsistence agriculture and livestock raising.	Sub-Saharan Africa, northern Mexico, Middle East, northeast Brazil, Australia, parts of China.
Natural and managed forests	Vulnerable to changes in precipitation, temperature and soil moisture, as well as pollution and development.	Southeastern United States, Amazon Basin, Indonesia, southern Siberia.
Inland lakes	Vulnerable to increased evapotranspiration and decreased streamflows.	Great Lakes, Lake Baykal, Aral Sea, Lake Victoria, Caspian Sea.
Water-stressed regions	Regions of periodic water shortage or where demand for water is increasing and threatening to exceed available supplies. Climate change may decrease the supply or worsen water quality and intensify conflicts among water users. Increases in demand may be caused by population growth or increased irrigation of agriculture.	California, Texas, parts of Mexico and China, the Middle East.
Very large metropolitan areas.	Megacities that are already at risk from heat stress, water scarcity, inadequate energy infrastructure, public health problems and conventional air and water pollution. Many of these problems could be exacerbated by climate change.	Calcutta, Bombay, Cairo, Jakata, São Paulo, Mexico City.
Industrial regions	Severe air and water pollution problems. Many feel that actions in response to climate change must wait until more urgent pollution problems are addressed. Integrated policies to address both sets of problems may be possible. Many regions lack the economic resources to respond to environmental issues.	Industrial regions in Eastern Europe, East Germany, India, Mexico and Brazil.

To justify comprehensive regional policies, identification of so-called tie-in measures holds promise. Such measures simultaneously address climate change and other, often longstanding, environmental, natural resource and economic problems. Several advantages are inherent in this approach. First, the time frame of climate change is too far into the future to lead to timely action. All political systems have great difficulties in acting on long-term threats. This may be different should the future threat help to solve a current problem. The city of Galveston, Texas, threatened by hurricanes, is planning to extend its seawall first built in response to a devastating hurricane that destroyed the city at the turn of the century. The design of the structure will be different and costs will be higher if sea-level rise is taken into account now. However, this additional effort will be less expensive than retrofitting the seawall several decades later. The Netherlands are already following this strategy in their century long battle against the sea.

The identification of current issues that have relevance to future climate change is another recurring theme in this volume. For developing countries, in particular, all of their available resources are being consumed by the problem of providing their impoverished citizens with the most basic needs. However, where water availability is already a problem, developing the appropriate infrastructure and improving water management practices could address current needs while preparing for an uncertain future. In another example, land-use planning that would discourage development in flood-prone areas would have obvious economic benefits now as well as in the future.

The regional strategies proposed throughout this volume will be important components of a pluralistic policy response based on a region's self-interest to secure its own future. This provides a more powerful motivation than the desire to preserve the global commons. But a well-designed regional strategy can contribute to the international goal of reducing greenhouse gas emissions. The condition is that regional mitigation can be accomplished at reasonable cost and with some benefits accruing to the region. Such a strategy will answer the following questions:

- What major threat or threats will this region be exposed to as a result of global warming?
- What can be done to minimize the threat?
- Can action on global warming help to resolve other environmental and resource problems in the region?
- How can the region reduce its greenhouse gas emissions, simultaneously reduce other sources of pollution, and not suffer undue economic losses in comparison to more complacent neighbors?

Institutions capable of addressing climate change, along with other issues, as part of development planning at the regional level do not yet exist. Some early institutional experiments are discussed by Moomaw and Toth in this volume. Many experiments all over the world are needed. In our view, new regional development institutions must address the entire agenda for sustainable development and actively involve both developmental and environmental interests. Such a grassroots effort, sustained over time and assisted by universities and state and national governments, holds promise for developing regional strategies for dealing with global climate change.

References

Glantz, M.H. (Ed.), 1988. *Societal Responses to Regional Climate Change: Forecasting by Analogy*. Westview Publishers, Boulder, Colo.

Grotch, S.L., 1988. *Regional Intercomparisons of General Circulation Model Predictions and Historic Climate Data*. Report prepared for the U.S. Department of Energy (DOE/NBB-0084). Lawrence Livermore National Laboratory, Livermore, Calif.

IPCC, 1990. *Climate Change: The IPCC Scientific Assessment*. J. T. Houghton, G. J. Jenkins and J.J. Ephraums (Eds.). Cambridge University Press, Cambridge, U.K.

Lashof, D. and E.L. Washburn, 1990. *The Statehouse Effect: State Policies to Cool the Greenhouse*. National Resources Defense Council, New York.

MacCracken, M.C., 1989. Scenarios for

future climate change: Results of GCM simulations. In: *Impacts of Climate Change on the Great Lakes Basin.* National Climate Program Office/NOAA and Canadian Climate Centre, pp. 43-48.

Roan, Sharon L., 1989. *Ozone Crisis: The 15-Year Evolution of a Sudden Global Emergency.* John Wiley, New York.

Schlesinger, M., 1990. Likely climate changes in the Western Hemisphere. Paper presented at the Annual Meeting of the American Aasociation for the Advancement of Science, New Orleans.

Schmandt, J., and G. Ward, 1992. Climate change and water resources in Texas. In: *Proceedings of the First National Conference on Climate Change and Water Resources Management.* T.M. Ballyntyne, E.Z. Stakhiv and J.B. Smith (Eds.). Institute for Water Resources, U.S. Army Corps of Engineers.

————, Hadden, S., and G. Ward, 1992. *Texas and Global Warming: Emissions, Surface Water Supplies and Sea Level Rise.* LBJ School of Public Affairs, University of Texas, Austin.

Waggoner, P.E., (Ed.), 1990. *Climate Change and U.S. Water Resources. Report of the AAAS Panel on Climate Variability, Climate Change, and the Planning and Management of U.S. Water Resources.* John Wiley, New York.

WMO-UNEP, 1988. *Developing Policies for Responding to Climatic Change. A Summary of the Discussions and Recommendations of the Workshops Held in Villach and Bellagio.* WCIP -1, WMO/TD-No.225. World Meteorological Organization, Geneva.

World Commission on Environment and Development, 1987. *Our Common Future.* Oxford University Press, Oxford and New York.

Part I

Identifying Regional Impacts of and Vulnerabilities to Climate Change

With an ever-increasing human population, the resources of the world are being stretched to their limits. Higher demand for food is resulting in the cultivation of increasingly marginal land, which is soon exhausted; widespread irrigation is causing salinization of soils in many areas. In addition, the clearing of forests for agricultural use and as a source of firewood adds another dimension to the environmental problems of the world, ultimately resulting in desertification of large areas.

Global climate change, although of somewhat uncertain consequence, is another facet of the problem of worldwide environmental stress. If populated areas become uninhabitable, because of harsher conditions or encroachment of the sea through sea level rise, this will increase the pressure on already-stressed areas, particularly in overpopulated regions. Some of these trends are already identifiable, such as increasing deforestation, and often act as a positive feedback mechanism in the process of global warming.

Twenty percent of the world's population lives in coastal areas. Encroachment of the sea, already a problem in some areas, will force the migration of people from some of the most productive regions. In developed countries this may simply result in the loss of structures that can be relocated; for overpopulated developing countries there may be limited areas for new immigrants. The first two chapters in Part I address the problems of sea level rise. Leatherman, in a historical study of the Chesapeake Bay area, shows how the encroachment of the sea has resulted in the progressive loss of large parts of off-shore islands. Historical records document the retreat of communities since the area was first settled.

For a nation like Bangladesh, the consequences are much greater, and the very survival of the population is at stake. Thus,

Mahtab describes the extreme vulnerability of much of the country, which is barely above sea level and experiences frequent natural calamities such as floods, drought, cyclones and tornadoes. Changing rainfall patterns could increase the frequency of these events, and if, as occurred in 1988, the peak flood occurs simultaneously on the Ganges and the Brahmaputra, devastating floods result.

The vulnerability of other regions stems from the overall increase in temperature and reduction in precipitation, aggravated by increases in evapotranspiration. A detailed account of such potential consequences in Mexico is given by Liverman. She describes how the existing social structure, including the allocation of land and water, contributes to the vulnerability of the majority of the population, and how, under warmer and drier conditions, the plight of the poor can only get worse.

A similar picture of vulnerability is painted by Ogbuagu for western and central Africa. In this case, flooding is a major problem in some areas, while other areas are drought-prone. Ogbuagu argues that the resources of the country are so limited that the government is unable to do much more than deal with the effects of immediate crises; long-term risk management strategies will require international assistance.

Some of the same themes emerge in the chapter by Pachauri on the effects of global warming in Southeast Asia. He stresses the need for research to obtain better information and for international assistance. In addition, a regional approach involving all of the nations influenced by the Ganges is emphasized. A cooperative effort to develop the upper reaches of the Ganges could provide India with hydroelectric power and Nepal with a source of income, while at the same time mitigating some of the most severe effects of major storms by controlling streamflow in the deltaic regions of Bangladesh.

Overall, these chapters emphasize the extreme vulnerability of areas of the globe that are already stressed. Those areas that are drought-prone are likely to have increasing problems with water availability; those that experience frequent floods can expect them to become more frequent and more severe. In addition, the extreme poverty of many of these areas reduces their ability to prepare for future events, because all of their resources are committed to problems of a more immediate nature.

2. COASTAL LAND LOSS IN THE CHESAPEAKE BAY REGION: AN HISTORICAL ANALOG APPROACH TO GLOBAL CHANGE ANALYSIS

Stephen P. Leatherman

Although the debate on global warming continues, it is clear that relative sea-level rise (SLR) is occurring along most of the world's coasts. Over the long term, global warming and rising sea level must be linked. Although some politicians and the public are unaware of or have largely ignored these warnings, the evidence for global change and its impacts is mounting.

A global rise in sea level appears to be the most dramatic and certain effect of all potential consequences of human-induced climate change. In fact, sea level can be thought of as the "dipstick" of climate change. Rising sea levels and growing coastal populations already place an increasing strain on the coastal zones of the world. Fortunately, societies can plan for the consequences of sea-level rise because at least we know the direction, if not the degree, of increase.

The Chesapeake Bay region (Figure 2.1) serves as an ideal laboratory to investigate the physical effects of accelerated SLR and explore the range of impacts and response strategies. This estuary, with over 3,200 km of shoreline, is subject to all SLR impacts: erosion, inundation of low-lying lands and wetlands loss, salt-water intrusion into aquifers and surface waters, higher water tables, and increased flooding and storm damage.

Coastal land loss in the Chesapeake Bay region is a major issue today, because more and more people are moving closer to the water's edge. It is projected that, by the year 2020, over 16 million people will live in the Chesapeake Bay region, twice the 1950 population (Kasowski, 1989, p.1). At the same time sea level has been rising at a rapid rate—approximately 28 cm during the past century (Hicks, 1978, p. 1378).

The Chesapeake Bay coastline responds to sea-level rise in two different ways— submergence and erosion. Submergence is the total inundation of the land by the rising water, and erosion is the physical removal of sediment, resulting in the diminishment of land area. Submergence and erosion are evident on the Chesapeake Bay's Eastern Shore, while bluff retreat through erosion is dominant on the Western Shore.

Fortunately, excellent historical data are available because of the early exploration and settlement of the Chesapeake Bay region. For instance, Kent Island, one of the first settlements, dates back to 1631, and the original Royal land charters and tax records of the Chesapeake Bay islands have been kept since that time. Maps, charts and air photos have been obtained from historical societies, museums, and the National Archives, and analysis of these voluminous historical data has been undertaken. Land loss information can also be obtained in a qualitative sense from newspaper accounts over the centuries. Even though the press has been recording incidents of land loss in response to sea level rise, as well as other factors, these incidents have not been distilled and analyzed in a scientific manner.

The use of historical analogs (of past changes and responses) to formulate present and future policies is only now being recognized as an invaluable approach to global change analysis where environmental problems have a long lead time. This study of coastal land loss in the Chesapeake Bay region documents the human adjustments to a 0.3 m rise in sea level. These types of studies can provide much insight into the response of other erodible and low-lying coastal areas worldwide. While communities have been forced to flee from the advancing water in the Chesapeake Bay, entire nations are at risk; elsewhere, in the Indian and Pacific oceans. Conceivably, the hard lessons learned by the Chesapeake Bay inhabitants in response to a relatively small

FIGURE 2.1. The Chesapeake Bay Region.

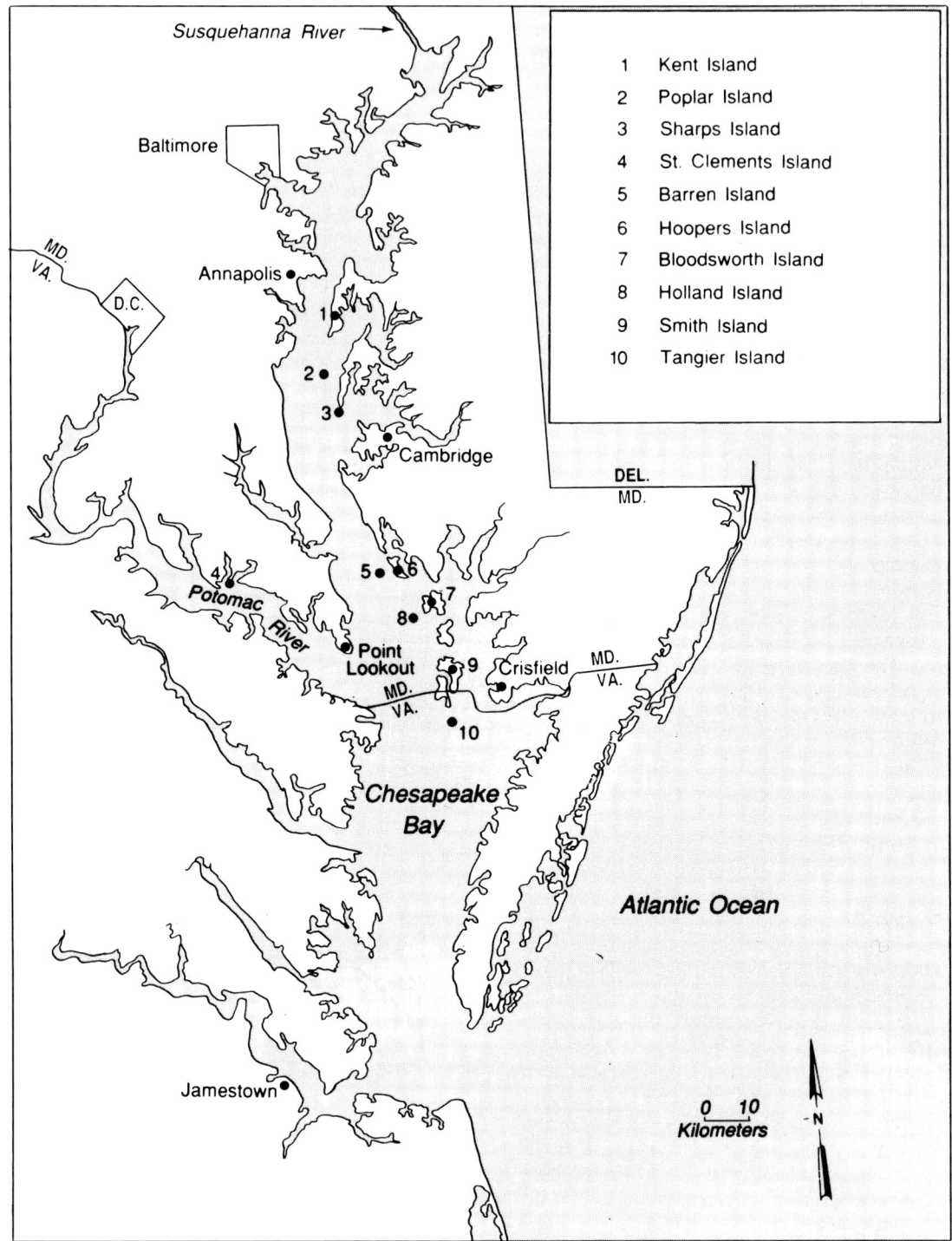

rise in sea level can be applied on a much broader scale, avoiding the prospect of millions of forced environmental refugees worldwide in the coming decades.

Chesapeake Bay Region

The Chesapeake Bay is the largest and probably the most important estuary along the U.S. Atlantic coast. It is by definition a drowned river valley, the product of SLR pushing salty water up the basin of the Susquehanna River and its tributaries during the last 10,000 years.

In a real sense, the Chesapeake Bay region can serve as a microcosm of coastal areas that are susceptible to SLR impacts elsewhere. Throughout the period of the bay's natural development, continual reshaping of the shoreline presented no human problems. The first inhabitants were hunter-gatherers known as the Clovis people. These early residents walked throughout the Chesapeake Bay region over the valley of the narrow Susquehanna River, at a time when sea level was over 60 m lower than at present.

The rates of postglacial SLR had dropped significantly some 6,000 years ago (National Research Council 1987, p. 12), and the vast salt marshes of the Chesapeake Bay began to form (Orson *et al.*, 1985, p. 30). During the last 100 years, the relative rise in sea level has been 0.3 m, about ten times faster than the average for the past 6,000 years.

Western Shore

The Western Shore, the cliffed shoreline of Maryland, is composed of sediments ranging from loose sand to blue marine clays. The cliffs tower up to 30 m above Chesapeake Bay. Historical maps reveal that these impressive cliffs, almost vertical in places, retreat landward 0.3 to 0.6 m per year as a result of constant wave action (Leatherman, 1986, p. 28; Singlewald and Slaughter, 1949).

The waves, particularly during storms, are literally eating away part of Maryland's history, as bits and chunks of sediment fall along the retreating cliff face. St. Clement's Island on the Potomac River (Figure 2.1), where the colonists first landed in 1634, is considered the Plymouth Rock of the State of Maryland. It once covered over 160 ha and was "thickly wooded with cedars, sassafras, and nut trees, with herbs and flowers everywhere" (Shultz, 1972, p. 7); today the island has shrunk, through cliff erosion, to about 16 hectares (ha) with few remaining trees.

The Native Americans lived in harmony with the changing shorelines, unlike the later Europeans. As the eroding edge threatened a campsite, the Piscataway tribe on St. Clements Island would move their tents and belongings farther inland. The European colonists, however, required more substantial housing, urbanized development, and those vital lighthouses to mark the navigable channels. Unfortunately, not enough attention was given to the soft, erodible sediments upon which these structures and sentinels to the sea were built, and no one realized at that time that sea levels were rising, mandating continued shore retreat.

Coastal land loss can be prevented, but the price can be exceedingly high, depending on the wave energy and currents. The historic towns and present-day ports of Annapolis and Baltimore have not been threatened by sea-level rise impacts. These harbors of refuge are protected from the larger waves along the main stem of the bay, and the value of the urban property mandates protection through the construction and maintenance of bulkheads and seawalls.

Eastern Shore

The flat and low-lying Eastern Shore of Maryland contrasts strongly with the high eroding bluffs of the Western Shore. While there is erosion along the bay edge, the primary response to rising sea levels is submergence. The biggest loser in the looming battle against SLR, however, may be the bay region's wetlands. Even as baywide policies and state legislative initiatives are put in place to protect wetlands, these vital natural habitats are disappearing before our very eyes.

The Eastern Shore is the site of the largest expanse of coastal wetlands along the U.S. mid-Atlantic coast. Blackwater Wildlife Refuge near Cambridge, Maryland, was established by the U.S. Fish and Wildlife Service because of the ecological importance of these undisturbed marshlands for migrating waterfowl. Unfortunately, protection of these indispensable wetlands has not been perfect, with over a third of the total marsh area

FIGURE 2.2. Wetlands loss is dramatically shown by historical aerial photography of Blackwater Wildlife Refuge in the Chesapeake Bay, Maryland.

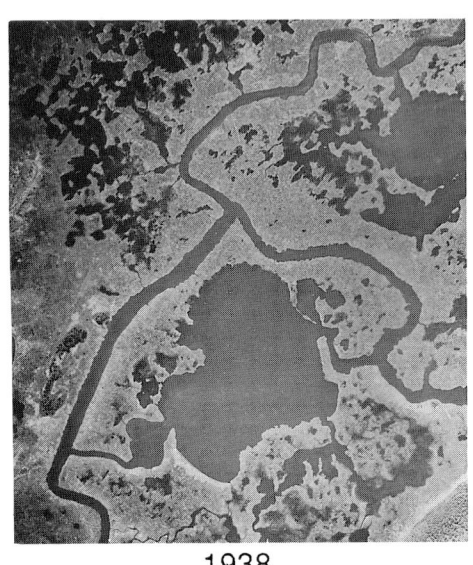

1938

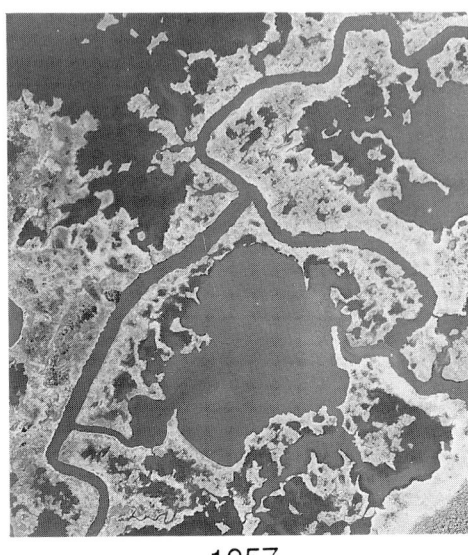

1957

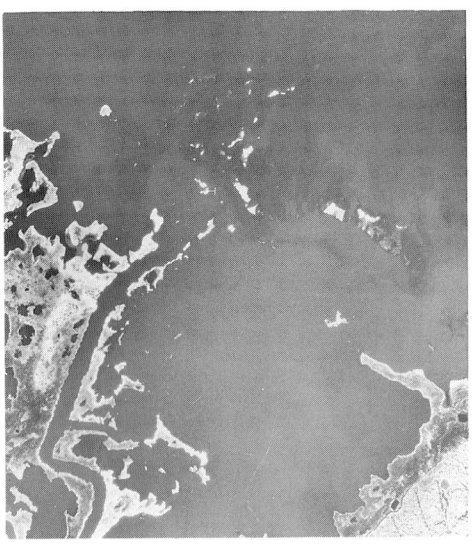

1972

1988

(about 2,000 ha) being lost between 1938 and 1979 (Leatherman, 1989, p. 46), and the trend is continuing (Figure 2.2). As sea levels continue to outpace the ability of the marsh to maintain elevation, the magnitude of the total loss will increase. The prospect for wetlands in the Chesapeake Bay region is indeed bleak.

Salt marshes, and their tropical equivalent mangroves, exist in a delicate balance with water levels. Sea level rise created the vast expanses of wetlands in the Chesapeake Bay region, and the same rise will destroy them, should the rate become too rapid (Gehrels and Leatherman, 1989, p. 10). About 6,000 years ago, the rate of relative sea-level rise in the Chesapeake Bay had slowed dramatically (about 30 cm per millennia; National Research Council, 1987, p. 12). At that time the marshes were a relatively narrow band along the interface between land and water. In the face of slowly rising water levels, this fringe marsh was able to keep pace with sea level rise. As the salty waters rose, the marshes continued to build vertically and the leading edge penetrated farther inland, up the gradually sloping coastal plain of the mainland.

While existing salt marshes are sustaining heavy losses, which will only accelerate through time, new marshes are now developing through upland conversion.

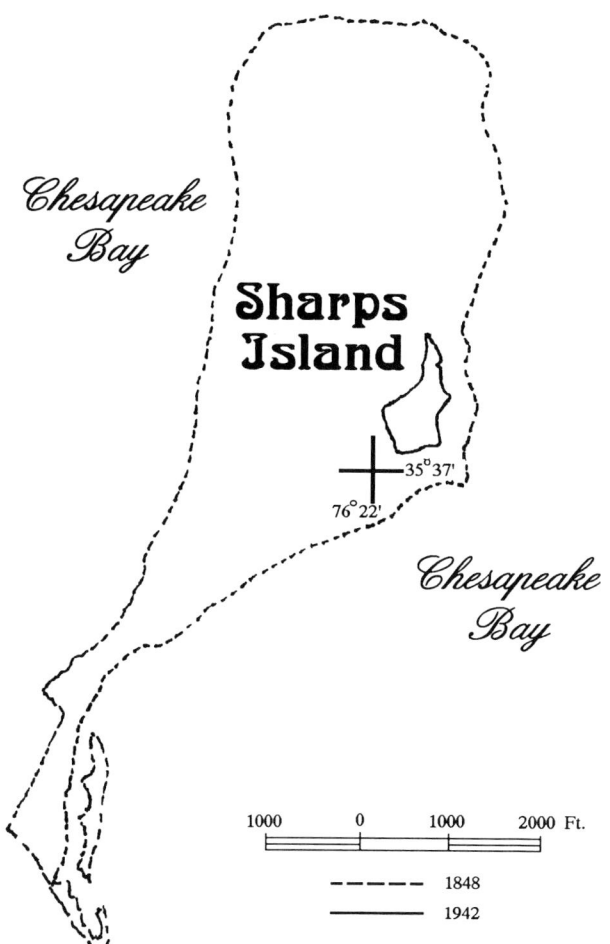

FIGURE 2.3. Shoreline changes of Sharp's Island, 1848-1942. (Source: Singlewald and Slaughter, 1949, p. 152.)

However, the landward penetration of brackish water plants is viewed by Eastern Shore farmers as a loss of good agricultural lands. The rate of landward intrusion of the salt water is about 2 to 3 m annually according to local observers (A. Eastman, 1990, personal communication). While the shrinking of the arable lands will undoubtedly reduce the farmer's yield of corn and other crops, the loss is not significant regionwide. Ample space is available elsewhere for agricultural activities, and defensive measures are not justified in terms of the relatively high cost of protecting this low-lying land. Even small towns, such as Crisfield on the Eastern Shore of Maryland (Figure 2.1), are wrestling with the question of sea defenses in order to prevent flooding during coastal storms. Dikes have been designed, but not yet constructed. The real fate of the new marshes is determined by increased development: as the bay's shore continues to be urbanized, the perceived benefits will justify the costs. As sea level rises, bulkheads will line the high ground, and the marshes will literally be squeezed out of existence.

Island Communities

The human dimension of SLR impacts is best exemplified by an examination of land loss on the Chesapeake Bay islands. Sizable islands have been substantially reduced in size or even drowned; their slow passing can be tracked on nineteenth and twentieth century maps and charts. Following sharp decreases in island area, settlements were abandoned on many of the islands in the first decades of this century. People left the islands as rising sea levels caused progressive erosion and/or submergence, especially in the wake of hurricanes.

There is a rich history of human occupation in the face of adversity (e.g., rising sea levels). The gradual "wearing away" of the land represents a case of an environmental problem with a long lead time. Recognition of the trend and human response to the SLR impacts are graphically portrayed in the Chesapeake Bay region as the land shrank and people were eventually forced to abandon their homes.

Sharp's Island

Sharp's Island was once a populous island with pine forests and cultivated land. Now it has been reduced through erosion to a sand shoal, totally submerged at high tide.

The earliest records show that when Lord Baltimore conveyed ownership of Sharp's Island in 1660 it consisted of 360 ha (Bryan, 1962, p. 10). Its average elevation above mean sea level was only about 1 m, as shown on historical maps. Located in the mouth of the Choptank River, Sharp's Island was subjected to both wave and tidal currents from nearly all directions, greatly contributing to its demise.

The original 360 ha shrank to 280 ha by 1775. By 1812, when the island was again sold, the deed stated the island had 252 ha, which translates to an annual loss of approximately 1 hectare. In 1838 the original Sharp's Island Lighthouse was built on this small, shrinking island. Only ten years later, the lighthouse was moved farther inland because of the encroachment of Chesapeake Bay. The Coast Guard then moved the lighthouse inland two more times in the following years for the same reason (Preston, 1982, p. 45).

Not only was the slowly rising sea level allowing the waves to attack higher on the land, causing severe erosion, but the island was also being submerged in place. In 1848 the island was said to consist of 175 ha, most of which was lowland grasses and marshland (Figure 2.3). The pine forest had been reduce to a few stands of trees.

Apparently oblivious to the island's vulnerability and recent history, Miller Creighton of Baltimore built the Sharp's Island Hotel in 1895 (Bryan, 1962, p. 11). Within a few years the eroding beach had undermined the new pier and was threatening the very foundation of the hotel. By 1900, Sharp's Island consisted of only 36 ha; about 80 percent of the island had been lost. Ten years later the island had shrunk to 21 ha, and the hotel was abandoned. The last sightings of the hotel were made in 1924, when at least half of the last remaining building was in the water. Finally, in the winter of 1962 the Chesapeake Bay consumed the last of Sharp's Island (Figure 2.3).

Barren Island

Barren Island was a thriving community in 1900, with a dozen families, stores, a doctor, a school, and a Methodist church; now it is deserted. A relatively large island in colonial times, this island of 344 ha was awarded by Lord Baltimore in 1664 to Richard Preston (Cronin, 1988, p. 66). Barren Island has been steadily eroding because of its exposed position in Chesapeake Bay. It is now broken up into two large and five small islands totaling about 100 ha, based on map calculations.

Beginning in the late 1890s, residents began to move away from Barren Island. Houses were jacked up and floated across the water to the nearby Hoopers Island, another low-lying bay island. The last family left in 1916. When Barren Island is finally consumed by the sea, the already eroding and submerging Hoopers Island will be fully exposed to the energies of the sea (Figure 2.1).

More recently, local hunting clubs bought the island because of the abundant game bird population. A lodge for the local gun club was built from the remains of a stately Baltimore hotel. Two bulkheads were built to stop the erosion and protect the lodge and remaining island area. Eventually, the bulkheads began to fail and the refusal of the U.S. Army Corps of Engineers in the 1970s to rebuild this structural defense doomed the building and the island (Figure 2.4).

Poplar Island

Poplar Island was one of the larger islands in the Chesapeake Bay group; the first recorded sale transaction in the late 1670s revealed 572 ha. By 1848 the first truly accurate map of this island was produced by the U.S. Coastal Survey (Figure 2.5). The island had been already greatly reduced in size, but still was roughly 320 ha. Erosion and submergence continued to take their toll, and this single

FIGURE 2.4. Rapid shore erosion has doomed this coastal resort on Barren Island.

FIGURE 2.5. Poplar Island has been greatly reduced in size and splintered to form several small islands by continuing erosion. Dashed line shows coastline in 1848; solid line shows coastline in 1987.

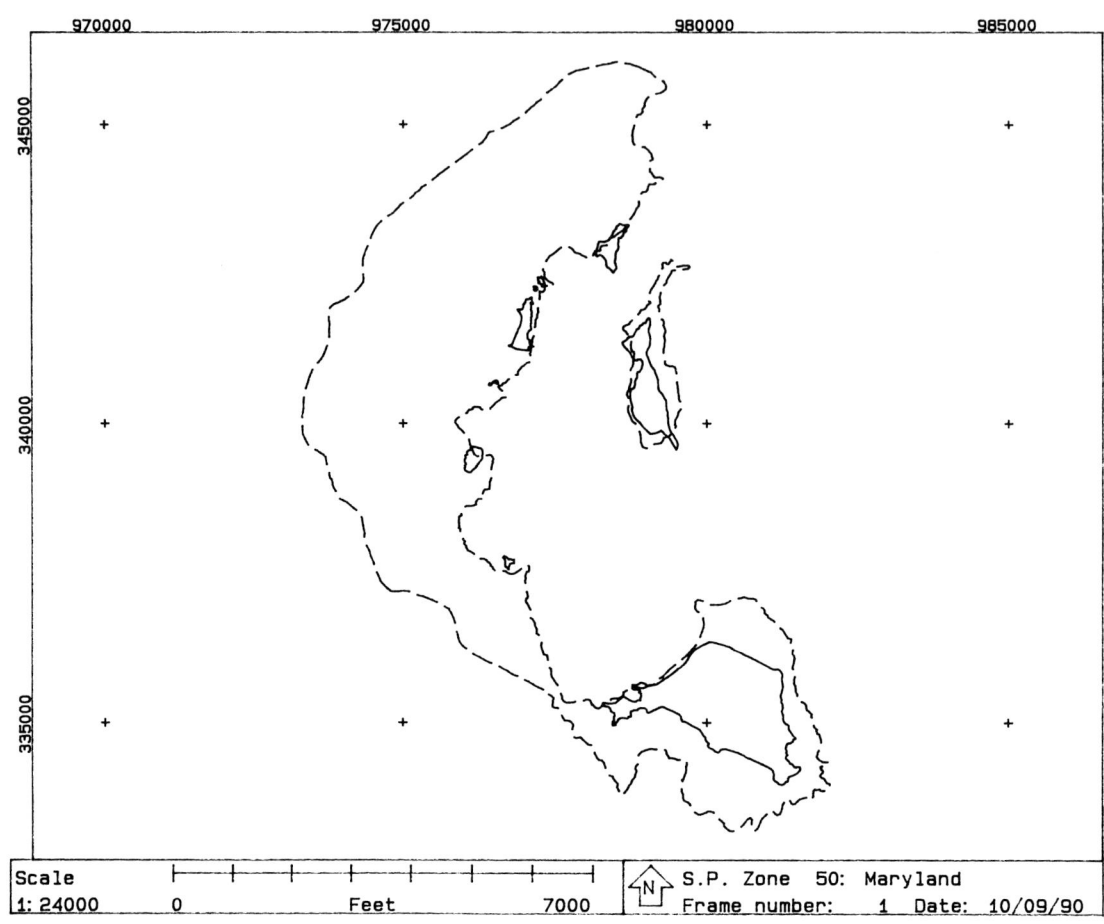

island was split into a main area and several smaller islands.

Around 1900, at the height of settlement, about 15 families, totaling 50 to 100 people, lived on the several Poplar islands. The main island was still large enough to support a small town known as Valiant. This community included a general store, post office, school, church, and even a sawmill. The town flourished until the 1920s, when the serious erosion problem became evident to all, and most islanders deserted their homes. By 1930 Poplar was completely deserted (Rummel Consultants, 1971, p. 15).

However, the following year the Poplar islands were purchased by a group of

Democratic politicians, who founded the Jefferson Islands Club in 1931. The club was officially established as a hideaway for "congenial" members and served as "a shrine for Democrats" (Abribat, 1988, p. 12). It was frequented by many famous Democrats, including Presidents Franklin D. Roosevelt and Harry S. Truman, until the complete destruction of the clubhouse by fire in 1946. This event, coupled with the fact that the island was continuing to erode and submerge at a rapid rate, forced the Democrats to relocate their club elsewhere.

Poplar's demise is in response to relative SLR and the resulting erosional agents. The island is slipping away at the alarming rate of over 6 m annually (Figure 2.5), and will probably disappear by the turn of the century at present rates of land loss. According to one observer, "What in colonial times had been a single, one thousand acre (400 ha) island was now in seven pieces, a chain of islands totalling no more than one hundred and twenty-five acres (50 ha) or so" (Meyer, 1986, p. 43).

The Maryland House of Delegates passed a resolution in 1968 to arrest the erosion at Poplar and help the Smithsonian Institution purchase this land. While an extensive feasibility study was conducted to place engineering structures around the western shoreline of Poplar, this proposed stabilization plan never materialized. In 1971 another study by federal and state governments was authorized with the goal of installing a breakwater along the bay shore. Again, this plan was never put into action, principally because of the cost of saving a rapidly eroding, small island. According to the director of the Chesapeake Bay Center, "This damn island has broken everyone's heart who has been on it" (Abribat, 1988, p. 21). Poplar Island—once called the playground of presidents—is fast on its way to becoming another sand shoal in the Chesapeake Bay; lost but not forgotten.

Discussion and Conclusions

The Chesapeake Bay regional study of coastal land loss and human adjustments to only a small rise in sea level demonstrates the potential for disaster in erodible and low-lying coastal areas worldwide in response to accelerated sea-level rise. The Chesapeake Bay area, with its high, eroding cliffs on the Western Shore, and disappearing islands and submerging coasts on the Eastern Shore, is actually a microcosm for such changes and problems elsewhere. Although there are no extensive river deltas in the Chesapeake Bay region, the low-lying coastal plain of the Eastern Shore manifests all the problems of people attempting to live and farm these areas near or at existing sea levels.

Continued rapid urbanization of the Chesapeake Bay region also reflects a worldwide trend. People in industrialized countries have a strong desire to build resort areas and retirement houses right at the sea's edge. In developing counties, the coastal fringe is often the most desirable because of the natural harvests of seafood and reclamation of intertidal areas for agriculture and aquaculture. Developing counties are also attempting to promote tourism (a major source of hard currency) by building seaside resorts.

Even the lower estimates of accelerated SLR will place many of the world's coastal inhabitants in jeopardy in the coming decades. The low-lying coastal fringe is subject to catastrophic events of flooding, particularly during hurricanes, typhoons, and cyclones; the intensity of these large tropical storms is likely to increase in the future as a result of global warming (Emanuel, 1987, p. 483).

The history of the Chesapeake Bay region provides fascinating stories of the fate of island communities that lost their land and livelihood in response to a small rise in sea level. While their struggle with coastal land loss can be seen as folly, in all fairness it must be realized that the islanders had little knowledge of coastal processes and less awareness of the long-term trend of sea-level rise. In this "age of information" hopefully we will not repeat the mistakes of the past in confronting the more severe and widespread consequences of accelerated SLR on island nations.

While the greenhouse effect is mainly caused by the industrialized counties at present, the developing nations will probably suffer the most from its impact. Adaptation measures will be required, and retreat is the most likely response in less developed countries (Vellinga and Leatherman, 1989). Even small changes in sea level can spell major coastal impacts, and smaller, poorer island counties are particularly vulnerable (Tickell, 1989). In fact, the IPCC (1990, p. 6-4) has concluded that the largest effects on humanity of climate change may be entire countries, such as the Maldives, Tuvalu, and Kirabati, imperiled by a rise of a few meters in sea levels. Populous deltaic plains—as in Egypt, Bangladesh, India, China, and Indonesia—are all threatened by inundation from even a moderate global sea-level rise.

Many small island countries would lose a significant part of their land area with a SLR of 1 meter. For instance, the 1,190 small islands that constitute the Republic of the Maldives have average elevations of 1 to

1.5 m above existing sea level (Small States Conference on Sea-Level Rise, 1989). Submergence and erosion can convert many smaller islands to sandbars and significantly reduce the usable dry land on the larger, more populated islands. In addition, salt-water intrusion and loss of fresh water supplies may be an equally severe limitation for human habitation. The Republic of the Maldives already has a high population density, as do many other island countries, so that retreat is not a viable alternative. Just as the islanders in the Chesapeake Bay before them, they must either protect themselves or evacuate and abandon their land. These imperiled people are growing more aware of the looming problems of global warming and SLR impacts. In November 1989 President Gayoom of the Maldives wrote to the Secretary General of the United Nations, expressing his concern about the problem:

> "...nor would any fisherman ever think that the sea which is the bountiful source of his livelihood could, in a matter of decades, become his eternal grave. But that, ladies and gentlemen, is precisely the prospect that we have to face today..."

Abandonment appears to be the only feasible option for coastal areas, like small island nations, followed by migration and resettlement. But where might the people go in a world of limited arable land and exploding populations? This human dimension of SLR-induced coastal erosion and inundation has far-reaching implications. There was plenty of open land in the Chesapeake Bay region for resettlement, and the inhabitants of the area were all of European descent and similar religion; this is obviously not the case in the Indian and Pacific ocean regions. India is less likely to welcome people from Bangladesh, with their distinct cultural and religious differences. The largest number of environmental refugees will most likely result from inundation of the deltaic plains of Bengal and the Nile, where the homes and livelihood of tens of million people are potentially threatened. Even a modest rise in global sea levels could produce millions of such refugees, and a rise of 1 meter in ocean levels worldwide would likely result in the creation of at least 50 million environmental refugees (Dennis *et al.,* 1990;

Jacobsen, 1988, p. 29; Tickell, 1989). Such refugees from sea-level rise impacts could become the single largest class of displaced persons in the world, far exceeding those fleeing famine and war.

Acknowledgements

This study was funded by the U.S. Fish and Wildlife Service Chesapeake Bay Estuary Program and the W. Alton Jones Foundation. The following graduate students are gratefully acknowledged: Ms. Rachel Donham and Mr. Greg French assisted with the graphics, and Ms. Kate Eldred provided editorial assistance.

References

Abribat, B., 1988. The Jefferson Islands Club. *The Weather Gauge*, 24(2):10-21.

Bryan, P., 1962. Sharp's Island Hotel doomed from the start. *The Star-Democrat Weekend Magazine*, November 10, 1962, pp. 10-11.

Cronin, W.B., 1988. Barren Island. *Chesapeake Bay*, (May), pp. 56-57, 66-68.

Dennis, K.C., R.J. Nicholls, and S.P. Leatherman, 1990. *Sea Level Rise Impacts in Developing Counties*. Report of the U.S. Environmental Protection Agency's International Project. USEPA, Washington D.C., pp. 5-14.

Emanuel, K.A., 1987. The dependency of hurricane intensity on climate. *Nature*, 326:483-85.

Gehrels, R., and S.P. Leatherman, 1989. *Sea Level Rise: Animator and Terminator of Coastal Marshes*. Vance Bibliography, Monticello, Ill.

Hicks, S.D., 1978. An average geopotential sea level series from the United States. *Journal of Geophysical Research*, 83:1377-79.

Intergovernmental Panel on Climate Change, 1990. *Potential Impacts of Climate Change*. Working Group II, various numbered pages.

Jacobsen, J.L., 1988. Environmental refugees: A yardstick of habitability. *Worldwatch Paper*, 86, 46 pp.

Kasowski, K., 1989. The rising Chesapeake.

Chesapeake Newsletter. Alliance for the Chesapeake Bay (July-August issue).

Leatherman, S.P., 1986. Cliff stability along western Chesapeake Bay, Maryland. *Marine Technology Society Journal*, **20**:28-36.

————, 1989. Impacts of accelerated sea level rise on beaches and coastal wetlands. In: *Global Climate Change Linkages*. J.C. White (Ed.), Elsevier Science Publ. Co., Amersterdam, pp. 43-57.

Meyer, E.L., 1986. *Maryland Lost and Found*. The Johns Hopkins University Press, Baltimore, Md.

National Research Council, 1987. *Responding to Changes in Sea Level: Engineering Implications*. National Academy Press, Washington D.C.

Orson, R., W. Panateogou, and S.P. Leatherman, 1985. Response of tidal salt marshes of the U.S. Atlantic and Gulf coasts to rising sea levels, *Journal of Coastal Research*, **1**:29-37.

Preston, D.J., 1982. Sharp's Island: Gone and forgotten. *Tidewater,* p. 45.

Rummel Consultants, 1971. *Report on Shore Erosion Control at Poplar Island, Talbot County, Md.* Baltimore, Md.

Shultz, M., 1972. State's "Plymouth Rock" is in trouble. *Baltimore Evening Sun* (July 17), p. 7.

Singlewald, J.T., and T. Slaughter, 1949. Shore erosion in Tidewater, Maryland. Maryland Department of Geology, *Mines and Water Resources Bulletin,* **6**, Baltimore, Md.

Small States Conference on Sea Level Rise, 1989. Male declaration on global warming and sea level rise. Male, Maldives, November 14-18, 1989.

Tickell, Crispin, 1989. Environmental refugees. National Environment Research Council Annual Lecture, The Royal Society, London, June 5, 1989.

Vellinga, P., and S.P. Leatherman, 1989. Sea-level rise: Consequences and policies. *Climatic Change,* **15**:175-89.

3. THE DELTA REGIONS AND GLOBAL WARMING: IMPACT AND RESPONSE STRATEGIES FOR BANGLADESH

Fasih Uddin Mahtab

Global change within the next century will be particularly hard-felt in densely populated deltaic areas like Bangladesh, where a substantial area is barely above sealevel and vulnerable to frequent natural calamities like droughts, floods, tropical cyclones and tornadoes. Unfortunately, our ability to deal with the impact of global change, particularly climate change due to global warming, will remain limited unless we can significantly extend our knowledge and broaden the range of practical policy responses.

People and Economy

Bangladesh is a heavily populated nation. Presently, more than 115 million people live huddled together in a territory of 144,000 square kilometers (km^2). By the year 2000 and 2030 the population of Bangladesh is expected to increase to 145 million and 232 million, respectively (World Bank, 1990). With a per capita income of only U.S.$170 per annum, Bangladesh is also one of the poorest nations on earth. Its limited resource base, large population and vulnerability to frequent natural disasters have all contributed to widespread and apparently increasing poverty.

Real GDP has increased only at an average rate of 3.19 percent annually since 1981, equivalent to an average growth rate of 0.79 percent per capita per annum (Mahtab, 1990). Moreover, this growth stemmed mainly from expansion of the service and construction sector; the contribution of productive sectors, such as agriculture and manufacturing, has been limited (World Bank, 1990). Agriculture constitutes about 50 percent of GDP and employs nearly 60 percent of the civilian labor force. Its growth, therefore, critically affects the other sectors. The manufacturing sector, though very small (only 10 percent of GDP), is equally important in creating employment opportunities for the growing work force and in the diversification of exports. The growth performance of the economy could be substantially improved if the potential of these two sectors were realized.

Physical Environment

Contrary to common belief, Bangladesh includes varying environmental conditions that provide a wide range of opportunities for, and limitations to, agricultural development. The physical environmental factors relevant to land use and agricultural potential are (1) landform, (2) soil, (3) inundated region and (4) climate.

Geography

Bangladesh consists primarily of a large alluvial basin lined with quaternary sediments deposited by three major rivers—the Ganges, Brahmaputra and Meghna—and their numerous associated streams and distributaries (see Figure 3.1). The country can be divided geomorphologically and physiographically into three major zones: floodplains, terraces and hills. In a recent study 34 physiographic units and subunits were identified (FAO 1988, p. 5).

The floodplain area, which comprises 79 percent of the land area, can be classified as piedmont plains, tidal and estuarine floodplains, and recent floodplains. These floodplains have very gently undulating relief with broad and narrow ridges alternating with depressions. Differences in elevation between adjoining ridges and depressions ranges from around 1 m on tidal floodplains near the coast to 2-4 m over much of the Ganges and Brahmaputra river flood- plains, and as much as 5 to 6 m in the Sylhet basin in the northeast.

Pleistocene terraces, which occupy about 8.3 percent of Bangladesh's land area, are restricted to two main regions—the Barind Tract in the northwestern region and the Modhupur Tract in the central region of Bangladesh. These terrace zones are formed on the oldest quaternary alluvium. Both tracts lie predominantly 1 to 5 m above the adjoining floodplain and less than 20 m above mean sea level (MSL). However, the western edge of the Barind Tract is 42 m above MSL at its highest point.

The hills, which occupy 12.7 percent of the land area, are located in the eastern part of Bangladesh, mainly in the Chittagong Hill Tract. The altitude of some of the ridges is about 1,000 m, with lower adjoining hills less than 100 m above MSL.

FIGURE 3.1. Map showing Bangladesh with generalized relief contours in meters. Insert shows the percentage of land area at different elevations relative to sea level.

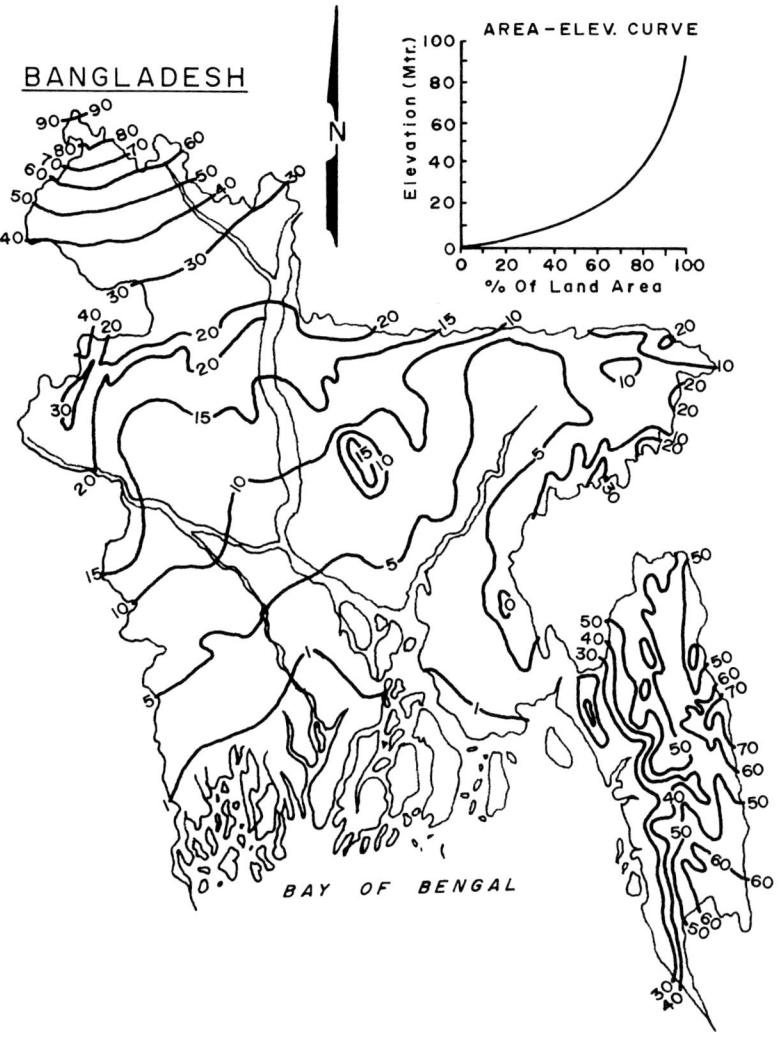

Subsidence is a common earth movement in Bangladesh, but there may also be some uplifting. Major areas of subsidence are the Surma Basin, Faridpur Trough, Chalan Beel, Dhaka Depression and the Khulna-Sundarbans area. Calculated values of subsidence range from 0.6 mm to 5.5 mm annually, but may exceed 20 mm annually in the central part of the Surma Basin (MPO Report 4, 1987).

A recent study by FAO identified 21 general soil types (FAO, 1988, pp. 29-33) by grouping soils that are broadly similar in appearance and characteristics as a result of development in response to similar environmental factors. The four main categories of general soil type account for 65 percent of Bangladesh's total area. They are: noncalcareous grey floodplain soil (27.52 percent), calcareous dark grey floodplain soil (11.65 percent), noncalcareous dark grey floodplain soil (13.00 percent) and brown hill soil (12.70 percent). There are six different types of terrace soil, of which shallow grey terrace soil (2.2 percent) and deep grey terrace soil are the main types (FAO 1988, p. 30).

Climate

Bangladesh lies in the tropical monsoon region to the south of the eastern Himalayas. These mountains act as a barrier providing protection from the rigors of the cold Siberian winds. The tropical monsoon climate is characterized by heavy precipitation during the rainy season and little or no precipitation during the dry winter months. The wind direction reverses from northeast in winter to southwest in summer.

Bangladesh has seasonal rainfall divided into four periods. During the pre-monsoon period (March-May) rainfall varies from 200 mm to 900 mm, or more. During the monsoon (June-September) and post-monsoon (October-November) periods the rainfall varies from 1,250 mm to 3,700 mm and 100 mm to 300 mm, respectively. During the winter months (December-February) rainfall accumulation is less than 90 mm (BBS, 1989, 30-34).

Like rainfall, temperature in Bangladesh also exhibits seasonal variation. Normal maximum temperature in the summer months varies between 30.5°C and 35°C, with temperatures occasionally exceeding 37.7°C during April and May; temperatures during the summer are seldom lower than 21.7°C. The winter months are characterized by a minimum temperature seldom falling below 10°C and then only in certain areas, such as Chittagong Hill Tracts, south Sylhet and North Bengal. Even in these areas, however, maximum temperatures generally exceed 23.9°C (BBS, 1989, pp. 30-34).

Length of Crop Growing Periods

Temperatures in Bangladesh are suitable for crop cultivation throughout the year. Therefore, the parameter that mainly determines the natural length of the growing period is soil moisture. There are three main rain-fed crop growing periods (FAO, 1988, pp. 71-80):

- Pre-kharif transition period in Bangladesh is characterized by unreliable rainfall, varying in timing, frequency and intensity from year to year. This period starts after the end of February, when precipitation first exceeds half of potential evapotranspiration (PET) and ends when precipitation continuously exceeds 0.5 PET.
- The kharif (monsoon crop) growing period begins when precipitation continuously exceeds 0.5 PET and ends on the date when the combination of precipitation plus an assumed 100 mm of soil moisture storage falls below 0.5 PET. The mean starting date ranges from April 1-10 in the north-east to May 20-31 in the extreme west and the mean ending date ranges from December 1-10 in the northwest to January 1-10 in the northeast.
- The Rabi growing period is the time between the end of the humid period (when rainfall is greater than PET) and the time when 250 mm of soil moisture has been exhausted by evapotranspiration. The starting date ranges from October 1-10 in the extreme west to November 1-10 in the northeast and in central and eastern coastal areas. The mean ending date ranges from February 1-10 in the extreme west to March 20-31 in the northeast.

On most floodplain and valley land, cropping patterns are primarily determined by seasonal flooding regimes. Farmers' traditional cropping patterns and practices are adapted to flooding regimes on a microtopographical scale: differences of only a few centimeters between neighboring fields may influence the choice of crops, varieties or management practices. Distributions of seasonal flooding regimes in terms of depth of flooding are given in Table 3.1.

Two kinds of thermal zones are important in agriculture. The five major zones are defined by the length of the cooler winter period. Subzones are defined by the number of days with extremely high summer temperatures (FAO, 1988, pp. 81-82).

River System and Land Accretion Potential

Bangladesh is a flat deltaic area of land formed at the confluence of three major rivers: Ganges-Padma, Brahmaputra-Jamuna and Meghna. The total area of the basin is 1,758,000 km^2, of which 62 percent is in India, 18 percent in China, 8 percent in Nepal, 4 percent in Bhutan, and only 8 percent in Bangladesh. The combined Ganges, Brahmaputra and Meghna rivers in the last 100 km of their run to the sea form a river 2.5 times the size of the Mississippi (Rogers, 1989, p. 57).

The Ganges River enters Bangladesh from the west near the town of Rajshahi, joins the Brahmaputra at Goalanda, and the combined flow travels southeast with the name Padma till it joins the Meghna River at Sureswar. The 20-year peak discharge in the Ganges at Hardinge Bridge and Goalanda is about 71,400 m^3/sec and 111,000 m^3/sec, respectively (MPO, 1987).

The Brahmaputra is one of the world's most turbulent and dynamic rivers. The Brahmaputra-Jamuna system gathers drainage water from the eastern Himalayas and Assam ranges—an area of exceedingly heavy rainfall. During the great floods of 1787, this river was divided into two channels—the Jamuna, flowing south, became the main channel, while the old Brahmaputra became the left bank distributary. The 20-year peak discharge at Bangladesh is 80,000 m^3/sec (MPO, 1987).

The Meghna River originates with the name of Barak and divides into two branches. The northern branch flows downstream with the name Meghna. At Sureswar it joins the Padma River, then flows and discharges into the Bay of Bengal, assuming the name of Lower Meghna. The 20-year peak discharge in the Meghna at Sureswar is about 18,000 m^3/sec (MPO, 1987).

Presently the Ganges, Brahmaputra and Meghna carry a huge amount of annual sediment (about 1.00 to 1.48 billion tons during wet years), which replenishes one of the world's largest deltas. Each year, during monsoon season, about 18 percent of the country is flooded, depositing a part of the sediment. It is believed that a delicate equilibrium exists between building of the floodplains with floodplain deposits and

TABLE 3.1. *Extent of Inundation Land Types*

Land type	Normal flood depth	Area (Ha)	Proportion (%)
Highland (H)	0 - 30 cm	4,199,952	29
Medium Highland (MH)	30 - 90 cm	4,039,724	35
Medium Lowland (ML)	90 - 180 cm	1,771,102	12
Lowland (L)	180 - 300 cm	1,101,560	8
Very Lowland (VL)	> 300 cm	193,243	1
Total soil area :		11,305,581	85
River, Urban, Homesteads, etc.		2,178,045	15
Total :		13,483,626	100

Source: FAO, 1988.

relative sea-level rise caused by exustatic sea-level rise (at present some 10 cm per century), compactions of the clayey deposits and tectonic subsidence of the area (MPO, Report 4, 1987).

Natural Hazards: Cyclones, Floods, Droughts and Tornadoes

Cyclones

Bangladesh's coastal regions are subjected to damaging cyclones almost every year. Because of the high density of population occupying a flat deltaic area north of the funnelling tidal estuaries, the loss of human life and property has at times been very great. The severe killer cyclones develop during spring and autumn, of which 75 percent occur between April 15 to June 15 and September 15 to December 15 (see Table 3.2). A comparative study of the monthly frequency of tropical cyclones for the period 1891-1960 and the subsequent 14 year period (1961-1974) clearly indicates that, in every month, the frequency of tropical cyclones during the latter period (i.e., 1961-1970) is significantly greater than that of the earlier period, as shown in Table 3.3 (Chowdhury, 1977). Although the tropical cyclone frequency during the 1975-1988 period is slightly less than that of the 1961-1974 period (except in the month of November), it is higher than the period between 1891 and 1960.

Storm surges associated with tropical cyclones are serious problems in the coastal areas of Bangladesh. Records show that the world's most pronounced storm surge disasters are observed in the Bay of Bengal. The impact of the cyclone of November 1970 was particularly severe. A wave of up to 9 m was produced by this cyclone and is thought to have killed over 300,000 people (BBS, 1989, p. 27).

Floods

In Bangladesh flooding is very much a part of the normal cycle of the seasons. A delta cannot develop physically without flooding, and it is to floodwaters bearing plant nutrients in the form of dissolved and suspended solids that much of Bangladesh's

TABLE 3.2. *Monthly Distribution of Tropical Cyclones in the Bay of Bengal*

	J	F	M	A	M	J	J	A	S	O	N	D	Total
Moderate	3	0	2	11	10	30	31	24	19	34	33	17	214
Severe	1	1	2	7	18	4	7	1	8	19	23	9	100
Total	4	1	4	18	28	34	38	25	27	53	56	26	314

Source: Ananthakrishnan and Rao (1964) as quoted in Chowdhury MHK (1977).

TABLE 3.3. *Frequency of Tropical Cyclones with Winds >/ 35Knots* (Number per Month)

Period	April	May	Sept.	Oct.	Nov.	Dec.	Total
1891-1960[a]	0.26	0.40	N/C	0.76	0.80	0.37	2.70
1961-1974[b]	0.14	0.86	0.79	1.07	1.29	0.64	4.00
1975-1988[b]	0.07	0.50	0.29	0.86	1.36	0.36	3.86

Sources: [a]Anathakrishnan and Rao, (1964) as quoted by Chowdhury (1977).
          [b]Calculated by Chowdhury, Director, Bangladesh Meteorological Dept., Dhaka.
Note: not comparable,

fertility can be attributed. Each year about 26,000 km^2 (18 percent of the country) is flooded. During severe floods, the affected area may exceed 52,000 km^2 (i.e., 36 percent of the country and nearly 60 percent of the net cultivable area). In an average year 775 billion cubic meters of water flow into the country from June to September through the three main rivers.

For two consecutive years, 1987 and 1988, Bangladesh was deluged by exceptionally severe floods. Both these catastrophic floods attracted worldwide attention and concern. The 1988 flood was generated by intensive rainfall extending over the northeast of the subcontinent. The flood peak of the Brahmaputra was the highest ever recorded. In an unusual coincidence, the flood peak of the Ganges also occurred, with devastating effects on the country. The size of the 1988 flood peak of the Brahmaputra at Bahadurabad was on the order of a 100-year event (FEC, 1989).

The effects of the 1988 flood were devastating. It inundated more than 90,000 km^2 of land area, affecting nearly half of the 110 million population, with 2,300 deaths. About 1.6 million tons of the standing monsoon rice crop was damaged. Many schools, houses, livestock, telecommunications, roads, railways and bridges were damaged or destroyed. Production in much of the country came to a standstill. Lines of communication were disrupted for over a month. Capital stock losses were well over U.S.$1 billion and GDP growth was set back severely.

The occurrence of catastrophic floods in two consecutive years necessitated close examination of the possibility of the simultaneous occurrence of major peaks in the two rivers involved. In a recent study (FEC, 1989), a detailed analysis involving the date of occurrence of the major yearly peak has been carried out for several stations of significant interest. Major findings of the study follow.

- A first flood event occurs frequently around August 8, typically associated with the normal occurrence of floods on the Brahmaputra.
- A second flood event happens classically around September 2, which corresponds to the period of the Ganges annual peak.
- The likelihood of having simultaneous

high stages on both of the rivers is quite high, and it depends on the date of occurrence of the Brahmaputra peak flow.

Drought

The consequences of a drought can be as far reaching and disastrous as the effect of a major flood. In 1975 Bangladesh had a major drought when about 47 percent of the area and 53 percent of the population were affected. Bangladesh also experienced severe drought conditions in 1951, 1957, 1958, 1961, 1966, 1979, 1981, 1982 and 1989 (Chowdhury and Hussain, 1982).

Although annual rainfall ranges from 1400 mm in the dry Rajshahi (northwest) region to over 5,000 mm in the wet Sylhet (northeast) region, about 80 percent of the precipitation generally occurs during the four months from June to September. Therefore, droughts of varying intensity occur in almost all parts of Bangladesh during the eight months from October to May. Droughts are particularly severe in the northwest and southwest regions of Bangladesh, where monsoon rain occurs for about three months, as compared to five months in the northeast. The areas suffering different intensities of drought for one in five years during the July-October period are shown in Figure 3.2.

The effect of soil-moisture on crop yield depends on:

- the stress day index (SDI), the ratio of number of stress days during a particular growth stage of the crop and duration of the growth stage;
- the crop susceptibility factor (CS), which indicates crop susceptibility to a given water deficit at each stage of development;
- normal achievable yields in the farmer's field; and
- the stress days (SD), the number of days when evapotranspiration is less than PET.

For the present study the SDI and stress index for the HYV transplanted Aman paddy rice (monsoon crop) were calculated using the actual rainfall days of 1980. Calculations were carried out for four different locations having different intensity of drought. Stress index values for HYV transplanted Aman were found to be highest for all locations during the milk to maturity period, when grain filling and starch accumulation occurs. The loss of yield for the HYV transplanted Aman paddy in Rajshahi, Jessore, Comilla and Sylhet districts was found to be 52.4 percent, 41.2 percent, 13.2 percent and 8.2 percent, respectively. In another study (Karim *et al.*, 1990), it was found that the effect of drought on HYV transplanted Aman (kharif crop) and wheat (Rabi crop) is greater than on Broadcast Aus (pre-kharif crop).

Tornadoes

A tornado is an intense atmospheric vortex, consisting of violently rotating, tall, narrow column of air that occasionally extends to the ground from a cumuliform cloud. Tornadoes are more common in the central part of Bangladesh, and 76 percent of them occur during the pre-monsoon period (see Table 3.4).

A severe tornado razed some parts of the Dhaka district on April 14, 1969. According to the official version, the tornado killed 922

FIGURE 3.2. Map of Bangladesh showing drought-prone areas. Reference crop T Amon; months July to October. (Source: Karim and Ibrahim, BARC, 1990).

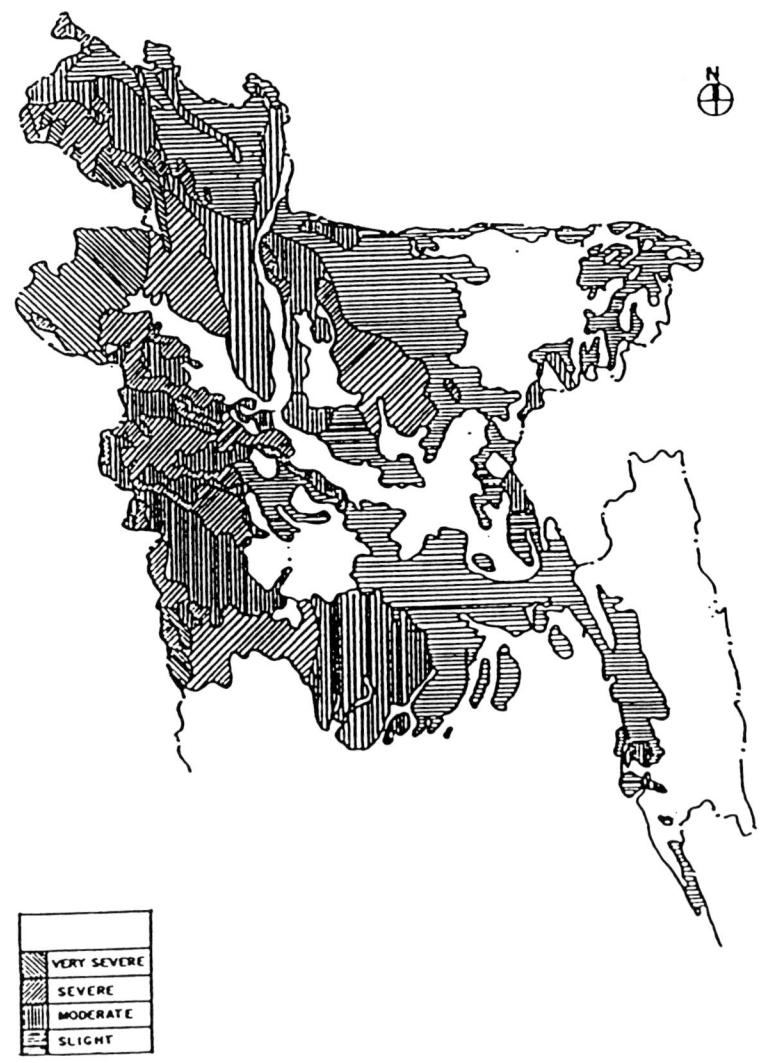

VERY SEVERE
SEVERE
MODERATE
SLIGHT

people; 16,511 were injured, 119,944 houses were completely destroyed, and 125,876 families were affected (Chowdhury, 1985). A more recent tornado (April 26, 1989) affected 6 upazilas of Manikganj, Dhaka and Tangail districts and left 800 people dead, over 100,000 people homeless, and more than 10,000 injured. The devastation wrought by this tornado was so colossal that, barring some tree skeletons, there were no signs of standing infrastructure anywhere in those areas most affected.

Impact of Greenhouse Effect on Bangladesh

Possible Scenario

The major hindrance in assessing the impact and possible response strategies to global warming is the lack of detail in the forecast of climatic change at the regional level. Under these circumstances, the predictions of experts can only be used as guidelines. According to one group of experts, who convened in Villach in 1987 (WMO, 1988), by the middle of the next century the greenhouse gases (GHG) will warm the humid tropical region by 0.3 to 5°C, causing the following changes.

- A rising water level along coasts and rivers, resulting from a combination of increasing sealevel, a greater chance of tropical storm surges and increasing peak runoff, will result in larger areas being subject to flooding and a greater risk of salinization.

TABLE 3.4. *Seasonal Distribution of Northwesters and Tornadoes (1975-1979)*

Season	Annual distribution	Percent
Winter	16	3
Pre-monsoon	397	76
Monsoon	89	17
Post-monsoon	21	4
Total	523	100

Source :   Chowdhury, 1985.

- Changing spatial and temporal distribution of temperature and precipitation will have an impact on industry, settlements, agriculture, grazing lands, fisheries and forests.
- An increase in rainfall in the range of 5 to 20 percent, largely through increases in rainfall intensity, will heighten the chances of major floods.
- Because warming will increase the potential evapotranspiration, there could be a tendency toward more drought stress.

Consequences of Sea-Level Rise

According to recent estimates of the effects of global warming, sealevel is expected to rise on the order of 30 cm, although this could be as much as 1.5 m by the middle of the next century (WMO, 1988), depending upon attempts to control GHG emissions. The sea-level rise along a specific coast will depend both on regional and local geological movement, and global sea-level rise. For the present study, a scenario with a 1 m rise in sealevel by the middle of next century is postulated as a result of a 90 cm (average of 30 cm and 1.5 m) overall rise in sealevel and about 10 cm due to subsidence (see earlier discussion).

A sea-level rise of 1 m will inundate about 5.608 million acres (22.889 km^2) of existing coastal land, about 15.8 percent of the total area of Bangladesh (see Figure 3.3). The area corresponds to 13.74 percent of net cropland and about 401,600 hectares of mangrove forest along with its wildlife. The inundation of 2.915 million acres of net cropland will cause a production loss of more than 2 million tons of rice, 13,000 tons of wheat, 214,000 tons of sugarcane, 405,000 tons of vegetables, 10,000 tons of jute and 97,000 tons of pulses. As can be seen in Figure 3.3, with a 1 m rise in sea level, the entire 401,600 hectares of mangrove forest (Sundarbans), as well as 36,000 hectares of newly established mangrove forests along the coast, will be gradually destroyed.

Already the western part of Sundarbans, which has been subjected to a progressive decline in freshwater supply (mainly due to diversion of substantial amount of Ganges water by Farraka Barrage), has seen an

FIGURE 3.3. Streamflow iso-saline map showing mean monthly maximum salinities. The shaded area shows the land that would be submerged with a 1 m rise in sea level. A: existing 1,000 μmhos isosaline line; a with 1 m sea level rise. B: existing 2,000 μmhos isosaline line; b with 1 m sea level rise. C: existing 10,000 μmhos isosaline line; c with 1 m sea level rise.

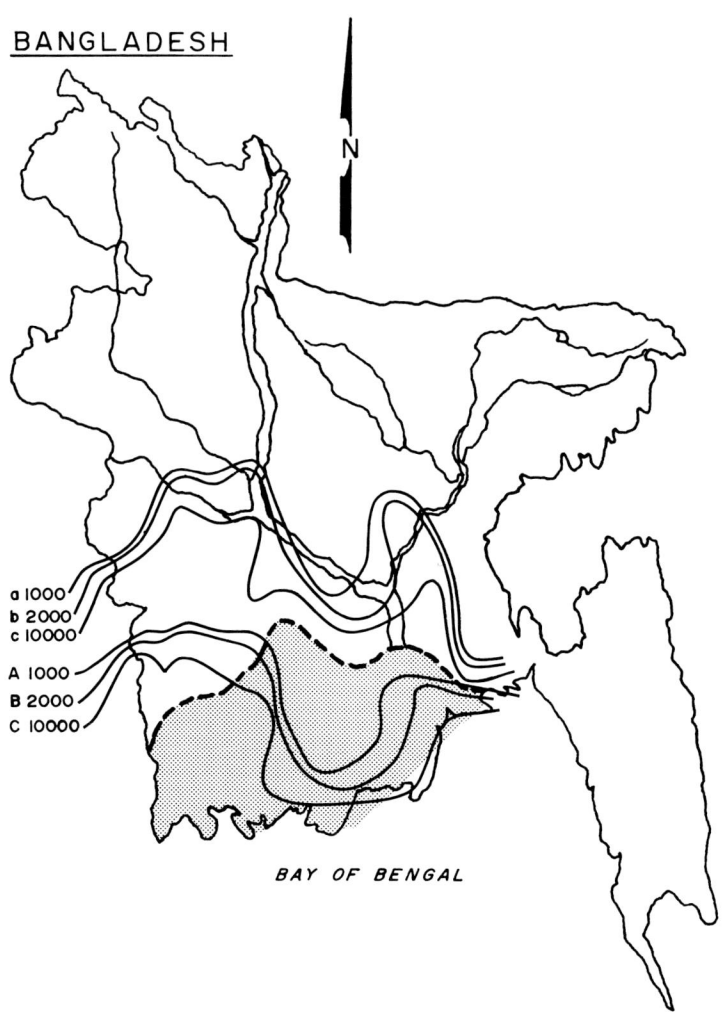

increase in salinity, which in turn reduced the regeneration rate of Sundari. As a habitat for wildlife, Sundarbans is unique. Its forests and numerous waterways support a wide range of mammals (including Royal Bengal tigers) birds, amphibia, fish, reptiles and crustacea. With a 1 m sea-level rise most of these animals will lose their habitat and may become extinct.

Monetarily, the total loss of assets and production in both small and cottage industries in takas (Tk) could be Tk 1,078 million (U.S.$33.5 million) and Tk 98.553 billion (U.S.$3 billion), respectively. About 10 percent of Bangladesh's 115 million population (about 2.05 million households) will be displaced and will not have any option but to migrate to unaffected urban areas, especially to major cities like Dhaka, Chittagong, Khulna and Rajshahi; they will live in perpetual poverty. The loss to housing and physical infrastructure will be extensive: about 1.9 million homes, 8,300 schools, 180 health centers, 1,470 km of railways, 10,300 bridges and culverts, 700 km of paved roads 19,800 km of unpaved roads, 375 food/fertilizer godowns (warehouses) and 1,760 markets. Output loss is estimated at about 13 percent of GDP and loss of assets about Tk 450 billion (U.S.$14 billion) at 1984-1985 prices.

During the dry season, when the streamflow of the three major rivers and their tributaries diminish greatly, there is substantial inland penetration of salinity in the coastal areas through the complex estuarine river systems. This salinity intrusion limits opportunities for irrigation of Rabi and pre-monsoon crops in the coastal freshwater areas and damages the same crops by flooding during high tides. Moreover, the large withdrawal of Ganges water by India during the dry season has caused considerable penetration of salinity inland.

With sea-level rise there will be further penetration of salinity inland. At this stage, it is extremely difficult to assess the extent of high salinity penetration with reasonable accuracy. However, a very rough assessment indicates that, with a 1 m sea-level rise, the area of high-salinity intrusion will increase from an existing 13 percent of Bangladesh land area to 32 percent (see Figure 3.3). In fact, the entire southern and southwestern part of Ganges-Padma-Lower Meghna river system will be affected by high salinity. This will reduce the crop-yield substantially in the affected areas. Salinity intrusion may affect power stations and industrial water use in the Siddhirganj and Narayanganj areas.

As a result of backwater effects, sea-level rise will cause an elevation in water levels of various rivers and streams. The surface water simulation study for a 1 m sea-level rise indicates that the water level of the Ganges and Jamuna rivers will be affected up to Aricha (confluence of the Jumuna and Ganges rivers) and up to the confluence of the Kangsa and Surma-Boulai rivers in the case of the Meghna rivers. As a result of this rise in water level, the floodplains and low-lying areas of another 20 percent of the land area of Bangladesh will be inundated, and considerable amounts of cropland now suitable for transplanted Aman rice will be lost. Moreover, both the duration and extent of flooding in these areas will increase.

The Coastal Embankment Project in Bangladesh is mainly a flood control and drainage project comprising 58 polders. The project features include embankments and drainage cum flushing sluices with vertical lift or flap gate arrangements. With sea-level rise, the water level outside the polders will rise proportionately and consume either part or whole of the freeboard. Strengthening and increasing the height of the embankments could require about Tk 18.26 billion (U.S.$562 million) at 1984-1985 prices.

Change in the Intensity and Frequency of Climatic Events

As mentioned earlier, in a humid climate like that of Bangladesh, there will be changes in spatial and temporal distribution of temperature and precipitation, which in turn will increase both the intensity and frequency of extreme events like droughts, floods, tropical cyclones and tornadoes.

The recent occurrence of a series of wet and dry years with climatic events of greater intensity raises the question as to whether identifiable trends with respect to changing rainfall patterns exist. Such fluctuations have implications for water resources and agriculture and, if such trends exist, it may be dangerous to rely on short term rainfall records in planning the development of water resources and agriculture. It may be noted that changes in runoff will not be directly proportional to changes in the amount of rainfall. Here, the rate of change is likely to be critical; rapid changes in rainfall due to global warming will have very serious implications.

For the present study, more than 85 years of rainfall data for six different stations (Dhaka, Sylhet, Kushtia, Barisal, Rajshahi and Rangpur) were analyzed. Results of the linear regression of rainfall data indicate that four (Dhaka, Sylhet, Barisal and Rangpur) out of six selected locations depict an upward trend, and the other two stations (Rajshahi and Kushtia) indicate a downward trend. It seems that rainfall is increasing in the wetter areas and decreasing in the drier parts of the country (see Figure 3.4). When the rainfall data were analyzed for seasonality, it was found that during the monsoon period, except for Kushtia, all stations indicated increasing trends. For dry and pre-monsoon seasons, a decreasing trend in two out of six stations and three out of six stations, respectively, was evident. The analysis of rainfall variation indicates that the coefficient of variation is more in the drier areas, as well as during pre-monsoon and post-monsoon periods. These changes have been more pronounced in the recent past .

Both the magnitude and the frequency of heavy rainfall were analyzed by assuming 43 mm or more rain in 24 hours to be heavy rainfall. Results of the analysis indicate that for Bangladesh both frequency and magnitude (mean monthly) are maximum during the months of June and July. Results of the linear regression analysis indicate that the annual frequency of heavy rainfall for Dhaka, Sylhet and Rangpur (wetter areas) depicts an increasing trend (see Figure 3.5), whereas for Rajshahi and Barisal (relatively drier areas) it depicts a decreasing trend (Mahtab, 1989).

The above results indicate that there is a change in spatial and temporal distribution of rainfall. However, it is not possible to prove, at this stage, whether these changes are due to global warming. But this change in spatial and temporal distribution of rainfall is likely to have a short-term effect on agricultural productivity through (1) variation in seasonal crop yield, (2) season-to-season yield variability, (3) winter season yield variability, (4) variations in yield quality, (5) variation in yield responsiveness to fertilizer, (6) carryover effects from a previous season and (7) combined impact on more than one system that may induce a nonlinear response in agricultural output.

Changes in climate are also likely to have long-term effects on agriculture through (1) changes in mean yields that are positively related to average seasonal precipitation, (2) changes in the level of crop yield dependability and (3) concurrent effects on soil fertility. In addition to the effect of changes in rainfall on agricultural

FIGURE 3.4. Annual rainfall at Sylhet and Dhaka for the period 1901-1988. Annual values are shown together with regression analyses for two selected periods, 1901-1988 (solid line) and 1948-1988 (dashed line).

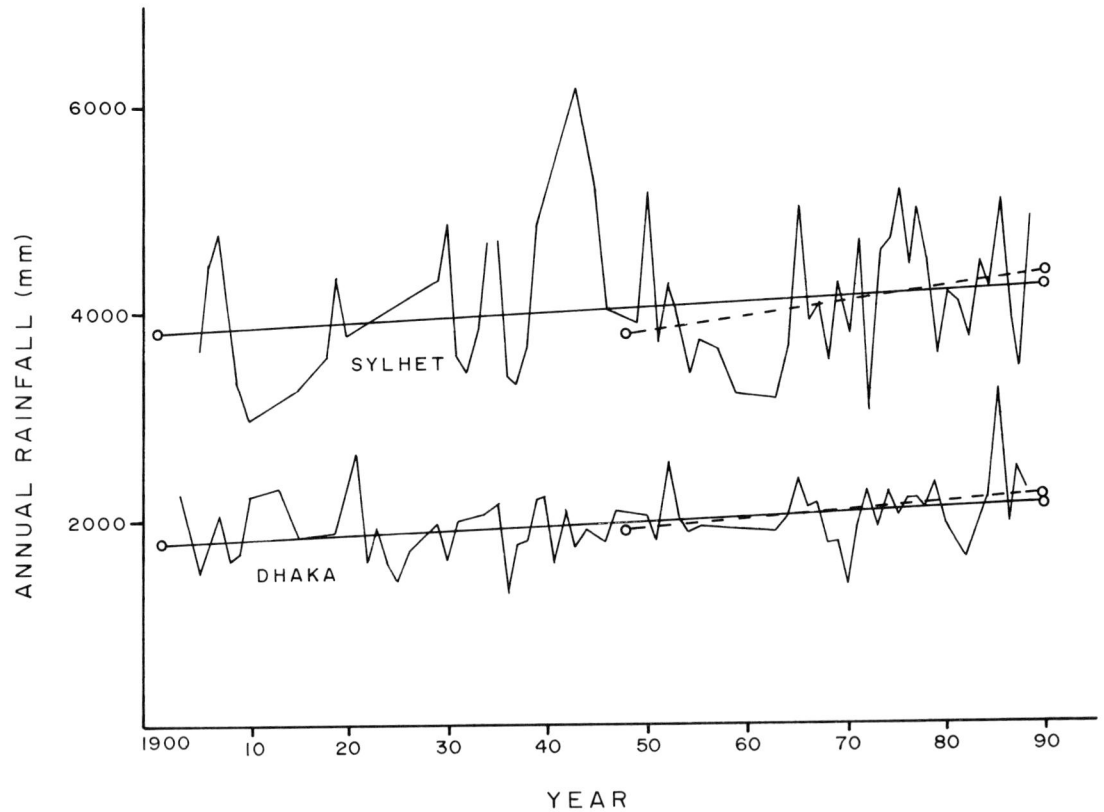

FIGURE 3.5. The annual frequency of heavy rainfall at Dhaka for the period 1953-1988.

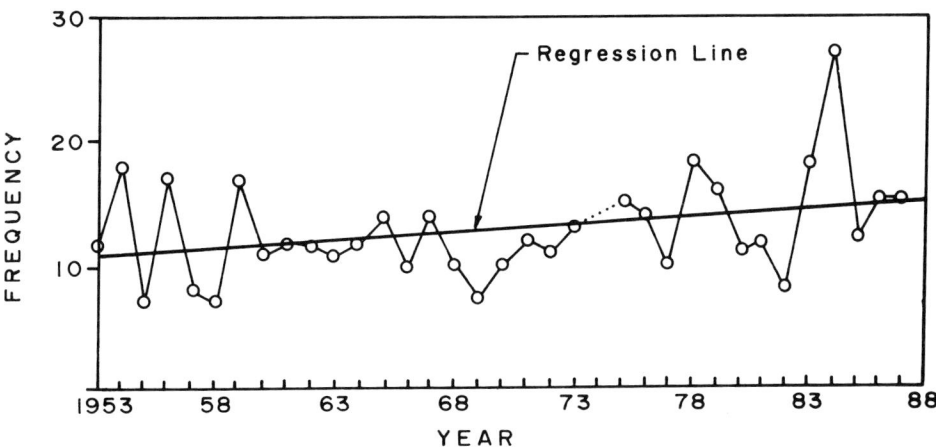

productivity at a given location, changes in spatial and temporal distribution of precipitation will cause both short-term and long-term shifts in the spatial pattern of agricultural potential and risk.

The above analysis of rainfall data indicates that:

- Annual rainfall in the western part of Bangladesh is about 1,400 mm and it is more than 5,000 mm in the northeast region.
- The amount of rainfall is increasing in the wetter areas of Bangladesh and decreasing in the drier areas.
- The coefficient of variation of rainfall is more in the drier areas and during the pre-monsoon and post-monsoon period.
- Changes in the magnitude of rainfall are mainly due to changes in intensity of rainfall (changes in frequency of heavy rainfall).
- On average, about 80 percent of the rainfall occurs during the four months from June to September.
- The duration of the monsoon period is shorter (about three months) in the western part of Bangladesh than in the northeast (about five months).

A decrease in rainfall, a higher coefficient of variation of rainfall, and a shorter monsoon period in the already drought-prone western part of Bangladesh are likely to increase both the frequency and intensity of drought stress. In all probability, drought-prone areas will also extend toward the southern and south-central region of Bangladesh and reduce the crop yield, especially transplanted Aman, wheat and Broadcast Aus, even more than that mentioned earlier.

Recent trends in Bangladesh indicate that the frequency of major floods is increasing. The area affected by major floods has increased from 50,500 km^2 in 1955 to 90,000 km^2 in 1988. Moreover, it seems that both the magnitude and the intensity of rainfall are increasing in Bangladesh's northeastern region. Although there is a lack of access to the rainfall data of the northeastern part of India, Nepal and Bhutan, it is quite probable that the rainfall in these areas has increased like that of northeastern Bangladesh. If that is the case, it is likely that both the intensity and the frequency of exceptionally high peaks in the Brahmaputra will increase. Moreover, as pointed out previously, because the likelihood of simultaneous peaking in both the Ganges and Brahmaputra rivers is high, the probability of having major floods in the future is also greatly increased.

Storm surges associated with tropical cyclones are serious problems in the coastal areas of Bangladesh. The average annual frequency of tropical cyclones in the Bay of Bengal ranges between 12 and 13, out of which 5 attain cyclonic strength (wind > 35 knots). As mentioned in the section on cyclones, a comparative study of the monthly frequency of tropical cyclones in the Bay of

Bengal for the period 1891 to 1960, as compared to the subsequent 14 year period (1961 to 1974), clearly indicates that the monthly frequency of tropical cyclones during the latter period (1961-1974) is significantly greater than that of the 1891-1960 period. Although the monthly frequency betwen 1975 and 1988 is slightly less than the 1961-1974 period (except for the month of November), it is higher than for the period between 1891 and 1960.

Response Strategies and Policy Recommendations

Key Issues

While formulating appropriate response strategies, it is necessary to consider the following key issues along with the possible impacts of global warming:

- Population pressure is already intense and will mount further. Thus, a dramatic increase in food production is required for which a more secure environment is essential.
- Enormous hydropower potential exists (Roger, 1989, pp. 10-15) in the mountains upstream of both the Ganges and Brahmaputra river systems, and this should be developed and used regionally.
- The Ganges and Brahmaputra rivers carry more than a billion tons of sediment every year and only a small part is now deposited in the floodplain. Experience shows that this natural sedimentation process can be speeded up with appropriate human intervention.
- The present practice of flood control by embankment limits the supply of water necessary to provide soil-moisture for the Rabi crops, recharge of beels and groundwater, supply of nutrients, and landbuildup in the low-lying floodplains and depressions.
- Recent studies indicate that only 39 percent of the land that is suitable for forestry has tree cover (Forestry, 1987). In Sundarbans, between 1959 and 1983, the standing volume of the two main species of trees, Sundari and Gewa, decreased on the order of 40 and 45 percent, respectively (ODA, 1985).

Moreover, village forest groves, which supply 80 percent of fuel-wood, 85 percent of timber and 90 percent of bamboo, are being gradually depleted.

The possible response strategies to global warming fall into two categories: adaptation strategies that will help to reduce the consequences of global warming, and limitation strategies that will enable us to control or stop growth of GHG emissions. Obviously, a prudent response would be to consider both types of strategies. Adaptation strategies can be subdivided into those requiring physical measures and those needing nonphysical measures. Most of the measures required for adaption to sea-level rise are physical in nature, whereas measures needed for adaption to climate change are of both the physical and nonphysical type.

Adaptation Strategies to Mitigate Sea-level Change

Three types of strategies could be employed to mitigate many of the consequences of sea level rise:

- Keeping high water out by constructing new or modifying existing embankments along the coast and river banks;
- Allowing the low-lying land to inundate and accelerate the siltation/land accretion process by appropriate intervention;
- A combination of the other two methods.

For the short and medium term, construction of embankments appears to be the most effective way to control flooding and prevent coastal inundation. However, embankments also limit the supply of silt-laden moisture required for Rabi crops, recharging beels and groundwater, supplying soil nutrients, and buildingup land in the low-lying floodplains and depressions. On the other hand, these disadvantages could be addressed by adopting the concept of controlled flooding. In essence, this implies that some amount of flooding may be allowed within the embankment in a controlled manner at an opportune time so as to allow the land buildup process to continue or even accelerate with appropriate interventions. For this, adequate control structures will be needed.

Almost all low-lying coastal floodplains, other than Sundarbans, are now protected against tidal flooding, saline water intrusion and river flooding. Prior to the Coastal Embankment Project (CEP), the river water used to enter into the numerous canals and creeks during high tides, and tidal water could enter into the low-lying floodplains. The silt carried by the river water could than ultimately settle on the low-lying floodplains. During lowtide, the tidal water free of silt came down the canals and creeks with tremendous force, clearing the river bed of excess silt. Since completion of the CEP, the silt-laden water is prevented from inundating the floodplains within polders. Consequently, silt is deposited in the river bed instead of the floodplains within polders. As a result, a substantial amount of land has now become uncultivable due to drainage problems.

Presently, the area where soil salinity exceeds 8,000 μmhos/cm remains fallow during the winter season. Usually only one crop of the local variety of transplanted Aman is grown and its yield is very low (about 1.5 mt/ha). On the other hand, at soil salinity levels exceeding 8,000 μmhos/cm there is a beneficial impact on shrimp cultivation, which is becoming more popular. Although available yield is only 100 kg/ha, a yield of 1 to 2 tons/ha is achievable with the adoption of improved practices. In the southwest, shrimp and rice can be grown as a two crop system, shrimp during winter and rice farming during the monsoon season. In fact, rice and shrimp are already grown as a two-crop system in the Satkhira, Khulna and Bagerhat districts.

Coastal areas outside the Coastal Embankment Project consist of Sundarbans and some islands. In these areas, the natural process of land accretion is proceeding. However, to mitigate loss of land due to sea-level rise, this process of land accretion should be accelerated by appropriate intervention.

As mentioned earlier, subsidence is a common occurrence in Bangladesh. Major areas of subsidence will be further inundated and the present natural process of siltation will not be enough to compensate for the combined effect of backwater and land subsidence. It will, therefore, be desirable to accelerate the process of land buildup in these areas by appropriate intervention. It will be unwise to undertake the usual flood protection measures of only constructing embankments in these areas.

To prevent further intrusion of salinity in the southern region of Bangladesh, it will be necessary to increase the fresh water supply in the Mathabanga and Gorai-Madhumati river systems. This will require implementation of the proposed Ganges and Brahmaputra barrages, which will (1) rejuvenate the southern distributaries of the Ganges-Padma River, (2) allow more freshwater supply in the southwest and south-central regions, (3) reduce and prevent further intrusion of salinity in the southwest and south-central regions, (4) increase irrigation facilities and (5) accelerate the land accretion process in the low-lying areas, including Sundarbans, thereby saving them and other low-lying coastal areas from inundation.

Adaptation Strategies to Mitigate Climatic Change

As discussed earlier, climatic change as a result of global warming will alter the spatial and temporal distribution of precipitation and temperature, and increase both the intensity and frequency of natural calamities.

Traditional farmers are well adapted to normal variations of climate. But the farmers may find it extremely difficult to adapt to the rapid change in climate that is likely to occur as a result of global warming. To mitigate the possible effects of this rapid climate change at the farm level, it will be necessary to diversify crop and employment opportunities and change cropping patterns, farming practices, irrigation regimes and household management practices.

At the national level it will be necessary to (1) adjust the food security system, (2) change resource allocation and input availability, (3) promote alternative employment, (4) develop new irrigation facilities, (5) reorganize agricultural infrastructure, (6) develop and introduce new crops suitable for changed or changing climate, (7) preserve genetic diversity and (8) provide improved information on agro-climatic potential. In addition to these measures, it will be necessary to undertake research in the field of agriculture, agro-climate, farming practice, etc., as well as socioeconomic aspects of climatic change.

Because the effects of floods, cyclones and tornadoes are more visible than droughts, we are normally more concerned about these three types of natural calamities. But the effects of drought on agriculture and water supply can be as great as those resulting from floods. To mitigate the effect of droughts on agriculture, more stress should be given to the development and introduction of drought resistant crops and the construction of more irrigation facilities.

The long-term approach for controlling floods in Bangladesh is building storage reservoirs behind dams on the mountains in the upstream reaches of the major rivers and their tributaries. This would not only provide water for winter irrigation and hydropower generation, but also reduce the flood stages in the river flows across the floodplains of Bangladesh. For the short and medium term, the construction of an embankment, with a provision for controlled flooding, seems to be the most effective way to achieve flood control. A recent report on flood policy also suggests a similar concept of controlled flooding (UNDP, 1989).

Regarding cyclones and tornadoes, nothing can be done at this stage other than the development and improvement of early warning systems and disaster preparedness plans and programs. For this, it will be necessary to develop appropriate nationwide telemetric networks, to elaborate remote sensing techniques, and to upgrade the capabilities of the manpower and institutional network.

Limitation Strategies

Although limitation of GHG emissions require a global effort, developed countries have to play a major role in reducing the emissions of CO_2 and other greenhouse gases. However, developing countries also have a role to play, because the growth in their energy demand in the future is likely to be very high.

In Bangladesh biomass fuel constitutes more than 70 percent of total energy use (157 kgOe), mostly for domestic cooking.

Domestic cookers presently in use in rural areas (where about 80 percent of the people live) are extremely inefficient. The use of efficient stoves/cookers could realize savings of up to 40 to 60 percent of existing biomass fuel. This conversion could result in a dramatic reduction in CO_2 emissions, both directly and by reducing the rate of deforestation. Moreover, this saving in biomass fuel (mostly agricultural residue) would enable farmers to use this as fodder and organic manure.

Finally, in Bangladesh only 39 percent of the forestland has tree cover, and village forest groves are being gradually depleted. The government should give more attention to afforestation and social forestry programs.

References

BBS, 1989. *Statistical yearbook of Bangladesh*. Bangladesh Bureau of Statistics.

Chowdhury, M.H.K, 1977. *Tropical Cyclone in the Bay of Bengal*, M.Sc. dissertation, Department of Geography, University of Reading, England, 1977, pp. 50-52.

———, and M.A. Hussain, 1982. *On the Aridity and Drought Condition*. Bangladesh Meteorological Department, Dhaka, July 1982.

———, 1985. *The Role of Meteorological Satellites in the Study of Local Severe Storms*. Proceedings of the Seminar on Local Severe Storms, Dhaka, Bangladesh.

FAO, 1988. Land resource appraisal of Bangladesh for agriculture development, Report 2, 1988.

FEC, 1989. Pre-feasibility study for flood control of Bangladesh. *French Engineering Consortium*, 2II:1-20.

Forestry, 1987. Chief Conservator of Forests, Government of Bangladesh, personal communication, 1987.

Karim, Z, *et al.*, 1990. *Drought Yield Estimate in Bangladesh* (in press).

Mahtab, F.U., 1989. *Effect of Climate Change and Sea-Level Rise on Bangladesh*. Commonwealth Secretariat, London.

———, 1990. State of development strategies and resource allocation.

Seminar organized by Supreme Court Bar Association, Dhaka, June 1990.

MPO, 1986. National water plan. Master Plan Organization. Ministry of Irrigation, Water Development and Flood Control, Government of Bangladesh, pp. 10-58.

MPO, 1987. *Geology of Bangladesh, Technical Report No. 4.* Master Plan Organization, Ministry of Irrigation, Water Development and Flood Control, Government of Bangladesh, pp. 4-17

ODA, 1985. *A Forest Inventory of the Sundarbans.* Main Report, Overseas Development Administration, Bangladesh, p. 14.

Rogers, P, *et al*, 1989. *Eastern Water Study: Strategies to Manage Flood and Drought in the Ganges-Brahmaputra Basin*, U.S. Aid for International Development, Washington D.C.

UNDP, 1989. *A Flood Policy for Bangladesh.* Joint Government of Bangladesh-UNDP Team, Ministry of Planning, Dhaka, Bangladesh.

WMO, 1988. *Developing Policies for Responding to Climate Change.* World Meteorological Organization and United Nations Environment Programme, WCIP-1, WMD/TD-No. 225, pp. 14-15.

World Bank, 1990. *Bangladesh: Managing the Adjustment Process—An Appraisal, Report, No.8344-BD.* Washington D.C., p. 164.

4. THE REGIONAL IMPACT OF GLOBAL WARMING IN MEXICO: UNCERTAINTY, VULNERABILITY AND RESPONSE

Diana Liverman

This chapter explores what global warming may signify for Mexico, a country with the eleventh largest population and thirteenth largest land area in the world. It will use the case of Mexico to argue that understanding and responding to the regional impacts of global warming requires not only the best possible estimates of how the climate will change, but also a detailed analysis of the vulnerability of society to these changes. The impacts of a climate change on environment and society are determined as much, if not more, by the characteristics of the regions and people affected as by the nature of the climate change itself. For example, rainfall declines of equal physical severity may have much less severe impacts on large commercial irrigated farms with insurance, good soils, and subsidized prices than on smaller, rainfed subsistence farms. This differential ecological, economic and technical vulnerability to climate variation and change is an important key to understanding the regional impacts of global warming.

Most climate impact studies use the results of complex climate models to examine the impacts of global warming. However, these general circulation models (GCMs) are an uncertain guide to the regional impacts of global warming in Mexico because of discrepancies among the models, their poor ability to reproduce observed climates, and the broad spatial and temporal scales for which model results are currently available. To address this uncertainty in the case of Mexico, various estimates of possible climate changes are used, and sensitivity analyses of how water resources and crop yields may respond to changes in key meteorological variables are undertaken. Such sensitivity analyses can indicate ways of reducing the impacts of global warming and can play a part in understanding the vulnerability of different sectors and people to climate change.

Even if immediate improvements in the regional projections of climate models are unlikely, we may still learn a lot about how global warming may affect Mexico by analyzing vulnerability to climate variation and change. In the second part of the chapter the vulnerability of different regions and people in Mexico to climatic change and variations in the past is described. Vulnerability studies, at regional and local levels, provide insights into the importance of demographic, political, economic, and technological changes in determining the impacts of climate on society in time and space. We can consider how such factors may change in the future and how they may determine the regional impacts of global warming. The chapter concludes with a discussion of policy options for responding to global warming in Mexico.

Why Is Mexico an Important and Useful Case Study?

The majority of studies on the regional impacts of global warming emphasize the direct effects on developed countries, paying less attention to developing countries. Climate models do project greater temperature increases in higher than lower latitudes, but tropical regions are nevertheless expected to experience significant temperature increases. Moreover, developing countries may be relatively more vulnerable to any climatic shift because of their economic situation and great reliance on both rainfed and irrigated agriculture (Gleick, 1989; Jodha, 1989).

Current climate models indicate significant increases in temperature and changes in precipitation in Mexico (e.g., Meehl and Washington, 1988). It has also been suggested that global warming will be accompanied by an increase in sea level of up to 1 m (USEPA, 1989). For Mexico, continually striving to support a growing population with an agricultural system that relies on relatively low and variable rainfall, any warmer, drier conditions could bring nutritional and economic disaster. More than one-third of Mexico's rapidly growing population works in agriculture, a sector whose prosperity is critical to the nation's debt-burdened economy. Although only one-fifth of Mexico's cropland is irrigated, this area accounts for half of the value of the country's agricultural production, including many export crops. Many irrigation districts rely on small reservoirs, and wells deplete rapidly in dry years. The remaining rainfed cropland supports many subsistence farmers and provides much of the domestic food supply. Frequent droughts already reduce harvests and increase hunger and poverty in much of Mexico (Liverman, 1990a).

Significant geographic mismatch exists between water and population in Mexico (Enge and Whiteford 1989).[1] Seven percent of the land, lying in the extreme southeast of the country, receives 40 percent of the rainfall. Only 12 percent of the nation's water is on the central plateau where 60 percent of the population and 51 percent of the cropland are located. Any reduction in water availability would also create problems for cities, industries and hydroelectric production. Mexico presently generates about one-fifth of its electricity from hydroelectric schemes and has developed less than a quarter of its potential (World Resources Institute, 1990). Water supply infrastructure has not been able to keep up with rapid urban and industrial development. These other sectors compete with the agricultural sector, which consumes more than 80 percent of total water supplies. Higher temperatures could increase air pollution and heat stress, and sea level rise could threaten agriculture, tourism, industry and natural ecosystems along Mexico's extensive coastlines. These impacts would likely create economic, political and demographic pressures that could aggravate tensions with neighboring countries, particularly the United States (Gleick, 1988; Liverman, 1990b).

Of course, many factors other than climate could affect Mexico's future. Economic crisis, social inequality, political instability

and population growth may well reduce living standards over the next 25 years (Bailey and Cohen, 1987; Castaneda, 1986; Pastor and Castaneda, 1988). Global warming is likely to occur over the same time as a doubling of the Mexican population to 160 million by the year 2025, and a corresponding increase in pressure on natural resources. Changes in agricultural technology and land reform to some extent allowed agriculture to keep up with growing populations in the first part of the twentieth century. But the economic and nutritional benefits of growth in agricultural production have been distributed unequally amongst regions and social groups (Sanderson, 1986). Further expansion of technology and land is unlikely in the next decade because of economic problems and land resource constraints. In any case, it is not clear that this type of change has reduced vulnerability to climate in the past, nor will do so in the future. Mexico provides a valuable opportunity to investigate such issues because primary and secondary data sources are relatively extensive and accessible.

Mexico is the thirteenth largest contributor to global greenhouse gases, producing 1.4 percent of the total net emissions (World Resources Institute, 1990). The Mexican government does perceive itself as an active participant, and leader among the Third World, in international discussions on environment and the atmosphere. For example, Mexico was an early signatory of the Montreal agreement to protect the ozone layer. In 1990, Mexico hosted World Environment Day for the UN Environment Programme (UNEP) during which President Carlos Salinas de Gortari emphasized his nation's commitment to dealing with climate change and environmental problems: "The environmental crisis affects the whole planet. Its most extreme consequences include the likely alteration of the climate...; we cannot elude our responsibility for the causes of this crisis. I would like to pledge improved effort to save energy. . . save and rationally use water. . . control pollution. . . recover the productivity of farmland. . . and protect our flora and fauna " (translation of portions of a speech by President Salinas de Gortari, 1990).

In summary, then, it is useful to assess the regional impacts of global warming in Mexico, not only because Mexico may be

---

[1] Many of the locations referred to in the text are shown in Figure 4.1.

FIGURE 4.1. Mexican states and major cities.

México
States and Major Cities

vulnerable to climatic change, but also because it is a significant greenhouse gas producer and may be willing to cooperate in international agreements.

Climate Models and Global Warming in Mexico

How might global warming alter climate conditions in different parts of Mexico? Studies that describe what may happen to the climate and resources of different regions if global warming occurs usually discuss the temperature and precipitation projections of general circulation models (GCMs) of the earth's climate. These models simulate climate across a relatively coarse geographic grid, using preindustrial levels of atmospheric carbon dioxide ($1xCO_2$) and the equivalent of double those levels ($2xCO_2$). Climate change scenarios are developed for places within the gridsquares by modifying observed average climates by the differences between the two simulations. These projected climate changes are then used to make estimates of

how crop yields, water supplies and other resources may change as a direct result of global warming (e.g., Parry, Carter and Konjin, 1988; USEPA, 1989).

Tremendous uncertainties are involved in using results from the various climate models to assess the regional impacts of global warming. Although most models produce an overall increase in temperatures, they differ in their regional projections of the magnitude of the warming, and often they do not agree on the direction and magnitude of precipitation changes. Many key processes and components such as cloudiness and ocean circulation are inadequately represented in the models (Kellogg and Zhao, 1988; Schneider, 1989a; USEPA, 1989). The coarse scale of the model grids makes it difficult to allocate climate changes to specific locations and tends to neglect some of the important subgrid scale weather patterns that can determine precipitation. In many cases the models do a poor job of reproducing contemporary climates and climate variability (Grotsch, 1988). My investigation of the regional impacts of global warming in Mexico begins by examining the ability of various GCMs to reproduce observed climates.[2]

Mexico lies between latitudes 15 and 32 degrees north, with a climate influenced by three seasonally shifting features of the general circulation: (1) the westerly winds, which bring winter rain to the extreme northwest of Mexico; (2) the subtropical high pressure belt, which brings stable, dry conditions to most of the country in winter, and to the north in summer; and, (3) the intertropical convergence zone and trade winds, which bring substantial rains as they move north in the summer. In winter cold polar air can surge southward over Mexico in "nortes," bringing rain and frost. In midsummer high pressure sometimes develops over the central plateau with a disruption of easterly flow, bringing a period of drought conditions called the *canicula* (Garcia, 1965; Metcalfe, 1987; Mosino, 1975; Wallen, 1955).

Meteorological records have been kept for some parts of Mexico since the nineteenth century, and 30-year normals are available for more than 300 stations (Servicio Meteorologico 1976). Lowland regions of Mexico have annual average temperatures around 20° Celsius. The highlands and the north have larger daily and annual temperature ranges, with annual averages decreasing with altitude. Annual rainfall totals range from below 100 millimeters (mm) a year in the desert northwest, to 2,000 mm a year in the south and east. Interannual variability in precipitation is very high, with coefficients of variation around 130 percent in the most arid regions, and approaching 80 percent in the seasonally dry areas of central Mexico. The low, seasonal and highly variable rainfall in much of Mexico means that the country experiences frequent droughts and can illafford any reduction in water availability.

Results from five major climate models for the gridsquares covering Mexico have been used. The models are the Geophysical Fluid Dynamics Laboratory (GFDL) model developed at Princeton and described by Manabe and Wetherald (1987), the Goddard Institute for Space Studies (GISS) model described by Hansen and others (1988), the National Center for Atmospheric Research (NCAR) model described by Washington and Meehl (1986), the Oregon State University (OSU or Oregon) model discussed by Schelsinger and Zhao (1989), and the United Kingdom Meteorological Office (UKMO) model described by Wilson and Mitchell (1987).

The models vary in their assumptions about clouds, ice feedbacks and oceans, and in their characterization of topography, geography, and soil depth, among other factors. Despite these differences, many similarities exist in structure and parameterizations among the models. The model characteristics and results have been compared by Bach and co-workers (1985), Schlesinger and Mitchell (1987), and Jenne (1989).

Model Simulations Compared to Observed Climates in Mexico

The models perform rather poorly in simulating the current climate of Mexico. Comparisons between model results for "current" or "1xCO$_2$" levels of carbon dioxide (approx. 300 ppm) and observed climates are complicated by differences in, and the large size of, the spatial grids of the models (Figure 4.2), as well as the uneven distribution of meteorological stations. One of our methods of comparison uses individual meteorological stations, comparing the most recent climatic normals (e.g., 1951-1980 average temperature and total precipitation) to the annual average temperature and total precipitation predicted by the 1xCO$_2$ simulation of the models, for the gridsquare in which the station is located. Table 4.1 shows the summary results of such a comparison for three representative stations—Chihuahua in the north of Mexico, Puebla in the central highlands, and Merida on the Yucatan peninsula. In warm, arid Chihuahua, the models estimate average temperatures ranging from -2°C to +6°C of the observed, with the UKMO model (UKMO) coming closest to observed conditions. This model also comes the closest to replicating observed annual rainfall at Chihuahua, in contrast to the GFDL and

---

[2]This work on the climate model results for Mexico was conducted in collaboration with Karen O'Brien, and a description of some of the work is found in her M.Sc. thesis (O'Brien K., University of Wisconsin, Land Resources Program 1990). We would like to thank the various GCM modelers who made their results available to us.

FIGURE 4.2. The GCM grids for Mexico for each of the five models.

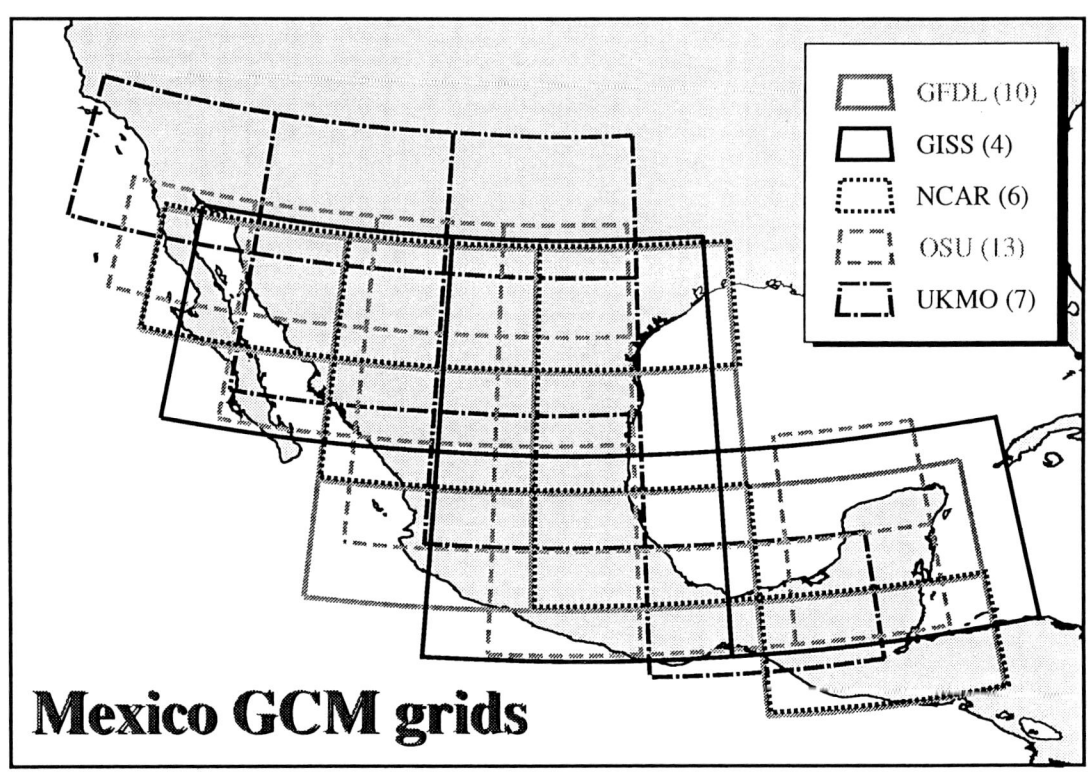

DEASY GEOGRAPHICS LAB

TABLE 4.1. *Comparison of Observed, Rand and 1xCO$_2$ Climates*

	Observed	RAND	GFDL	GISS	NCAR	OSU	UKMO
Chihuahua							
Avg. Temp. (°C)	18.4	18.9	20.8	15.6	24.3	16.3	18.7
Avg. Precip. (mm)	327	384	1,442	720	1,098	434	304
Puebla							
Avg. Temp. (°C)	16.7	20.3	19.6	22.1	21.9	26.2	16.7
Avg. Precip. (mm)	833	1,016	3,198	1,710	444	467	2,462
Merida							
Avg. Temp. (°C)	25.9	25.2	20.8	24.4	24.5	24.0	22.6
Avg. Precip. (mm)	987	1,086	1,681	1,234	447	500	2,578

Key: Observed, Current climate (1951-1980 normals); RAND, Rand Corporation 4° x 5° average observed climatology; GFDL, Geophysical Fluid Dynamic Laboratory Model; GISS, Goddard Institutute for Space Studies Model; NCAR, National Center for Atmospheric Research Model; OSU, Oregon State University Model; UKMO, U.K. Meteorological Office Model

NCAR models, which seriously overestimate it.

All models, except for the British, overestimate temperatures at Puebla, by almost 10°C in the case of the Oregon (OSU) model. This probably results from the lack of topographical detail in the models in that Puebla's cooler observed temperatures result from its 2,166 meter (m) altitude. The UKMO model may have more accurate temperatures because it sets the Puebla gridsquare at 1,351 m, whereas the GISS and GFDL models have much lower elevations at 785 m and 616 m with correspondingly warmer temperatures. For rainfall, the NCAR and OSU models only produce about half of the observed annual amount at Puebla, whereas the other models overestimate rainfall by up to four times. In Merida, the models all slightly underestimate temperature. Again, the NCAR and OSU models have rainfall that is too low whereas that for the UKMO model is too high.

Further discrepancies can be revealed in comparing monthly and seasonal results of $1xCO_2$ simulations to observed data. The most serious problem is the inability of the models to reproduce the seasonality of rainfall in northern Mexico. All models except GFDL show a winter maximum of rainfall in Chihuahua, whereas observed maximum rainfall occurs in the summer. Possible explanations for this discrepancy include: (1) the models are generating the westerlies (which bring winter rainfall in the far northwest of Mexico) much farther south than actually occurs; or (2) the lack of topography in the models is removing the observed rainshadow effect of the Sierra Madre Mountains in winter. Three of the models correctly replicate the summer rainfall maximum in Puebla, but the NCAR and OSU models produce low rainfall all year. In Merida, only the UKMO model's rainfall seasonality comes close to observed patterns.

Individual meteorological station conditions may not be representative of the regional climates that correspond to the scale of the GCM grids. The heterogeneity of the Mexican environment means that meteorological observations within a model gridsquare may vary widely. This was investigated using regional averages of observed climates developed by the Rand Corporation, in which temperature and precipitation values are averaged for meteorological stations within a 5-by-4 degree geographic grid. As can be seen from Table 4.1, the Rand conditions are a little wetter than the individual station values in all three regions, and slightly warmer in the north and central regions. Compared to the Rand values, the UKMO model seems to replicate the climate of northern Mexico slightly better than do the other models, and the GISS model seems to do the best in the central and southern regions of Mexico.

These results demonstrate some of the inadequacies in the ability of climate models to reproduce observed climates, and can lead to distrust in the changes projected by the models, particularly those for precipitation. Further caution in using model results for climate impact studies is indicated in the chapter on climate variability in the USEPA (1989) report, which suggests that the models are unable to reproduce the variability of observed climate.

Differences Between the Model Projections of Climate Change in Mexico

Keeping in mind the models' problems in reproducing observed climates, it may still be possible to use their output to construct relative estimates of how global warming may change the climate. Some authors claim that although regional estimates of $1xCO_2$ climates may be unrealistic, the way in which the model estimates the change from $1xCO_2$ to $2xCO_2$ conditions is assumed to be relatively consistent (Cohen, 1986; Gleick, 1987; Parry *et al.*, 1988). Rather than relying on $2xCO_2$ results alone as a guide to future climates, the difference between the model representation of the present and the future is used to generate a measure of climate change.

Thus, the temperature change is calculated by subtracting the $1xCO_2$ from the $2xCO_2$ temperatures for each gridsquare. In the case of rainfall, especially in an arid region, the absolute change in precipitation from $1xCO_2$ to $2xCO_2$ is sometimes much greater than the total observed rainfall. In these cases, using the absolute change could give an unrealistically high, or even below zero, rainfall. Many researchers have used this ratio or percentage method to estimate precipitation changes (Adams *et al.*, 1990; Parry *et al.*, 1988).

What changes do the models project for Mexico, and do these projections show any agreement? The models each have a different number of gridsquares covering Mexico (Figure 4.2). Averaging the changes projected by each model for the whole of Mexico gives a general sense of the similarities and differences among them. The UKMO model forecasts the greatest regional temperature increase of 5.44°C and the NCAR model the least at 2.38°C (Table 4.2). The NCAR model predicts a 23 percent decrease in rainfall, and the GISS model a 3 percent increase. Both the magnitude and, in the case of rainfall, the direction of change vary depending on the spatial and temporal scale of analysis. When we look at average changes for the whole of Mexico, but for seasonal and monthly rather than annual time scales, we see much greater differences among the models (Figure 4.3). For example, we find that the OSU model has its greatest temperature increases in the summer, whereas the GISS and NCAR models have their greatest increases in the fall and winter. The differences among the model predictions for precipitation are very dramatic on a seasonal basis. For example, in the NCAR model autumn rainfall decreases by almost 50 percent with doubled CO_2 and that for the OSU model increases by 15 percent. Monthly comparisons show even more extreme differences in the magnitude and direction of projected changes.

These comparisons are for annual changes averaged across the whole of Mexico, and the variations among the models are greater at the more local level. Table 4.3 presents summary statistics for the change in rainfall at different spatial scales for four of the models. In almost all cases, the range and the standard deviation of monthly precipitation changes produced by the models increase from the national, to the regional, to the gridsquare scale. In some cases, the direction of the average change shifts as the spatial scale decreases. The UKMO model shows a 1 percent decrease in rainfall at the Mexican national scale, a 6 percent increase for the northern region, and a 39 percent increase for the one northwestern gridsquare. Many of these changes are clearly an outcome of temporal and spatial statistical averaging procedures. The results suggest that averaging gridsquares and using annual conditions to facilitate the comparison among the models will obscure many of the differences among them.

Monthly results for individual gridsquares were used to investigate climate change in Mexico. This highlights the differences among the models and maximizes the representation of seasonal variability. Daily data, not available for this study, would permit an even better examination and representation of variability within and among the models. By using the results from all five models in generating climate change scenarios for Mexico, some of the uncertainties in the projections for global warming are captured.

The Nature and Implications of Climate Changes Associated with Global Warming

The most common approach to analyzing the regional impacts of global warming is to take the type of changes estimated by the models, as discussed above, and modify current climates to represent doubled CO_2 conditions. Temperature changes are added or subtracted from observed conditions, and observed precipitation is multiplied by the estimated percentage change.

For four meteorological stations in northern Mexico all of the models project increases in temperature in every month of the year, resulting in more months of extremely high summer temperatures (Figure 4.4). For

TABLE 4.2. *Change Projected by GCMs for the Whole of Mexico*

	GFDL	GISS	NCAR	OSU	UKMO
Temperature (°C)	3.11	3.92	2.38	3.15	5.44
Precipitation (%)	-1.75	2.74	-23.01	-1.05	-0.09

FIGURE 4.3. Monthly and seasonal temperature and precipitation changes projected by climate models.

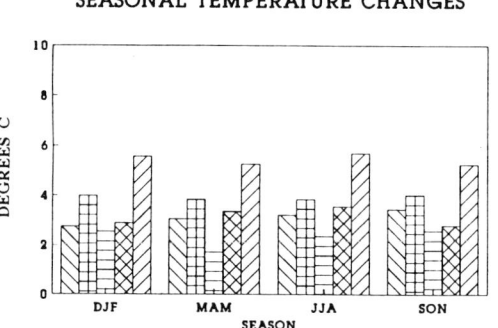

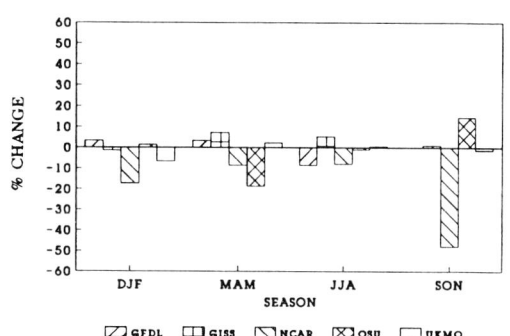

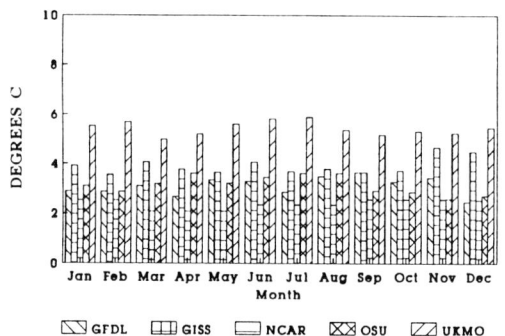

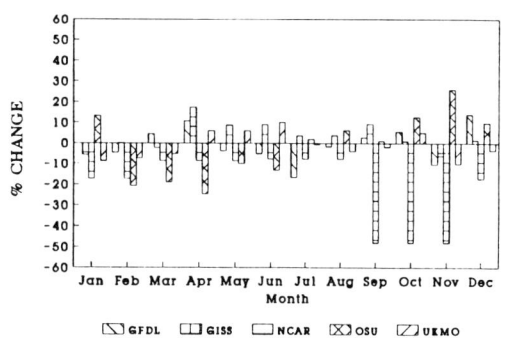

TABLE 4.3. *Differences in the Rainfall Changes Estimated at Different Scales*

	A	S	A	S	A	S	
GFDL	0	8	-7	11	3	42	North
			9	19	-2	26	Central
			3	23	3	24	South
GISS	3	7	5	14	38	53	North
			-4	7	-4	7	Central
			15	30	15	30	South
OSU	-1	15	-11	13	-16	12	North
			0	17	5	23	Central
			25	23	28	20	South
UKMO	-1	6	6	12	40	47	North
			-4	15	-4	15	Central
			10	19	-10	19	South

Key: A = Average percent change in annual total; S = Standard deviation of percent change in annual total

example, mean monthly temperatures of about 25°C in Chihuahua could increase to almost 30°C, with some models predicting at least one month above 35°C . In Ciudad Obregon, which already experiences summer temperatures near 30°C, several months could approach 35°C. These temperature increases in the already hot summers of northern Mexico are likely to bring great stress to humans, plants and animals, especially if these monthly averages imply even higher daily extremes. On the other hand, warmer temperatures may extend the growing season at higher elevations, such as Chihuahua, which sometimes has cold temperatures and

frosts in winter months. In Tijuana, a water-limited and rapidly growing border city, several models suggest that winter rainfall may be higher. The UKMO model also projects higher summer rainfall, but the NCAR and GFDL models produce a lower annual rainfall total. In Ciudad Obregon, located in one of the most important irrigated agricultural regions of Mexico, the GISS and UKMO models indicate that rainfall may increase in July and August, whereas other models bring lower rainfall throughout the year. Chihuahua's climate scenarios follow a similar pattern, but with distinct drops in OSU and GFDL estimates for summer rainfall.

FIGURE 4.4. Monthly temperature and precipitation for various locations in Mexico under conditions of global warming. The graphs were compiled using data from the five climate models; vertical lines show the range of projections.

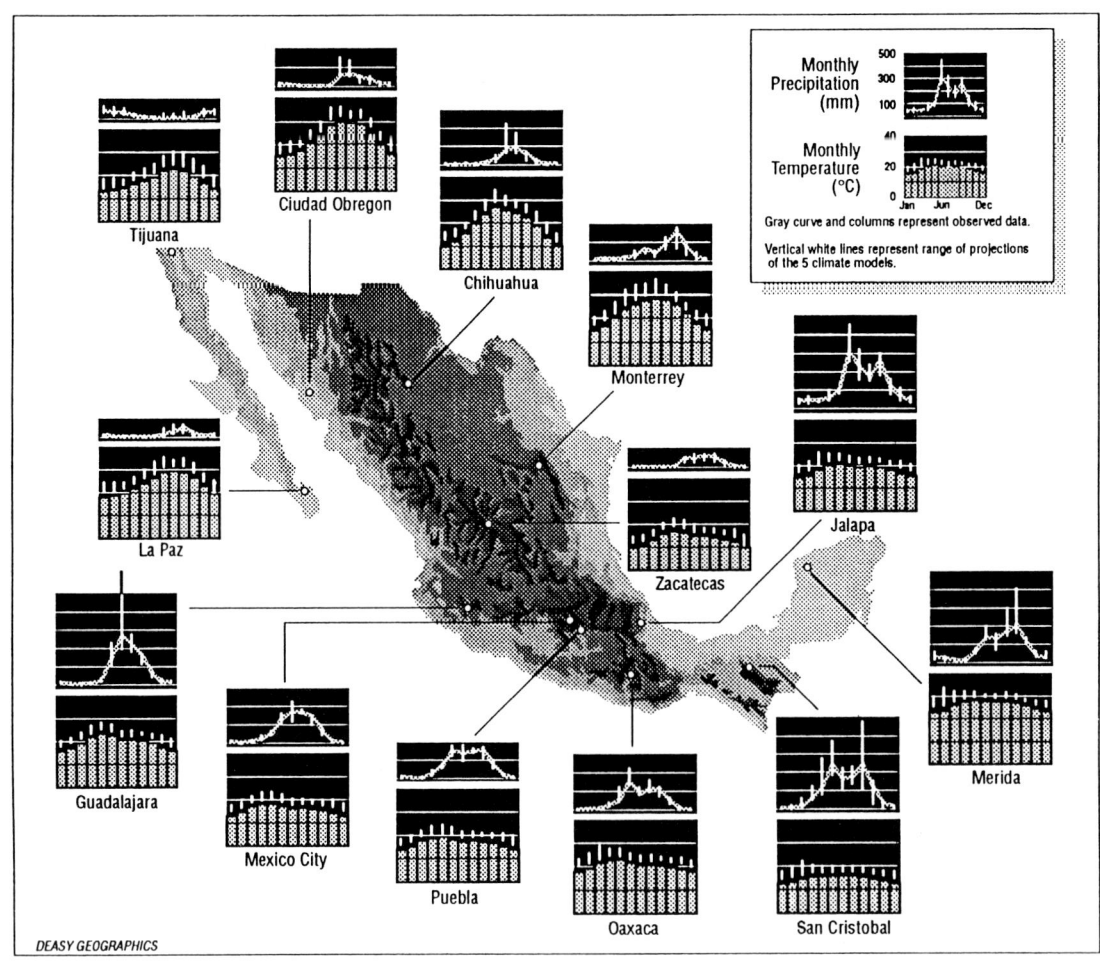

Monterrey will have lower rainfall in June according to all of the models, as well as slightly lower annual totals for four of the models.

In central Mexico temperature increases are not quite as high as in the north but are still projected to increase by all models in all months. For example, Mexico City's current winter temperatures of around 12°C could increase to as high as 18°C. Warmer temperatures could be important in reducing frost risks in the highland agricultural regions around Mexico City and Puebla. On the other hand, higher temperatures in Mexico City and, to a lesser extent Guadalajara, could lead to an increase in the frequency and severity of air pollution. In Guadalajara, all models with the exception of GFDL produce decreases in summer rainfall. In Mexico City and Puebla, rainfall may be lower in spring according to three models, and may increase in the summer in the UKMO and NCAR scenarios. Overall, the GISS and UKMO models project an annual precipitation increase, whereas the others suggest that decreases are more likely. For Jalapa, all models except one suggest lower annual rainfall.

In southern Mexico, significant temperature increases are forecast by all models, and these could be particularly important in Merida where high summer temperatures and humidity currently combine to make conditions stressful. The slight frost risk in areas of Oaxaca might be reduced under warmer conditions. Summer rainfall increases are projected for Oaxaca by two models, but the UKMO model shows a late summer decrease. At San Cristobal, in the highlands of Chiapas, all models produce a rainfall increase in June, which could bring an important boost to soil moisture early in the growing season. However, some of the models show lower rainfall in late summer and winter. All the models except the UKMO model project increased annual rainfall at Merida, especially in the fall. Under current conditions, autumn rainfall in Merida is often produced by tropical storms. Some authors have speculated that the intensity and frequency of such storms would increase with global warming (USEPA, 1989).

A limitation of these climate change scenarios for Mexico is that we have no reliable information on how the variability of rainfall may be altered. The ratio method of constructing scenarios for rainfall change means that months with very low observed rainfall tend to also have low rainfall under global warming. The use of average monthly output rather than several years of daily data in our studies means that we cannot fully investigate all-important changes in the beginning of the rainy season or in the frequency of frosts. One approach to understanding the implications of these uncertainties is to examine the sensitivity of various impact assessments to a range of possible conditions.

Sensitivity Analyses of Global Warming Scenarios for Mexican Water Resources and Agriculture

For Mexican agriculture, the most important impacts of these climate changes would be changes in the length of the growing season and in moisture availability. Assessing changes in moisture availability, particularly in locations where both temperature and rainfall are projected to increase, requires the analysis of evaporation. We need to know how any increase in rainfall or growing season length is likely to interact with increased evaporation associated with higher temperatures. Most of the models calculate evaporation and soil moisture internally, but the surface energy and water budgets of the models tend to be rather simple. For example, all but the GISS model represent soil water using a 15 cm deep "bucket," which when full produces runoff. Moisture is lost through evaporation, at slower rates as the bucket empties (Kellogg and Zhao, 1988). The problem arises in dry regions where the $1xCO_2$ simulation of soil moisture is low. Even if potential evaporation in the model rises under the higher temperatures of $2xCO_2$ there is little or no water to evaporate in the bucket, and, as a result, the estimated change in soil moisture will be very small. In reality, deeper soils can often contain more water, and soil moisture could be reduced significantly as temperatures rise. Information obtained for the water balance of the GISS and OSU models suggests that potential evaporation will increase in all months because higher air temperatures increase the atmospheric demand for moisture in air that is relatively dry. The GISS model internally estimates a 15 percent increase in potential evaporation for the Chihuahua gridsquare and a 2 percent increase in Puebla, The OSU model projects a 12 percent decrease in potential evaporation at Chihuahua and a 6 percent increase in Puebla.

Because of the limitations of modeling soil moisture, model projections of precipitation, temperature and solar radiation changes were used to estimate exogenous evaporation and moisture availability using the Penman method. Table 4.4 shows the exogenously estimated changes in Penman potential evaporation for the stations of Puebla in central Mexico, and Ciudad Obregon in the northwest. Current Penman potential evaporation totals about 1,528 mm per year under current (1951-1980) climate in Puebla. With global warming, this could increase from 7 percent to 16 percent, depending on the climate model. These results have serious implications for water resources and irrigation in central Mexico, because they imply increased losses from open water surfaces such as reservoirs. In Ciudad Obregon, current potential evaporation in this desert like environment is 2,438 mm per annum. Higher temperatures associated with global warming could increase this by 8 percent to 27 percent. In northwest Mexico, the economy and livelihoods depend on adequate supplies of water to cities and irrigated agriculture. Three-quarters of the water needs in this region depend on limited supplies from reservoirs and rivers. In 1987, a severe drought

year in much of Mexico, reservoir levels in the northwest dropped on average to only 30 percent of their capacity with a corresponding drop in production of wheat and other important crops. The northwest is also a region where international conflicts over water from the Colorado River could be exacerbated by increases in potential evaporation (Gleick, 1988).

The relationship between increased potential evaporation and changes in rainfall can be approximated over a year by subtracting total annual potential evaporation from annual rainfall to produce an estimate of moisture surplus or deficit. Under current climate, such a calculation for Puebla produces a moisture deficit of 694 mm. That is, potential evaporation is about double precipitation each year. Global warming could increase this deficit by 17 percent to 45 percent depending on the climate model. A 45 percent moisture deficit would occur with the GFDL scenario, because this model combines a temperature increase and a large precipitation decrease at Puebla. Although the UKMO model projects the largest potential evaporation increase for Puebla, an increase in precipitation partly offsets this for a deficit increase of 30 percent. Similar scenarios occur in the northwest with moisture deficits increasing in Ciudad Obregon by between 15 percent and 23 percent. This reduction in water availability would impact both rainfed and irrigated agriculture in the region.

Rainfed agriculture and economic devel-

TABLE 4.4. *Change in Potential Evaporation and Moisture Deficits*

| | Puebla | | Ciudad Obregon | |
	Potential evaporation	Moisture deficit	Potential evaporation	Moisture deficit
Observed	1,528 mm	-694 mm	2,438 mm	-2,182 mm
GFDL	14.2%	-44.7%	16.0%	-21.2%
GISS	13.3%	-33.1%	20.0%	-16.3%
NCAR	7.1%	-22.8%	8.0%	-12.3%
OSU	10.0%	-17.5%	12.7%	-14.6%
UKMO	16.0%	-29.5%	26.6%	-22.7%

Key: See Table 4.1.

opment are already limited by annual water deficits in central Mexico. Harvests often fail through a combination of high temperatures and lower than expected rainfall. The Puebla region is an important supplier of food to Mexico City. The capital has a growing demand for water, which is currently met by bringing in water from the surrounding area (Kandell, 1988). If water supplies shrink by 17 percent to 45 percent in central Mexico, there likely will be serious implications for economic growth and living standards in Mexico City.

The implications of possible errors in the model projections for precipitation and solar radiation can be investigated through a sensitivity analysis of the Penman calculations. These indicate that it would require significant decreases in wind speeds (-50 percent) or solar radiation (-20 percent) to prevent an increase in potential evaporation with higher temperatures. Moisture deficits could only be reduced by much larger increases in rainfall than currently projected by the models. Only if the climate models are greatly overestimating temperature increases, or if they are very wrong regarding rainfall and solar radiation, does the future seem optimistic for soil moisture in Mexico.

The principal conclusion of the analysis of what climate models say about the regional impacts of global warming is that Mexico is likely to be warmer and drier. Whichever model we use it seems that potential evaporation will increase, and moisture availability will decrease, even in those cases where the model projects an increase in precipitation. Sensitivity analyses of our evaporation and moisture deficit calculations indicate that water availability could increase only if the model results were to produce much higher rainfall and relative humidities, or significantly less solar radiation and wind. Such changes are, of course, possible as the modeling of clouds and synoptic conditions improves, but at present it seems that a moisture decrease is more likely.

The Implication of these Results for Crop Production

The possible impacts of global warming on Mexican crop production are being explored using the Ceres-maize crop yield model. The Ceres model uses daily weather data to estimate the effects of climate on yields of maize under water management regimes ranging from rainfed to fully irrigated. The model is physiologically based, simulating the influence of solar radiation, degree days, nutrients and water availability on the development of maize through major phenological stages (Adams *et al.* 1990; Jones and Kiniry 1986). It has been suggested that many crops will benefit from a higher level of atmospheric carbon dioxide and that higher yields may result. The Ceres model has been modified to include the direct effects of elevated CO_2 levels on crop growth and water use efficiency, but does not account for losses from, and changes in, pests and diseases.

Yields of maize, the major staple in Mexico, have been simulated for agricultural regions near Mexico City. The Ceres model has been validated for a site near Cuernavaca using agricultural experiment data provided by the International Center for Wheat and Maize Improvement (CIMMYT).[3] The model has then been used with observed weather data from 1973 to 1989 to simulate 17 years of rainfed and irrigated yields of maize under typical soil and management conditions.

Climate change scenarios are generated using results of the GFDL, GISS and UKMO climate models concerning likely changes in temperature, precipitation and solar radiation with a doubling of CO_2. Rainfed and irrigated yields are then simulated under these changed climates and compared to the yields for the years of observed weather. Results of these experiments, which initially assumed no alterations in planting dates or other input conditions as the climate changes, are shown in Table 4.5. On average, rainfed yields decline by 53 percent for the GISS climate projections, 85 percent for GFDL, and 55 percent for the UKMO model. Under observed conditions, only one year, 1989, produces yields below 1,000 kilograms per hectare (kg/ha). The GISS and UKMO models increase this frequency to 7 years and GFDL to 2 years.

[3]Research on the impact of global warming on maize yields in Mexico is being done in collaboration with K. O'Brien, M. Dilley, L. Menchaca and G. Edmeades, with the support of the EPA project on international agriculture and global warming.

These low yields occur in spite of overall rainfall increases projected by some models, because higher temperatures create elevated evaporation early in the growing seasons, and in some cases high rainfall leaches nutrients out of the soil.

Irrigated yields are relatively high under current climate, averaging 6,200 kg/ha, but are limited by poor soil fertility and a lack of fertilizer. Irrigated scenarios for global warming produce less variable yields, but do not exceed, on average, the current yields. The demand for irrigation water with global warming increases by 15 percent to 23 percent depending on the model, compared to irrigation needs under observed conditions. Projections for moisture deficits, indicated by the Penman calculations above, suggest that this extra irrigation water may not be available in a warmer climate.

Sensitivity analyses indicate the importance of some of the current uncertainties in estimating the impacts of future climatic changes. For example, the simulations described above use the ratio method and assume that any increases in rainfall with global warming only occur on those days that precipitation falls under observed conditions (i.e., if there is no rain under observed conditions, then any change is multiplied by zero). If we also increase the number of days on which precipitation occurs, then yields can change significantly. Increasing the number of rainy days in the British model's rainfed simulation produces yields that are 5 percent higher, on average, than when just the original ratio method is used.

As with estimates of Penman evapotranspiration, the yields are sensitive to solar radiation changes. The simulations described above use changes in solar radiation based on the GCMs. If we assume that solar radiation does not change with global warming then average yields are 62 percent less than observed for the GISS model, 78 percent less for the GFDL model, and 32 percent less for the UKMO model. The results reported above also assume no change in the average date of planting, with seeds planted on May 1 in both observed and GCM conditions. May 1 is an average planting date under current climate, which allows farmers to balance the risk of late spring rains and early fall frosts. If, assuming warmer temperatures, we adjust planting dates to one month earlier with global warming, rainfed yields decrease compared to a May 1 planting date for any of the model scenarios. However, if we plant one month later, rainfed yields are slightly higher than with a May 1 planting date in the GISS and UKMO scenarios, because the crop grows once the rains have started and can continue to grow into the fall without frost risk. However, the yields are still lower than a May 1 crop planted under current climate.

We have also left the crop variety unchanged in this initial experiments. When we change the variety, yields of both rainfed and irrigated maize under global warming in Mexico can increase by up to 20 percent compared to the yields obtained if the variety is unchanged. However, we have been unable to find a variety that compensates for global warming under rainfed conditions. It is possible to design hypothetical genetic coefficients within the Ceres model that maintain high yields under global warming, but it is not clear that these coefficients are biologically possible. The sensitivity of the model results to these factors indicates possibilities for adjustment to global warming by altering the dates of planting and the genetic type or

TABLE 4.5. *Ceres-Maize Model Yields (1973-1989 Base Period) at Tlaltizapan, Mexico*

	Rainfed average yield (kg/ha)	Irrigated average yield (kg/ha)	Irrigation water (mm)
Observed	3,600	6,200	230
GISS	1,900	4,900	276
GFDL	3,000	5,100	265
UKMO	2,000	4,200	282

Key: See Table 4.1.

characteristics of the seed.

The amount of fertilizer and irrigation is critically important to the maize yields predicted by the Ceres model. The experiments above all assume relatively low fertilizer use because many Mexican producers can only afford to use 50 to 100 kilograms (kg) of nitrogen fertilizer at planting. If more fertilizer becomes available to more farmers as the climate changes, then some of the yield reductions might be offset. Only with adequate water and fertilization can the direct effects of CO_2 be of significant advantage to C-3 plants such as maize. When we run observed and model scenarios with no water or nutrient limitations we find that current climate produces yields averaging 6,800 kg/ha. The Goddard scenario produces 5,200 kg/ha, the GFDL produces 5,900 kg/ha and the UKMO model 5,000 kg/ha. Given the environmental and economic constraints and trends in agricultural inputs in Mexico, unlimited water and nutrients are extremely unlikely. The experiment illustrates, however, the sensitivity of maize yields to assumptions about future resource availability.

It is important to note that, in terms of export-oriented agriculture, and price incentives to producers, it will be Mexico's comparative advantages and disadvantages in agricultural production in a warmer world that may determine the national and regional economic impacts of climate changes. We have not yet undertaken any detailed estimates on other crops or the areas of land that may be threatened by sea level rises. It is clear that a number of low lying coastal agricultural areas in Mexico would be threatened by such sea level rises. In some cases the threat might be of direct flooding of land, or of increased storm damages (e.g., along the Gulf Coast), but in others, such as Sonora and Sinaloa, the threat would be to irrigation systems and soil fertility through rising water tables and salinization.

Again, in spite of differences between the climate model results, the general direction of changes in Mexican maize yields with global warming is a decrease, whatever the model, and whether irrigation is used or not. Sensitivity analysis of the Ceres model indicates that decreases in crop yields will be severe under global warming unless irrigation expands, fertilizer use increases, or new varieties of crops are developed.

Vulnerability to Climate Change in Mexico

As noted at the beginning of the chapter, we should and need not wait for improvements in the climate models to develop methods for assessing the regional impacts of global warming; we can acquire a general sense of possible changes in climate by examining a range of model results and sensitivities. More importantly, we can approach the study of the impacts of global warming from a different perspective by looking at vulnerability to climate change. The impacts of global warming will depend much less on the exact amount of the temperature or moisture change than on the characteristics of the regions and people who experience the changes.

In Mexico, we gain considerable insights into the possible regional impacts of global warming by analyzing the vulnerability of agriculture and other resources to current variations in temperature and precipitation. The advantage of vulnerability studies is that they do not depend on climate model results, yet they point to many ways in which we might reduce the negative impacts of climate variation both today as well as in the case of a warmer future. Vulnerability analysis shows us that some people and countries will suffer more severely, and more quickly, than others from global warming (and, of course, that some will benefit). It also expands the range of choices in responding to global warming by demonstrating that we can reduce impacts of global warming, not only by slowing the rate of climate change, but also by reducing the vulnerability of populations and economies to these changes. It may also focus our responses by highlighting regions and peoples who are most vulnerable, or whose vulnerability can be most easily reduced.

In many cases, the most vulnerable people are considered to be those living in the most precarious physical environments. Drought vulnerability would be associated, for example, with low or variable rainfall and sandy soils. This kind of physical vulnerability—which I have termed "biophysical vulnerability" (Liverman, 1990b)—is often combined with demographic analyses that examine population pressures in physically marginal areas. For example, the FAO estimates the agricultural production potential of lands in the developing world based on climate and soil constraints (Harrison, 1984). A number of regions and countries are identified as "critical zones" whose current or future population demands for food already exceed agricultural production potential. From this perspective, if biophysical and demographic conditions define vulnerability, then vulnerability can be reduced by modifying these conditions, or by moving people away from biophysically marginal areas.

A very different framework criticizes both physical and demographic characterization of vulnerability. The "political economy" approach has become increasingly important in climate impact studies. In this framework, vulnerability is defined by the political, social, and economic conditions of a society. One definition is provided in Hewitt (1984) where vulnerability "is the degree to which different classes in society are differentially at risk." It is suggested that underdevelopment (flows of resources out of a region, land expropriations, exploitative labor conditions, political oppression, and other processes associated with colonialism and capitalism) has made people, especially the poor, more vulnerable to disaster, and has forced them to degrade their environment. Similar arguments are made in R. V. Garcia's studies of the 1972 drought in the Sahel and other regions, where a "structural" approach for diagnosing the impact of climate anomalies was proposed (Garcia, 1981). By analyzing the historical evolution of social systems in various regions, the study sought to demonstrate how certain groups become so disadvantaged and exploited that they are unable to cope with drought, or struggle for the resources to overcome environmental stress.

In some cases, technology can be seen to reduce vulnerability to environmental change, as irrigation, improved seed varieties and fertilizers are designed to mitigate the impacts of rainfall deficits and infertile soils. However, technology does not always reduce biophysical vulnerability. Irrigation can cause the salinization and waterlogging of productive land and, in severe drought years, an overdependence on irrigation can mean greater vulnerability when water storage is eventually depleted (Wionczek 1982). In a political economy framework, technology is frequently tendered as increasing social inequality and vulnerability. The Green Revolution has provided the archetypical case of the social impacts of a set of technologies (improved seeds, irrigation, chemicals) that were partly designed to reduce vulnerability to environmental variation. Many authors have documented the social marginalization of small farmers and the poor associated with the introduction of Green Revolution technologies, (Barkin and Suarez, 1983; Hewitt de Alcantara, 1976; Pearse, 1980; Sanderson, 1986).

These ideas are important in thinking about the regional impacts of global warming in Mexico and raise a number of key questions. Who and where are the most climate-vulnerable regions, sectors and social groups? How have, and will, changes in technology and economy alter vulnerability to climate? The climate model results and sensitivity analyses discussed thus far suggest that it is not unreasonable to assume that global warming may bring warmer, drier conditions to Mexico. Therefore, the following discussion of vulnerability focuses on the factors that influence the impacts of warm, dry weather periods—droughts—in Mexico

Historical Context of Climate Vulnerability in Mexico

Paleoclimatic and instrumental records suggest that there have been significant shifts in climate in Mexico with long periods of drought (Metcalfe, 1987). Such changes have been implicated, together with changes in social organization, in both the rise and collapse of powerful pre-Hispanic civilizations such as those in the Yucatan and in the valley of Mexico. Historians have also blamed climate for famine and social unrest in the colonial period. Florescano (1969) has linked variations in the price of maize in the sixteenth and seventeenth centuries to droughts and other climatic events. Florescano claims that the majority of price rises were preceded by a severe drought, but he also notes the role of speculation and economic arrangements in triggering price rises and associated famines. He suggests that the economic and land tenure relations imposed by the Spanish Crown created a tremendous vulnerability to drought among the poorer and indigenous campesino populations. The colonial political economy allowed the larger landholders and merchants to manipulate the price of staples in drought years to the disadvantage of poor consumers and small producers. In postrevolutionary Mexico he suggests that differential vulnerabilities to drought, inherited from the colonial land tenure systems, are very evident. Florescano claims that:

> the most disastrous effects of drought, as in earlier times, are concentrated in the rainfed agriculture practiced by the poorest ejidatarios and campesinos, lacking credit, irrigation, fertilizers, and improved seeds. (Florescano, 1980, p. 17)

In this century, a steady and unvarying expansion of Mexican agricultural production has been necessary to meet the demands of a rapidly growing population and agricultural export market (Wellhausen, 1976). In the drive to modernize and expand production, the Mexican agricultural system has incorporated some techniques (such as irrigation, improved seed varieties, and chemical inputs) that may reduce hazard losses, (Venezian and Gamble 1969; Yates 1981). There is very little research on the role of the Green Revolution in reducing or increasing vulnerability to climatic variation. Michaels (1979) and Hazell (1984) have suggested that improved wheat varieties in Mexico and other countries may be more sensitive to drought and climate variability than traditional seeds. Some believe that the large regions of single variety high yielding crops established through the Green Revolution are much more vulnerable to pests and diseases than traditional mixed cropping (Pearse, 1980). Others suggest that these new techniques have replaced some traditional hazard prevention strategies, such as mixed cropping and microclimate modification (Wilken, 1987), and have allowed agriculture to expand into areas of high hazard risk such as deserts, mountains, coastal regions, and the disease-susceptible humid tropics.

Another question in Mexican agriculture concerns land reform, particularly the performance of the ejido sector.[4] Although some

authors claim that the ejidos are inefficient and unproductive (Wellhausen 1976; Whetten 1948), others suggest that, in terms of input use, they are relatively efficient and produce yields equal to the private sector (Dovring, 1970; Mueller, 1970; Nguyen, 1979). The problems of the ejido sector have been explained in terms of their lack of political power, difficulties in obtaining access to credit and inputs, bad management, and poor land resource endowment (Coll-Hurtado, 1982; Hewitt de Alcantara, 1976).

Agricultural Vulnerability to Drought in Mexican States

The differential vulnerability to drought and other natural hazards at the regional level in twentieth century Mexico has been analyzed in terms of physical geography, access to technology, and land tenure were analyzed. To achieve this objective, questions and hypotheses were raised and evaluated using data for the states of Sonora and Puebla in 1970, and for the whole of Mexico from 1930 to 1980:

- What is the pattern and severity of drought loss, measured by the area reported in the census as planted but not harvested due to natural hazards?
- What is the relationship between reported drought losses and the physical pattern of climate, as characterized by various indices of physical drought calculated from monthly rainfall and temperature data?
- What variables can be found to document vulnerability to drought and thus to explain patterns of reported drought losses? Data on technology, economy, and land tenure are used to answer the following questions:
  - Does irrigation reduce drought loss?
  - Do improved seeds or chemical fertilizer reduce drought losses ?
  - Are certain crops more vulnerable to drought than others ?
  - Are ejidos more vulnerable to drought than private farms ?

Meteorological and census data have been

_____

[4]The edijo land tenure system was established in postrevolutionary Mexico (McBride 1923; Yates 1981). Land is held communally in usufruct by a group of families and, although some plots are farmed collectively, most land is

farmed individually.

obtained from government offices in Mexico City and has identified no major systematic errors in the data (Liverman, 1990a).

In a case study of drought in the two agriculturally important states of Sonora and Puebla, the physical pattern of drought, expressed in terms of rainfall deficits and evapotranspiration, did not correlate strongly with the pattern of drought loss reported in the census. In both states there were *municipios* (local administrative districts) with high drought losses in areas of low rainfall, and some with low losses where there were severe rainfall deficits. This indicates that vulnerability to drought loss was not directly linked to physical climate conditions. In general, lower drought losses seemed to be associated with the use of irrigation, high yielding varieties, and fertilizer, suggesting that these technologies may reduce drought vulnerability. However, there were municipios with high reported drought losses where irrigation did not seem to buffer the agricultural system against rainfall deficits. The ejido sector in both states was, on average, reporting twice the drought losses of the large private landholdings.

Overall natural hazard losses in agriculture from 1930 to 1980 were also analyzed at the national level in Mexico. Hazards include drought, flood, hail and frosts, with detailed losses from drought only reported for 1950, 1960 and 1970. Based on these censuses it appears that, on average, more than 90 percent of hazard losses in Mexican agriculture are from drought. The area of total hazard losses increased in Mexico from 1930 to 1970. This partly reflects a change in the total crop area, which increased from 7.8 million hectares (m ha) in 1930 to 11.8 m ha in 1960. Relative (percentage) losses also increased, indicating that some of the land expansion may have included more hazard vulnerable land. However, both absolute and percentage hazard losses increased dramatically in 1970 when land area dropped to 10.2 m ha. There is no indication in the meteorological record of an increasing severity of weather events in the census years or from 1960 to 1970. The increase may, therefore, support a hypothesis that hazard losses have been increasing irrespective of weather severity, because of increases in vulnerability to natural disasters.

High drought losses tend to occur in the arid northern region of Mexico, where precipitation is low and highly variable. Figure 4.5 shows that in parts of northern Mexico more than 35 percent of the area planted was lost to drought in the summer of 1970. High flood losses tend to occur in states such as Tabasco and Veracruz along the Atlantic Gulf Coast, or in the southern regions of the Pacific coast such as Sinaloa, Colima, Nayarit, or Oaxaca. These are all regions that can receive heavy cyclonic summer rainfall and with major rivers that often overflow their banks. These geographic variations in hazard vulnerability indicate that any global changes that bring drier conditions in northern Mexico, or increased storminess along coasts, could significantly exacerbate existing problems.

Is there any evidence that irrigation buffers agriculture against drought at the national level? States with high levels of irrigation, such as Sonora and Sinaloa, do have much lower drought losses than many of the states with smaller proportions of irrigated land. However, in 1970, Aguascalientes and San Luis Potosi, with about 20 percent of their land irrigated, reported drought losses of 70 percent and 50 percent, respectively, indicating that irrigation may not always buffer Mexican states against drought. As noted earlier, some people suggest that the use of improved seeds may be associated with changes in natural hazard vulnerability. The correlation between expenditure on improved seeds and drought losses is negative but weak (-0.18 in 1970), suggesting that higher expenditures on improved seeds may be associated with lower drought losses.

As noted earlier, the relative economic vulnerability of the ejidos has become a matter of considerable academic and political interest. Table 4.6 shows the average losses, at the state level, for the two main land tenure sectors in Mexico. In every census year the average losses in the private sector are less than those in the ejido sector. In 1970, losses are almost double on ejido lands. Differences between private and ejido losses are significant for each hazard, with ejidos generally reporting higher drought, flood, and frost losses. Pest losses seem to be similar, possibly because of vulnerability of monoculture in the private sector.

FIGURE 4.5. Drought losses sustained in Mexico in 1970, shown as percent planted but not harvested.

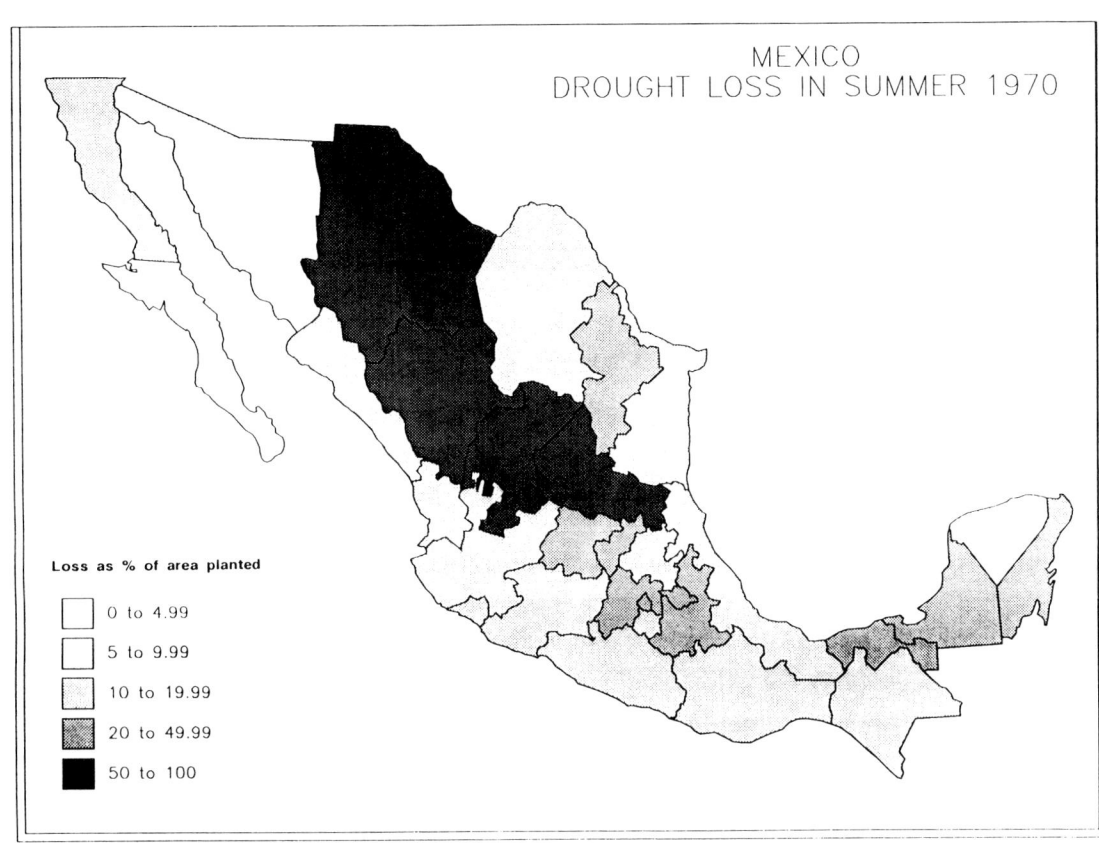

These results seem to confirm the ideas of Florescano about the relative vulnerability of the ejido sector to drought, and supports my findings of drought losses at the municipio level in Sonora and Puebla in 1970. This increased vulnerability appears to be be-

cause more biophysically marginal land was given to ejidos in the land reform and that ejidos are socially vulnerable as a result of reduced access to irrigation, credit, improved seeds and other resources.

The results suggest that agricultural modernization and land reform affect vulnerability to drought and climate variations. The expansion of irrigation, the adoption of new seed varieties, the use of chemical fertilizers, and the modification of land tenure systems will often reduce, but sometimes increase, vulnerability. Clearly we must consider such factors in assessing the possible impacts of global warming in Mexico and elsewhere. It is clear that rainfed and ejido land may be the most severely affected by global warming. However, if global warming brings more intense droughts, then even commercial agriculture, with its dependence on irrigation and improved crop varieties, may also suffer.

TABLE 4.6. *Hazard Losses and Land Tenure*

	Percentage of land sown but not harvested	
	Private	Ejido
1930	14.9	23.2
1940	12.0	16.8
1950	10.2	20.2
1960	11.8	17.7
1970	26.7	28.0
1980	12.3	14.9

Vulnerability at the Local Scale

The tremendous heterogeneity of the Mexican landscape means that even these regional assessments smooth over the real temporal, spatial and social variations in vulnerability to climate. Concrete insights into climate vulnerability are provided by the many studies of Mexican communities that emphasize the local use of land and water in different parts of Mexico. These studies, undertaken by geographers and anthropologists in the cultural-ecology tradition, reveal the meaning and experience of climatic change and variability in rural Mexico.

For example, Johnson's study of resource use in an Otomi community in the Mezquital valley, just north of Mexico City, reveals a long history of conflict over land and water in this drought-prone region of the central highlands (Johnson, 1975). Traditional livelihoods of hunting and gathering and rainfed maize production were transformed, first by the Aztec through their systems of tribute, then by the Spanish through forceful land takeovers and the introduction of livestock and cash crops, and, more recently, the Mexican government through land reform and irrigation. Johnson shows how these processes have resulted in inequalities in access to land and water, and an increase in vulnerability to drought. For example, she indicates that the Spanish colonists took over the better land in the relatively well-watered valley and planted cash crops that local people could not consume in dry years. Extensive livestock grazing introduced onto the drier land during the colonial period led to the denudation of hillsides and reductions in the moisture-holding capacity of vegetation and soil. She believes that land distribution and agricultural development have been biased toward the more powerful groups in the valley, and have continued to limit the access of subsistence farmers to favorable moisture conditions. Social stratification and vulnerability are documented using data on ownership of rainfed and irrigated land, and on indicators of wealth such as livestock ownership. Johnson finds that 83 percent of the households in the community have so little good land that they produce food for only half their needs in an average year. Almost a half of the households can provide only a quarter of their food needs. With limited opportunities for these people to earn extra income from crafts or from wage labor on adjacent commercial farms or in Mexico City, Johnson notes that vulnerability to frequent (1 in 4) drought years is extreme.

Similar inequities in access to land and water have been documented by DeWalt (1979) in the municipio of Temasalcingo, in the Rio Lerma valley, northwest of Mexico City. DeWalt stresses the extraordinary heterogeneity of Mexican landscape and societies. Like Johnson, he documents the impact of Spanish colonialism and postrevolutionary government intervention in this productive valley. Although land distribution is more equal than in many other parts of Mexico, he notes that microecological differences in topography, water availability and soil type create wide variations in the level and variability of yields. Low-lying fields in the valley are at risk from fall frosts brought by cold air flowing down the hillsides, and they may also become waterlogged. More sandy soils are vulnerable to drought. He found that poor farmers with smaller plots of this less favorable land are at economic and nutritional risk from droughts, frosts, and floods in the valley. Wealthier farmers may have several plots and alternative income sources in bad years. They can also afford to purchase fertilizer and machines to increase their production. Many farmers do not make full or efficient use of irrigation. Some will not water after planting, even in dry years, for fear that the water will pond and destroy young plants in the uneven fields. Population increases and inheritance in the community are also resulting in smaller, less viable plots, competition for land and water, and environmental degradation.

To the southeast of Mexico City, Enge and Whiteford (1989) have studied the use of land and water in the Tehuacan valley, which has been cultivated for over 2,000 years. They describe how drought occurs three out of every ten years and agriculture depends on a variety of water-management systems. They provided a detailed historical background to the politics of water in several communities. Some farmers depend on water from the springs in the valley. These springs, which have been privately owned in the past, were fiercely contested during reforms that reallocated land without the accompanying water rights. Some ejidos have been unable to farm their land because they could not buy or obtain water. Some farmers form cooperatives to construct *galerias* - tunnels dug back into hillsides to meet the water table. Other, wealthier farmers have drilled deep private wells that have already started to run dry and that sap water from the galerias and springs. The depletion of water resources in the valley has also been associated with population growth (3.4 percent annually) and with the expansion of commercial springwater bottling at cities farther along the valley. From 1976 to 1983, Enge and Whiteford report that spring output dropped between 10 percent and 70 percent, and flow from galerias declined by an average of 38 percent. Water is actually sold on the market in the Tehuacan valley, unlike most other parts of Mexico where it is state owned or privately owned and rarely priced. Enge and Whiteford show how shifts from subsistence to cash crops have raised the price of water and left many people unable to farm their lands except in wetter years. Some farmers "sharecrop" their land in order to gain access to water from the "waterlords" of the valley. The amount and distribution of water in the Tehuacan valley has created conflict and resulted in certain people accumulating wealth and influence in local politics.

Water rights in the southwestern state of Oaxaca, as Kirkby (1973) and Lees (1976) have shown, are somewhat different. The scanty rainfall of the Oaxaca valley is managed through small floodwater dams, canals and hand-operated wells. Water rights are associated with the land granted to ejidos and are obtained through membership of a community rather than ownership or size of landholding. Although the Oaxaca valley contains numerous small holdings of less than 2 ha, much of the best irrigated land is still owned by the large haciendas stemming from the colonial period. Yields, and often food availability, are determined by the timing of the rains, irrigation water supplies, and, in some higher parts of the valley, by frosts. Interestingly, Kirkby suggests that yields are not always as high as they could be in the valley. She suggests that farmers are "satisfiers" rather than "optimizers," planting less land when rainfall is favorable and they know yields may be higher. In low rainfall years, however, the limitations of landholding size and labor mean that adequate land cannot be planted to support household sub-

sistence needs.

The desert-like conditions of northwest Mexico bring even greater risks and conflicts to agriculture and water management. In interior regions of Sonora, communities still rely on intermittent streams and land along valley bottoms. Sheridan (1988), in his study of the municipio of Cucurpe, on the Rio Salado, conveys the pivotal role of land distribution and water availability in community structure, conflict and cooperation. He finds that six percent of the households control 56 percent of the irrigated land. Thirty-five percent of the irrigated land belongs to non-residents who can rely on outside resources and incomes when water is scarce. Fifty-eight percent of the households have no access to irrigated land. Many of the latter group are extremely poor, living in adobe houses without electricity and growing a little rainfed corn and beans.

Economic conditions and government incentives have led many farmers to switch from growing corn and wheat to producing alfalfa, barley and sorghum for livestock forage. Although alfalfa requires a lot of water, barley and sorghum are seen as an economic hedge in low rainfall years because of their drought resistance. They cannot, however, be easily consumed by local hungry households. Livestock raising is important on the higher land in the municipio. Again, the unequal distribution of land among private owners and communities results in the overgrazing and desertification of some areas. Sheridan (1988) notes that most of the ranchers think that land quality has declined, not because of overgrazing, but because of a decline in rainfall since 1943. Water allocation to fields and canals in Cucurpe is decided by an elected water judge, although, in general, upstream users have priority over those downstream. Sheridan notes that farmers can become so desperate for water in dry years that they steal water allocations and violence can erupt.

Coastal Sonora, where large-scale irrigation supports an export-oriented agriculture, has also seen great changes and conflicts in the management and distribution of water resources. Hewitt de Alcantara (1976) describes how massive government investment in irrigation in northwest Mexico during the first half of the twentieth century brought dramatic increases in crop production and benefited the families of those revolutionary generals who received irrigable land. These irrigated lands are now sold for prohibitive amounts. She describes the problems now being encountered on much of the irrigated land because of overpumping of wells and inefficient application of free or subsidized water. Vast areas of land are becoming salinized, especially around Hermosillo. Her study of the Yaqui valley shows the difficulties encountered by indigenous communities in obtaining access to land, water, credit and other inputs. She indicates that ejiditarios, frustrated by unequal access to resources, and frightened by the imbalance between their input costs and crop yields in dry years, are increasingly renting their land to large private farmers.

What do these studies tell us about changing vulnerabilities to water shortages in rainfed and irrigated agriculture in Mexico? They all indicate that climatic hazards are indeed frequent and serious at the local level, often bringing hunger, poverty and migration. Drought, flooding and frosts are the main problems, sometimes combining in years where the rains come late, followed by devastating floods and early frosts. In these years, only those with outside resources or non-farm income are resilient. Climatic hazards may, by impoverishing some, also create opportunities for others to obtain their land. These local studies demonstrate the exceptional variety of the Mexican landscape with microecological differences in soils, slope, and aspect that often define why some farmers are vulnerable to drought and others are not. The studies show how differences in land quality have given rise to competition for land in many communities. Historically, these land conflicts have been resolved by force or by government policy, which some claim is biased toward the more powerful groups. Many Mexicans have rainfed plots of land that are too small to provide food or an income, even in average years. In most communities, successful agriculture requires irrigation, and this has created complex social organizations and politics for irrigation management. Irrigation is a source of wealth and power in rural Mexico because it buffers people against variations in the climate. In most communities, access to irrigation and good land is unequally distributed and this inequality is felt hardest when rains are limited or water resources become depleted. Further irrigation development may benefit

some, but it has been demonstrated in the past to have increased the vulnerability of others, and to have fostered both intercommunity and intracommunity conflict. The studies also demonstrate how irrigation is exhausting water resources and leading to the salinization of fertile land.

In many cases the availability of fertilizer, hybrid seeds and machines has permitted some farmers to take advantage of irrigation and stabilize their yields. But some communities have found that fertilizer applications in dry years harm the crops and incur costs that cannot be recuperated. New seeds are costly and are sometimes too tightly attuned to average conditions, failing more easily than traditional open pollinated varieties in extreme years. Economic crisis means that many farmers can no longer afford the inputs that they were persuaded to use in more affluent times.

Local communities in Mexico have many traditional technologies for coping with environmental variation and limitations. In addition to the floodwater, galeria and well-irrigation systems mentioned already, others have also described ingenious ways of modifying microclimate through raised fields, terraces, and intercropping (Nahmad, Gonzalez and Rees, 1988; Johnson, 1975; Wilken, 1987). In drought years some communities rely on traditional sources of food and liquid such as the maguey and nopal cacti.

Market integration, with the production of crops for sale and export, has benefited some people and communities. Some of the cash crops are more drought resistant than traditional food crops and they receive higher prices. But forage and fiber crops grown for sale cannot be easily converted to subsistence food sources for local people when markets or rains fail.

Finally, both population growth and redistribution have clearly increased vulnerability to drought in some communities by fostering land subdivisions into unsustainable parcels and increasing pressure on land and water resources. In some cases, this population pressure results from high local birthrates, but in other cases it results from migration from other areas where land concentration and agricultural modernization have occurred.

These community studies can provide useful insights into the distribution and dynamics of climate vulnerability in Mexico. They illustrate how local politics, local knowledge, land tenure, water rights, government subsidies, agricultural technology and microenvironments define the risks of producing different crops under varying environmental conditions. This chapter suggests that they also provide insights into how global warming might be locally experienced and adapted to in Mexico.

For example, if global warming means more frequent droughts in Mexico, then conflict over land and water, land concentration and rural migration may increase. If government responds to water shortages by expanding irrigation or intensifying yields, then it may be important to avoid some of the problems associated with previous efforts. Local communities may exploit traditional resources and technologies to cope with a changing climate. Population growth, market integration, technological change, and political development will alter vulnerability to climate in Mexico's future as they have in Mexico's past.

Policy Implications

What are the policy options and sustainable development strategies for Mexico in the context of global warming? This chapter has established that global warming may have many serious impacts on Mexican agriculture and water resources. Despite inconsistencies among the different climate model results, and among model simulations and observed climate, it is evident that, whichever model is used, the general direction of global change in Mexico is toward hotter, drier conditions. The chapter also demonstrates the profound vulnerability of Mexico to climatic change and variation, demonstrated at the local level by intense conflicts over access to water, and at the regional level by high drought losses and differential vulnerability. Overall, it may be a rise in the number and severity of natural disasters such as floods, hurricanes, and droughts that may pose the greatest risks to Mexico's food and economic security in a changed climate.

Global warming has significant implications for a wide range of agricultural, energy, and conservation policies in Mexico and for many development projects planned or al-

ready underway. Any future hydroelectric or irrigation project should probably take account of possible declines in water supply and increased conflicts over access to water, which may occur with global warming. The results also lend support to the calls for sustainable development, with less deforestation, greater care of soil and water resources, and stewardship of biodiversity. Groups concerned with such issues in Mexico should probably consider what a warmer, drier, and possibly more variable and extreme climate will mean for their programs, and adjust them accordingly. Despite the urgency of other problems in Mexico, the possible impacts of climate change merit more attention. Hydroelectric energy, irrigation systems, and limited urban water resources could be threatened by any decline in water supplies in major river basins. A rise in sea level could threaten coastal agriculture, industry, and tourist developments, and higher temperatures could directly affect human comfort and labor productivity. Climatic changes could alter natural ecosystems, affect biosphere reserves, important species, and pests and diseases. Changes in resources along borders with the United States and Central America could create political tensions, and any change in Mexican agricultural production will have some impacts on regional and global trade in key products such as maize, wheat, cotton, winter fruit and vegetables, and beef. In planning irrigation, energy, and other developments, in thinking about food security, and in joining the planned international negotiations to control global warming, Mexico needs to be fully aware of the implications of climate change for national security. For, in addition to the direct impacts of climate change on agricultural production and other sectors, global climate change may be accompanied by major shifts in the agricultural comparative advantage and population movements of both Mexico and other nations (Crosson, 1989). The contributions of Mexico to greenhouse gas concentrations also need to be assessed, and the feasibility of limiting fossil fuel burning and deforestation discussed, so that Mexico can decide its position in any international negotiations. Mexico has so far made only a small contribution to such increases in carbon dioxide and methane, but future expansion of population, industry and agriculture could bring much greater responsibility.

Because Mexico has many problems that may be more urgent, it is important to find ways to tie the response to global warming with other programs and policies and to confront the considerable uncertainties in predictions of regional climate change and its impacts. The Mexican government is already attempting to do this by linking local air pollution control programs and improvements in energy efficiency to reductions in greenhouse gas emissions.

International economic and political relations are also important in considering the response to global warming. Mexico is likely to come under international pressure as a significant source of greenhouse gas emissions from fossil fuel burning, livestock production, and deforestation. Although Mexico was an early signatory of the Montreal ozone protocol, and is eager to participate in discussions about a greenhouse gas convention, the ability to control fossil fuel emissions or deforestation may be limited because of economic crises and enforcement problems. Foreign debt and trade imbalances hamper the ability of the Mexican government to respond to economic and ecological crises and to manage the risks of climatic change.

Finally, the climate model results and vulnerability analysis contained in this chapter have been used to describe some of possible regional impacts of global warming in Mexico. In doing so, some of the uncertainties, conceptual frameworks and methods that can be used to understand the regional impacts of global warming have been illustrated, and some responses designed to manage the risks of climate change were outlined.

References

Adams R.M., C. Rosenzweig, R.M. Peart, *et al.*, 1990. Global climate change and U.S. agriculture. *Nature*, **345**:219-24.

Bach W., H.J. Jung, and H. Knottenburg, 1985. *Modeling the Influence of Carbon Dioxide on the Global and Regional Climate: Methodology and Results.* Ferdinand Schoeningh, Paderborn, Germany.

Bailey N.A., and R. Cohen R, 1987. *The Mexican Timebomb.* Priority Press, New York.

Barkin D. and B. Suarez, 1983. *El Fin de Principio: Las Semillas y la Seguridad Alimentaria.* Ediciones Oceano, Mexico.

Castaneda J.G., 1986. Mexico's coming challenges. *Foreign Policy*, **64**:120-39.

Cohen S.J., 1986. Impacts of CO_2-induced climatic change on water resources in the Great Lakes basin. *Climatic Change*, **8**:135-53.

Coll-Hurtado, A., 1982. *Es Mexico un Pais Agricola?: Un analisis geografico.* Siglo Veintiuno, Mexico, D.F.

Crosson P., 1989. Greenhouse warming and climate change: Why should we care? *Food Policy* (May): 107-18.

DeWalt B.R., 1979. *Modernization in a Mexican Ejido.* Cambridge University Press, New York.

Dovring, F., 1970. Land reform and productivity in Mexico. *Land Economics*, **46**:264-74.

Enge K.I., and S. Whiteford, 1989. *The Keepers of Water and Earth: Mexican Rural Social Organization and Irrigation.* University of Texas Press, Austin.

Florescano, E., 1969. *Precios del Maiz y Crisis Agricolas en Mexico 1708-1810.* Ediciones Era., Mexico, D.F.

———, 1980. Una historia olvidada: La sequia en Mexico. *Nexos*, **32**:9-18.

Garcia, E. de Miranda, 1965. *Distribucion de Precipitacion en la Republica Mexicana.* Instituto de Geografia de Universidad Nacional Autonomia Mexicana (UNAM), Mexico, D.F.

Garcia, R.V., 1981. *Nature Pleads Not Guilty.* Oxford, U.K., Pergamon Press.

Gleick P.H., 1987. Regional hydrological consequences of increases in atmospheric CO_2 and other trace gases. *Climatic Change*, **10**:137-60.

———, 1988. The effects of future climatic changes on international water resources: The Colorado River, the United States, and Mexico. *Policy Sciences*, **21**:23-39.

———, 1989. Climate change and international politics: Problems facing developing countries. *Ambio*, **18**:333-39.

Grotsch S., 1988. *Regional Intercomparison of GCM Predictions and Historical Climate Data.* U.S. Department of Energy, Washington D.C.

Hansen J., I. Fung, A. Lacis, *et al.*, 1988. Global climate changes as forecast by the

GISS three-dimensional climate model. *Journal of Geophysical Research*, **93**:9341-64.

Harrison P., 1984. *Land, Food, and People.* FAO, Rome.

Hazell, P.B.R., 1984. Sources of increased instability in Indian and U.S. cereal production. *American Journal of Agricultural Economics*, **66**:302-11.

Hewitt de Alcantara, C.,1976. *Modernizing Mexican Agriculture.* United Nations Institute for Research on Society and Development (UNIRSD), Geneva.

Hewitt, K. (Ed.), 1984. *Interpretations of Calamity.* Allen and Unwin, Boston.

Jenne, R., 1989. Data from climate models. Unpublished manuscript. National Center for Atmsopheric Research, Boulder, Colo.

Jodha N.S., 1989. Potential strategies for adpating to greenhouse warming: Perspectives from the developing world. In: *Greenhouse Warming: Abatement or Adaption?* N.J. Rosenberg, W.E. Easterling, P.R. Crosson, *et al.*.(Eds.), Resources for the Future, Washington D.C., pp. 147-58.

Johnson, K., 1975. Do as the land bids: Otomi resource use on the eve of irrigation. Doctoral dissertation, Clark University, Worcester, Mass.

Jones C.A., and J.R. Kiniry, 1986. *CERES-Maize: A Simulation Model of Maize Growth and Development.* Texas A&M Press, College Station.

Kandell, J., 1988. *La Capital: The Biography of Mexico City.* Random House, New York.

Kellogg W.W., and Z. Zhao, 1988. Sensitivity of soil moisture to doubling of carbon dioxide in climate experiments. Part 1: North America. *Journal of Climate*, **1**:348-66.

Kirkby, A.V.T., 1973. The use of land and water resources in the past and present valley of Oaxaca, Mexico. *Memoirs of the Museum of Anthropology No 5.* University of Michigan, Ann Arbor.

Lees, S.H., 1976. Hydraulic development and political response in the Valley of Oaxaca, Mexico. *Anthropological Quarterly*, **49**:107-210.

Liverman, D.M., 1990a. Vulnerability to drought in Mexico: The cases of Sonora and Puebla in 1970. *Annals of the Association of American Geographers*, **80**:49-72.

———, 1990b. Seguridad y Medioambiente en Mexico. In: *Temas de Seguridad Mexicana.* S. Aguayo and B.Bagley (Eds.), Siglo Veintiuno, Mexico.

Manabe S. and R.T. Wetherald, 1987. Large-scale changes in soil wetness induced by an increase in atmospheric carbon dioxide. *Journal of the Atmospheric Sciences*, **44**:1211-35.

McBride, G.M., 1923. *The Land Systems of Mexico.* Conde Nast, New York.

Meehl G.A., and W.M. Washington, 1988. A comparison of soil moisture sensitivity in two global climate models. *Journal of the Atmospheric Sciences*, **45**:1476-92.

Metcalfe, S.E., 1987. Historical data and climate change in Mexico: A review. *The Geographical Journal*, **153**: 211-22.

Michaels, P.J., 1979. The response of the Green Revolution to climatic variability. *Climatic Change*, **5**:255-79.

Mosino, P.A., 1975. Los climas de la Republica Mexicana. In: *El Escenario Geografico.* Z. Cserna (Ed.). UNAM, Mexico, pp. 57-171.

Mueller, M.W., 1970. Changing patterns of agricultural output and productivity in the private and land reform sectors in Mexico, 1940-1960. *Economic Development and Cultural Change*, **18**:262-66.

Nahmad S., A. Gonzalez and M. Rees, 1988. Tecnologias indigenas y medio ambiente. Centro de Ecodesarrollo, Mexico.

Nguyen, D.T., 1979. The effects of land reform on agricultural production, employment and income distribution: A statistical study of Mexican states, 1959-1969. *Economic Journal*, **89**:624-35.

Parry, M.L., T.R. Carter and N.T. Konjin (Eds.), 1988. *The Impact of Climatic Variations on Agriculture. Vol. 1 and 2.* Kluwer, Dordrecht, The Netherlands.

Pastor R.A., and J. G. Castaneda, 1988. *The Limits to Friendship: The United States and Mexico.* Alfred Knopf, New York.

Pearse, A., 1980. *Seeds of Plenty, Seeds of Change.* Clarendon Press, Oxford, U.K.

Sanderson, S. 1986. *The Transformation of Mexican Agriculture: International Structure and the Politics of Rural Change.* Princeton University Press,

Princeton, N.J.

Schneider, S.H., 1989. The changing climate. *Scientific American*, **261**:70-79.

Schlesinger, M.E., and J.F.B. Mitchell, 1987. Model projections of the equilibrium response to increased carbon dioxide. *Review of Geophysics*, **25**:760-98.

————, and Z. Zhao, 1989. Seasonal climatic changes induced by doubled CO_2 as simulated by the OSU atmospheric GCM/mixed-layer ocean model. *Journal of Climate*, **2**:459-95.

Servicio Meteorologico, 1976. *Normales Climatologicas*. Servicio Meteorologico, Mexico.

Sheridan, T.E., 1988. *Where the Dove Calls: The Political Ecology of a Peasant Corporate Community in Northwestern Mexico*. University of Arizona Press,Tucson.

United States Environmental Protection Agency, 1989. *The Potential Effects of Global Climate Change on the United States*. USEPA, Washington, D.C.

Venezian, E.L., and W.K. Gamble, 1969. *The Agricultural Development of Mexico: Its Structure and Growth Since 1950*. Praeger, New York.

Wallen, C.C., 1955. Some characteristics of precipitation in Mexico. *Geografiska Annaler*, **37**:51-85.

Washington, W.M., and G.A. Meehl, 1986. Climate sensitivity due to increased CO_2: Experiments with a coupled atmosphere and ocean GCM. *Climate Dynamics*, **4**:1-38.

Wellhausen, E., 1976. The agriculture of Mexico. *Scientific American*, **235**:128-50.

Whetten, N.L., 1948. *Rural Mexico*. University of Chicago Press, Chicago.

Wilken, G.C., 1987. *The Good Farmers: Traditional Agricultural Resource Management in Mexico and Central America*. University of California Press, Berkeley.

Wilson, C.A., and J.F.B. Mitchell, 1987. A doubled CO_2 climate sensitivity experiment with a global climate model including a simple ocean. *Journal of Geophysical Research*, **92(D11)**:13315-43.

Wionczek, M., 1982. The roots of the Mexican agricultural crisis: Water resources development policies, 1920-1970. *Development and Change*, **13**:365-99.

World Resources Institute, 1990. *World Resources, 1990-91*. Oxford University Press, New York.

Yates, P.L., 1981. *Mexico's Agricultural Dilemma*. University of Arizona Press, Tucson.

5. POTENTIAL IMPACTS OF CLIMATE CHANGE ON HUMAN SETTLEMENT PATTERNS IN THE WEST AND CENTRAL AFRICAN (WACAF) REGION: RESPONSE STRATEGIES

Stella C. Ogbuagu

Climatic conditions tend to dictate the pattern of human settlements, such that extremes of climate generally inhibit huge concentrations of human populations. For example, the polar and desert regions of the world attract the sparcest of human settlements. Furthermore, when people settle communities and initiate other social, economic and political activities, some kind of equilibrium is established between them and the environment/climate. It follows that, when such a balance is altered as a consequence of, say, climate change, serious readjustment problems can be anticipated. It is in this regard that the issue of the impacts of climate change on human settlements in the West and Central African (WACAF) region is examined.

Climate change and one of its manifestations, sea level rise, have come into sharp focus in recent decades, largely because of the anticipated impacts on the human race. Although the debates cover different aspects of the problem, attention is, essentially, drawn to reducing the rate of emission of greenhouse gases (GHGs) by the various countries that make up the international community. The differing perspectives—"tradeable permits" and "comprehensive approach"—indicate the lack of consensus on what should be done to give humanity a cleaner space to inhabit.

In the negotiations that currently prevail, the Third World countries seem to be in a weak position as they argue for the development of their societies with whatever resources are available (Orago, 1991). While the developed world argues for what could be termed "luxury" emissions, developing countries are producing "survival" emissions. While the debate or dialogue goes on, human activities continue to degrade the environment with severe implications for human comfort and survival. This is particularly threatening to the poor countries that are faced with meeting the diverse, but basic, socioeconomic needs of their citizens. Pressured by mounting "other" problems, it may appear injudicious to devote their meager resources to dealing with an issue as uncertain as climate change.

The WACAF region, which stretches from Mauritania in the north to Angola in the south (see Figure 5.1), falls squarely in the financially strapped group of nations. Recent estimates show that the region carries a big chunk of Africa's US$240 billion debt burden. For example, two of its members - Nigeria and Cote d'Ivoire belong to the world's 15 most heavily indebted countries (Uduebo, 1990). This heavy debt load has devastating implications for their people's standard of living (which is estimated to have fallen by 25 percent in the last few years) and economic and political stability of their countries.

To compound the region's problems, it is currently witnessing a phenomenal population growth rate. This is occurring as a result of lowered mortality rates associated with improved health facilities. With governments in the region already financially handicapped to provide, among other things, adequate housing and employment for all who are available and willing to work, individuals have no choice but to engage themselves in the struggle for survival. Indeed, most of them resort to the unrestrained and unplanned use of whatever resources are available such as land, trees, forests and space. There is, therefore, no gainsaying the fact that this high rate of population growth poses additional serious constraints to governments' capacity for achieving sustainable growth.

This lack of adequate funds and facilities compounds the problems already evident in the WACAF region, where many people live dangerously close to coastlines or creeks, with houses built on rafters that could easily be destroyed. It is also a region where many important and expensive structures are located either too close to seashores or on floodplains. In the deserts of the region, many citizens live in arid zones where precipitation is low and evapotranspiration is high. People living in such high-risk locations are vulnerable to major disasters, and these are likely to increase in frequency if the predicted climate change occurs.

It is the concern over what could happen to these high risk settlements and groups that has prompted this study. It is primarily di-rected at looking more closely and critically at their situation and suggesting potential intervention options. There is no doubt that many of the residents of the region have, in several ingenious ways, adjusted to their environments and life-styles, but these are likely to be disrupted by climate change. It is, therefore, important to anticipate such disruptions in an attempt to mitigate the most serious consequences.

To handle the topic adequately, the remainder of the chapter is organized under eight general headings: the aim of the study,

FIGURE 5.1. Countries and zones of the WACAF region (after Ibe, 1989).
Northern Zone: 1. Mauritania, 2. Cape Verde, 3. Senegal, 4. Gambia, 5. Guinea Bissau, 6. Guinea, 7. Sierra Leone, 8. Liberia.
Middle Zone: 9. Ivory Coast, 10. Ghana, 11. Togo, 12. Benin, 13. Nigeria, 14. Cameroon, 15. Equatorial Guinea, 16. Sao Tome and Principe, 17. Gabon.
Southern Zone: 18. Congo, 19. Zaire, 20. Angola, 21. Namibia.

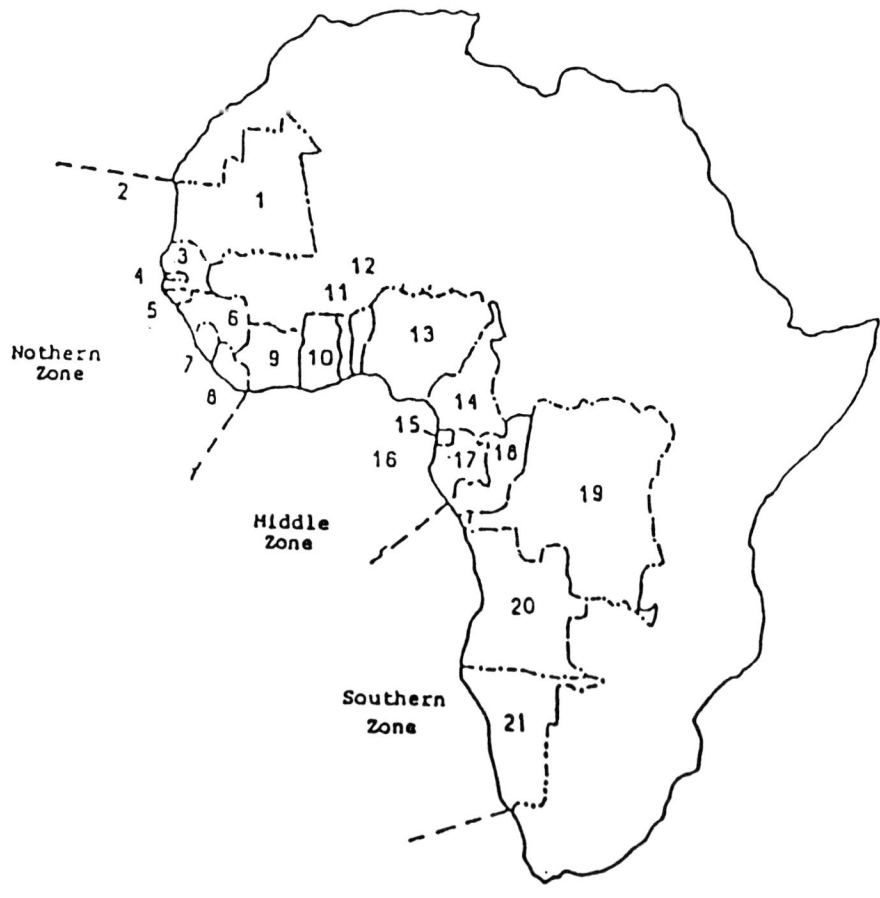

the rationale for choice of region, sources of data, temperature increase and migration in the region, sea level rise and possible relocation of some coastal peoples, current risk-management strategies, suggested risk management options and conclusion.

Aim of the Study

As indicated earlier, the chief aim of the study is to gain an objective insight into the problem of high risk populations in the region. This is in order to demonstrate concretely that, in the event of temperature increase and sea-level rise (SLR), significant proportions of these people would be seriously threatened to the extent that they might require relocation. Second, it is the aim of the study to provide policymakers and implementors with data to enable them to initiate further studies of the areas and their needs, so as to plan accordingly. A third major aim of the work is to create an awareness of the need to establish a multidisciplinary team of experts to monitor, assess and produce problem-solving recommendations with regard to the expected impacts of climate change. The primary thesis is that, if policymakers, planners and implementors are not sufficiently motivated to examine seriously the issues of climate change and how it relates to the lives of people, they might wave it off as "not of immediate concern."

These aims are stated with the awareness that people who have adjusted to their settlements generally feel attached to them. They are usually unwilling to make a change. To uproot them for resettlement could be a Herculean job. If they have to be moved, there is the need to prepare their minds for the move prior to the actual occurrence. It follows that time is required to do the groundwork of educating and helping them to accept the facts of the situation and, thus, make alternative plans for their lives. Without doubt, if this is done well ahead of time, lives and property would be saved. But if left until the changes take place, there would be little or no time for action, with disastrous consequences.

From the scientific predictions and projections already available, it does not appear that there is much time to waste. This is especially so if consideration is given to the

view that the scale of the anticipated climate change would have widespread adverse effects globally.

Rationale for Constituting WACAF as a Region

The countries of the WACAF region (see Figure 5.1) share a number of similarities—poverty, rapid population growth, a coastal location for most of their big cities, and a large proportion of their citizenry. Consequently, they are preoccupied with finding solutions to their individual developmental problems. For this reason, it would be difficult for them individually to establish effective research, monitoring and assessment centers to handle the challenging impacts of climate change reasonably.

Another important consideration is the fact that the United Nations Environment Programme (UNEP) has earmarked WACAF as a region for climate change assessment studies. Thus, relevant literature already exists on the region. For example, the UNEP expert team set up in 1989 wrote an overview of the impact of climate change on the region. Earlier, Ibé and Qué lennec (1989) had written for UNEP on the "Methodology for assessment and control of coastal erosion in West and Central Africa." These and several other sources provide some basic materials for additional work.

It is further reasoned that the expert team already established could serve as a resource and form a strong base for networking in the region. The region's manageable size is an added advantage. It is neither too small for pulling resources together nor too large to render cooperation impossible. In all, it is a convenient unit of analysis.

Sources of Data

The primary source of data derives from actual visits to some high risk sites. Given resource constraints, visits were limited to areas in Nigeria and the Meteorological Office in Ghana. Information from these two countries (which seems fairly representative of the WACAF region) would form the bulk of the materials used. Secondary data from works already done in the area, as well as media reports, are brought in at relevant points in the presentation to supplement site-specific information.

It must be pointed out that the relative newness of climate change concern and study in the region shows up in the lack of hard data on many of the issues. Several existing works are couched in tentative statements, just as the scenarios provided represent little more than "mere" predictions. This situation is even more pronounced in the developing countries, where meteorological centers are not adequately provided with necessary facilities. Limitations of this nature impose constraints on data buildup and level of analysis. Furthermore, in many of the countries of the region, very little assessment of the potential risks and management strategies has been done. Nevertheless, a lot of interest has been generated, and many scholars are now busy studying different aspects of climate change and its potential impacts. Such studies are likely to provide more concrete evidence of what to anticipate and should inform planners on how best to proceed.

Temperature Increase and Migration

Variations in climate in the WACAF region wreck havoc on the inhabitants because of their lack of preparedness to deal with the impacts. The drought of the early 1980s in the Sahel effectively dramatizes the vulnerability of poor countries to the vagaries of weather. Writing about the effect of droughts, Mazawase (1985) notes that drought often devastates several of the Sahelian countries and renders their governments' efforts at achieving food self-sufficiency ineffective.

The drought of the 1980s is not the first of its kind in the region's history. On the contrary, droughts have occurred repeatedly. The sad part is the unpreparedness of the governments and peoples of the affected areas to mitigate the impacts and to forestall future occurrences. Instead, each event is like a new experience in its devastation. It usually leaves in its trail wrecked homes, farms, livestock and casualties each time. In Nigeria, for instance, Mazawase (1985) observes that from the most severely affected areas—Borno, Sokoto, Kano, Gongola, and Kaduna states—people were moving south on a daily basis, having lost their "lands and pastures" to the encroaching desert. In Ghana and other countries that are similarly affected, the story is the same.

If the governments of these countries had sustained plans to rehabilitate the migrants, perhaps their sufferings could be mitigated. Usually, however, what is observed is ad hoc efforts. Thus, when the events occur, immediate relief of food and some clothing may be offered by the government. After that, the migrants are on their own. Given this situation, the migrants usually move to other less affected areas. In the new places, they swell the resident population, exert additional pressure on the already overpopulated and overused environment, and as a result cause further environmental degradation, as well as increased societal misery.

To illustrate, a significant number of the victims of the drought in the 1980s can still be seen begging for alms in many streets in Nigeria and Ghana. This experience is often reported in other parts of the region. Generally, drought-rendered refugees have lost virtually all they owned in life.

It is paradoxical that humans, who often are the worst affected by such disasters, can be said to be the "architect of their own misfortune." For example, humans engage in indiscriminate tree destruction. On this score alone, it is not surprising that many scientists estimate that at the current rate of destruction, much of the tropical forests will have been destroyed before the end of the next century. By UN estimates, about "11 million hectares of tropical forests are lost every year" (Hinrichsen 1990). In another case, a World Bank official expresses horror at the devastation being caused to a Liberian forest reserve by a foreign mining company, in addition to the bush burning and tree destruction by refugees fleeing from the war areas (Sawyer, 1990). What is true for Liberia (a WACAF nation), except for the war effect, is not unique but can be said of almost all the countries in the region.

To underscore this point, it is necessary to state that, in an effort to provide data for writing this chapter, some known sites of "natural forests" in Imo, Rivers, Akwa Ibom and Anambra States (all in the tropical forest zone) of Nigeria were revisited. Out of the 13 sites, eight have been converted to farmland, three are now industrial locations, while two have been transformed into residential areas. Two of the forests were of particular interest, as the writer had lived close to them (one in Umuagwo and another in Mbieri—both locations in the state of Imo). These two forests had huge, ancient trees, big animals, snakes, tortoise and so on. The one at Mbieri grew umbrella-like mushrooms that children loved to pick. Indeed, it was called "Nneohia," that is, "Mother of forests." Both of these places are now farmlands looking forlorn, with hardly any big standing trees. Viewing them makes one muse nostalgically over what used to be but is no more.

Unfortunately, trees destroyed in the process described above are not usually replanted. The resultant effect is further desertification, enhancement of the greenhouse effect, and possible depletion of the ozone layer. This cycle of environmental degradation is likely to continue unless massive actions are taken to reverse it.

Conditions as described above are bad enough. However, there is no doubt that should temperatures rise by the predicted 0.3°C per decade, the situation would be aggravated by increased human migration to cooler and wetter places. The receiving areas would most likely record greater overcrowding, overuse of resources and facilities, further depletion of the quality of the environment, with the attendant negative effects on already endangered human settlements.

This is illustrated by the situation in Calabar in the southeastern part of Nigeria, which experienced a large influx of people who had left areas of severe drought conditions. Many of these individuals squatted with kind-hearted people and had to resort to begging. To the general public, they constitute a nuisance, given the fact that they pester their victims until they are given alms. To the migrants, it is no doubt a dehumanizing experience to be reduced to "mere" beggars. Many more persons would be so humiliated if climate change impacts are ignored until they occur.

To farmers even in the wetter regions of the south, increases in temperature would most certainly cause untold hardships. For instance, farmers were distraught with worry when the rains they expected in the first few months of 1990 failed to materialize. It was obvious that they depended entirely on fate to counter the dryness and high temperatures. If, for example, the rains did not fall, many of the farmers would have been forced to move to more favorable locations, at least as an interim measure.

The question is, for how long can people live precariously in the hope that nature will remain clement? Already, many scientists are hard at work in efforts to provide realistic scenarios of climate change in order to plan for appropriate intervention. The considerable evidence that they have provided should be seriously considered in order to devise protective measures for areas most at risk.

Sea Level Rise and Possible Relocation of Some Peoples in High-Risk Zones

It can be argued that many areas of the WACAF region suffer from the menace of floods. This view is shared by Ibé and Qué lennec (1989) who, reviewing the situation of erosion in the region, observed that: "erosion is a prevalent phenomenon along the Atlantic coast of West and Central Africa." Indeed, floods along the low lying coastal areas are commonplace. Even further inland, the rainy season (which normally lasts from April to October in many instances) is usually a nightmare to residents of the flood-prone zones. For example, several parts of southern Nigeria, eastern Ghana, as well as parts of Sierra Leone and Benin, to mention a few, get heavily inundated during the rains. In this period, the rains are particularly heavy, and in some months up to 660.5 millimeters of rainfall may be recorded. In those months, particularly June and July, the city of Calabar is usually clogged with water. As a result, many of its sections remain inaccessible with clogged gutters, broken roads, flooded streets, and damaged physical installations and drainage systems. Such periods remain a nightmare to motorists. Indeed, many road users stay at home or literally swim through gushing torrents of water.

Calabar is not alone in this experience. In the Lagos areas (another site visited), floods threaten many structures. In a recent incident, the Victoria beach in Lagos was so badly flooded that some of its most posh buildings and hotels were seriously in danger of being overtaken by the floods. In other sections of the city, many residents literally could not get into their homes because of flooding.

In other parts of the country—for example, Nanka, Oko, Agulueze Chukwu, Umuezechukwu and Umuchu in Anambra state—erosion is so ferocious that some houses are known to have fallen into erosion gullies. The threat to human habitation is so great that relocation of various parts of these towns is under serious consideration. In Ikwuano/Umuahia, Uzuakoli and other towns in the state of Imo, the major roads have nearly all been rendered impassable because of mile deep gullies created by flood erosions. Many of these problems are far beyond the capacity of state governments and require external assistance. Similar stories are told about parts of Ghana, especially the Takoradi-Sekondi region where the sea is fast encroaching on the land. In fact, some areas along the Cape Coast are so close to being submerged that the residents are considering moving out.

Even the relatively drier northern areas of the region do not escape flooding during the few months of rainfall. For instance, Hart (1989) writes under the heading, "Weathermen predict thunderstorms, flood in the North." He notes also "the loss of lives and property consequent upon intense rains in the South." He further highlights the warning of the "Weathermen" that "the worst is not over yet." This report affirms the common observation that the problem of floods and erosions are here to stay because of heavy rains and very poor and ineffective drainage systems. The concern is that they will remain as long as lessons are not learnt from previous experiences and adequate plans are not made to counter the impact of this virtually annual occurrence.

In some of the islands the threat of the sea is foreboding. Before dikes were constructed on the islands of Bonny and Opobo in Nigeria, land structures such as churches, playing fields, tombstones, commercial houses, etc., were almost washed off. In the creek areas of the region, it is a common sight to see small settlements dangerously located on tiny islands. The huts literally hang on the edge of the water. One cannot but wonder what would be the fate of such citizens in the event of a small rise in sea level.

It has been necessary to examine the flood and erosion situation in the region in order to prepare the ground for assessing what might happen if the predicted 6 cm per decade rise in sea level occurs over the next several decades. In this case, the low lying areas would most probably be flooded, people's lives would be severely disrupted, their economies altered, and their residences threatened. In Nigeria and Ghana, as well as in other countries of the region, thousands of people do lose their homes to floods each year. If there is a further increase in flooding due to SLR, this problem will no doubt be accentuated.

It stands to reason that, when people's settlements are destroyed, alternative arrangements will have to be made for them. From experiments carried out in several places, especially in Nigeria, it is obvious that resettlement is not easy. Recently, one of the worst slums in Lagos (Maroko) was forcibly demolished by the government and most residents were either poorly relocated or not relocated at all. Affected people found it extremely difficult to settle in their new locations. The great difficulties experienced from this exercise show that, if there is the need to relocate people in the event of climate change, the time to start educating them and creating awareness of the implications of such anticipated changes is now. Areas of need should be mapped out as a matter of immediate concern, and the people prepared for relocation.

Current Risk-Management Strategy

Effective risk management should be based on sound information and proper assessment of the situation. Although some records exist in government offices on flooding and drought, they are grossly inadequate for meaningful determination of the causes, intensity of the problem, future prospects of recurrence, and possible preventive measures. Mainly because of the inadequacy of these data in the WACAF region, most actions of government and other bodies are hardly pre-planned. In other words, much of the risk management in the region has been carried out on an ad hoc and reactive basis. Often, too, response strategies have different meanings to different administrators. An example of the reactions of two military governors to the issue of flooding in their states will help to underscore this point. One of the governors is reported to have stated that his government had set up an agency to "cater" victims of the 1988 flood, as well as handle other flood related problems in the state. To prevent future flood menace to his people, he expressed the willingness of his government to relocate the people living in "likely target areas."

In contrast to this governor's reactions, his counterpart in another "victim" state reportedly argued that there was no need to make any contingency plans for fighting floods. Rather, he considered it appropriate for relevant government agencies to be contacted whenever floods occur (Fadason, 1989). This perspective represents the view of most government planners. In effect, adequate forward-looking programs are rarely put into operation in preparation for a possible recurrence.

Although this is the general pattern, it needs to be stated that governments and their agencies have sometimes taken some limited steps to prevent further repetitions, especially with regard to SLR. Specifically, governments have sometimes found it expedient to build embankments in some high risk areas in the region to forestall the sea rising beyond containable limits. Such are the cases in Opobo, Calabar and Lagos in Nigeria, for example. In the case of Lagos—Victoria beach—flood-prevention constructions were recently washed out and, as a result, the sea water threatened lives and property.

Risk management in the WACAF region has been anything but adequate. As has been pointed out, usually relief materials are rushed to the affected areas and victims are given temporary shelters. When the flood dries up, the people return to their old locations, only to suffer a similar fate in the next season of flooding.

Governments in the region are becoming aware of the hazards posed by environmental destruction caused by human activities. It seems that the enhanced tempo of global activities in this regard has greatly aided this awareness. They now realize that if the environment is not adequately protected, even their development efforts will be jeopardized. To this end, some of the societies have, in response to the call for better and cleaner environments, initiated policies directed at addressing these problems. In this spirit, Nigeria has put in place a policy on the National Environment. This policy is generally designed to:

> secure for all Nigerians a quality of environment adequate for their health and well being, conserve and use the environment and national resources for the benefit of present and future generations, raise public awareness and promote understanding of essential linkages between environment and development (Odeyinka, 1989, p. 9).

It must be stated that the articulation of this policy is the culmination of various efforts of the federal government, pressured by nongovernmental agencies, to bring issues of environment and development to the forefront of the public agenda. It should be seen as a big step in the right direction. If it is followed up with concrete programs and actions, Nigeria would have started a move away from the haphazard risk management strategy that had hitherto characterized its reactions to climate variations, fluctuations and change.

Since the national policy on environment is quite recent, it is still too early to assess its impact. From all indications, it is still business as usual for citizens with regard to deforestation, indiscriminate dumping of refuse, industrial pollution, gullies that threaten lives and property, and other environmental problems.

Some Risk-Management Options

Before putting forward any suggestions about risk management, it is necessary to restate some of the factors that contribute to accumulation of greenhouse gases, depletion of the ozone layer and acceleration of the pace of climate change.

Unrestricted and indiscriminate use of natural resources, especially land, forests, trees and water, constitutes the major source of environmental "abuse." For example, natural tropical forests that were the home to varieties of flora and fauna have been turned into farmlands with hardly any big trees still standing. Trees have been ravaged for construction purposes and for firewood. Industrial and human wastes are improperly disposed of, water and air pollution is high, and the population pressures, especially in the urban areas, result in excessive use of available meager socioeconomic facilities. Put together, these factors create conditions that are environmentally unhealthy.

In addressing the issues associated with resource depletion and environmental degradation, it must be realized that vulnerable areas and peoples that might require, for instance, relocation or resettlement in the event of climate change need the most immediate attention. In this regard, three categories can be identified as follows:

- coastal and creek peoples who live precariously close to the coast and on the edges of rivers and lagoons,
- those in drought prone zones and
- the inland populations that suffer annual flooding.

With these three broad categories as special cases, it is necessary to suggest some risk-management strategies. Aina (1989) in his discussion of "Current Trends in Sustainable Development," highlights some of the main issues in climate change in Nigeria. They include "control of desertification, erosion and flood control, emergency event response, establishment of environmental monitoring programs, central and zonal laboratories, international relations . . . legislative and regulatory measures."

For effective action on these major concerns, basic research to provide benchmark data for proper analysis is called for. Local studies of this nature highlight the unique problems of specific sites and guide relevant policy options for them.

So far, such basic data are rarely available in the few meteorological stations in each country, as well as in the region. Even where some data exist, they are largely uncoordinated. It is, therefore, vitally important that the advocated research should cover both local areas and the region as a whole. It should be detailed enough to be meaningfully applied for intervention purposes.

Perhaps of even greater importance is the creation of awareness about man's role in environmental degradation and the buildup of greenhouse gases. Given the region's high rate of illiteracy (which is still above 50 percent of the population), determined efforts should be made to publicize and inform people of the dangers of abusing the environment. It is only when people fully understand the negative consequences of their poor agricultural habits and other practices on the environment that they will begin to appreciate the importance of preserving their habitat.

At the same time as the public is being sufficiently sensitized about environmental matters (emissions and other factors), governments should proceed to put into place supporting legislation that demand certain sound environmental behaviors. Quite often in Third World countries, such legislation is made without public support and the means to enforce it. It usually remains nothing more than a mere adornment on the statute books.

For proper monitoring and assessment of the impacts of climate change generally and on human settlements in particular, it is suggested that regional, national, state and local government agencies be set up to deal with the issues of global warning. Already in many WACAF countries national agencies in charge of environmental protection are in operation. In Nigeria, for example, there is the Federal Environmental Protection Agency (FEPA), which has omnibus responsibility. Such agencies should be expanded, improved and empowered with appropriate facilities to deal with climate change.

At the regional level, a corresponding organization should be built up and sustained to a level that would ensure adequate and reliable risk assessment, risk management and the accumulation of a hard and useful database. When established, this regional body of experts should network with other regional and international organizations for information and technology sharing.

It needs to be mentioned that this is no time for the developing world to continue arguing that it has contributed to less than 25 percent of the problem of global warming. Although this position is true, it does not help the situation. What is needed, and what will help, are problem-solving actions.

From all indications, it seems likely that countries in the developing world will be the worst hit when the anticipated changes occur, largely because of their weak technological and financial status. They need to wholeheartedly join in global activities to improve the quality of the world environment.

Since climate change is a global issue, it can be reasonably argued that what affects one part would somehow affect the other parts and upset the world ecosystems. Nations in the industrialized world cannot, therefore, rest on their laurels with the belief that all will be well for them. They should assist the WACAF region, and others like it, financially and technically. This help cannot be delayed for too long.

Conclusions

An attempt has been made in this chapter to discuss possible impacts of climate change on the WACAF region, with special reference to human settlements. It was noted that those individuals who "perch" on shorelines or live in floodplains or drought regions are most likely to require resettlement. Since relocation is not an easy task, it was suggested that early preparation in the form of education and training in dealing with possible natural hazards should be started without further delay.

It was noted, however, that to carry out this plan effectively and realistically both international and regional financial and technical assistance need to be made available to the WACAF region.

Furthermore, it was stressed that for any of the proposed strategies to work, there must be efficient ways of monitoring and assessing the impacts of climate change. Of particular importance is the need for area research to determine firsthand the problems of each high-risk location or group so as to make meaningful suggestions on how to reduce the impacts.

When such data are available, it is anticipated that plans and programs on human settlements (in the event of climate change) would be based on sound information. Given the fact that gaps in knowledge of the exact degree and magnitude of the expected change exist, emphasis should be placed on the need

to set up, in the WACAF region, national and regional agencies made up of experts with varied disciplinary backgrounds to provide the necessary leadership role.

Finally, it is argued that, although the already available data are limited, it seems obvious that there is no time to "stand and stare." Instead, intensive studies should be carried out to determine priority areas for immediate intervention. Furthermore, it is observed that, because of the poor technological and financial condition of the WACAF region, any meaningful intervention would require great assistance from the developed world.

The choice seems clear. Either the region is assisted in taking well informed actions to avert the anticipated adverse impacts of climate change, or it does little or nothing and faces the possibility of severe global consequences. There is no doubt that our planet—Earth—would be a happier and safer place to live in if the former option, as stated above, is chosen.

References

Abrahamson, D. E. (Ed.), 1989. *The Challenge of Global Warming*. Island Press, Washington, D.C.

Aina, E. O., 1989. Current trends in sustainable development and the future of the Nigerian environment. *African Peace Research Institute (APRI) Newsletter*, **IV(5)**:10-15.

Cutter Information Corp., 1991. Global Environmental Change Report **III**, No. 15.

Fadason, V., 1989. Northern states take measures to combat floods. *The Guardian*, **16 (4444)**: 1.

Greenpeace, N. D. Climate Convention Commentary. Issue I.

Hart, B., 1989. Weathermen predict thunderstorms, flood in the north. *The Guardian*, **6 (4232)**: 1-2.

Hinrichsen, D., 1990. People and forests: What future? *Earthwatch*, **39**:1-3.

Ibé , A. C., and R. E. Qué lennec, 1989. Methodology for assessment and control of coastal erosion in West and Central Africa. UNEP Regional Seas Reports and Studies No. 1. UNEP, Nairobi.

Krause, F., W. Bach, and J. Koomey, 1990.

Energy Policy in the Green House. Earthscan Publications Limited, London.

Mazawase, S., 1985. Menace of desert encroachment. *Daily Times*, June 7, p. 5.

Odeyinka, G., 1989. FG's policy on Environment Soon. *National Concord*, p. 9.

Ogbuagu, S. C., 1991. Climate change and spatial planning. In: *A Change in the Weather: African Perspectives on Climate Change.* E.S.H. Ominde and C. Juma (Eds.), ACTS Press, Nairobi, pp. 151-63.

Orago, F., 1991. Promoting a healthy climate. *Impact*, **1**:2-11.

Osusuluwa, M. W., 1989. The question of third world debt burden: What role for the non-aligned movement? *African Peace Research Institute (APRI) Newsletter,* **5**:5-9.

Sawyer, J., 1990. Tropical forests—Multiple use, not multiple abuse. *Earthwatch*, **39**:4-6.

Uduebo, M.A., 1990. Debt crisis in the West African subregion: Which way out? *Bullion*, **1**:2-20.

6. GLOBAL WARMING: IMPACTS AND IMPLICATIONS FOR SOUTH ASIA

R. K. Pachauri

The Development Scene

The South Asian subcontinent provides a unique range of political, social, economic, geographic and ecological conditions in the context of which the complex impacts and implications of global warming assume a significant dimension. For instance, if one considers diversity of climate, the countries of the region include areas that experience cold, subzero temperatures perennially, deserts that attain high temperatures matching the hottest regions of the world, and precipitation levels that are the highest anywhere on the planet. Further, not only does the region have the highest mountain peaks on the globe, it also has some of the lowest islands inhabited by humans, islands that rise barely a meter above the sea that surrounds them. In economic terms, South Asia is a poor region, with a large population that ekes out a living at bare subsistence level from land which is degraded and, in several areas, is slipping gradually from the marginal category into an uncultivable condition. On the other hand, the region has centers of significant industrial activity, and a science and technology infrastructure that compares, in some cases, with the best. In several respects, therefore, not only is South Asia a varied and rich microcosm of regions that would experience the full range of impacts that global warming would have on different parts of the globe, but it also provides a cross-section of capabilities and human capacities for coping with or adapting to these impacts.

There is, as yet, relatively meager research in South Asia on the impacts and implications of global warming. Indeed, the first major event to focus attention on this subject in South Asia was the International Conference on Global Warming and Climate Change: Perspectives from Developing Countries, which was held in New Delhi in February 1989. This conference, the proceedings (Gupta and Pachauri 1990) of which are an important contribution to the literature in this field, was attended by outstanding scientists and researchers from all over the world, including those from the countries of South Asia, and led to several policy initiatives on the subject, at least within the government of India. Subsequently, several research projects were launched to study both the likely impacts of global warming in the region and possible responses that at least some of the countries of South Asia can develop and implement. But knowledge in this field specific to the countries in the region remains scant and imperfect, and this chapter, is, therefore, based on extensive analogies drawn from other parts of the world. To unfold a backdrop of likely demographic and economic developments against which the effects of global warming would have to be evaluated, we first review briefly the economic scenario pertaining to South Asia in the coming decades.

A major departure in the thinking of multilateral development organizations has been initiated recently in the UN Development Programme (UNDP) publication on human development (UNDP 1990) and the most recent World Development Report of the World Bank (1990). The UNDP report makes a pertinent point, stating "Life does not begin at $11,000, the average per capita income in the industrial world." The report establishes and reiterates the possibility and importance of human development independent (and in support) of efforts to improve material standards of living. The World Bank report of 1990, cited above, also makes this point in a somewhat different way, emphasizing the importance of poverty removal. This requires provision of adequate social services for health, education and family planning. Human development and the elimination of poverty are critical, not only for minimizing the impacts of global warming on human activities and actions, but also for ensuring that abatement measures and adaptive responses can be launched to the full potential of the human capacity that should be developed in the region. To this end, the prognosis for South Asia does not appear heartening. This is conveyed by the projections made in Tables 6.1 and 6.2.

While these projections portend a substantial decline in undesirable indicators such as illiteracy and child mortality, as well as a major dent on the absolute poverty that

exists in South Asia, this region would still remain weak and deficient in infrastructure to deal with crises and adverse impacts that may have to be faced as a result of global warming. The decade of the 1990s is, therefore, crucial for the people of South Asia, and much will depend on whether the tempo of economic growth of the 1980s can be maintained or improved on and whether the countries of the region will make a greater commitment to the provision and widespread supply of a package of social services.

TABLE 6.1. *Social Indicators by Developing Region, 1985 and 2000*

Region	Net primary school enrollment ratio		Under 5 mortality	
	1985	2000	1985	2000
Sub-Saharan Africa	56	86	185	136
East Asia	96	100	54	31
China	93	95	44	25
South Asia	74	88	150	98
India	81	96	148	94
Eastern Europe	90	92	25	16
Middle East and North Africa	75	94	119	71
Latin America and the Caribbean	92	100	75	52
Total	84	91	102	67

Source: World Development Report, 1990.

TABLE 6.2. *Poverty in the Year 2000 by Developing Region*

Region	Incidence of poverty (%)		Number of poor (millions)	
	1985	2000	1985	2000
Sub-Saharan Africa	46.8	43.1	180	265
East Asia	20.4	4.0	280	70
China	20.0	2.9	210	35
South Asia	50.9	26.0	525	365
India	55.0	25.4	420	255
Eastern Europe	7.8	7.9	5	5
Middle East, N. Africa and other Europe	31.0	22.6	60	60
Latin America and the Caribbean	19.1	11.4	75	60
Total	32.7	18.0	1,125	825

Source: World Development Report, 1990.

Global Warming

The Intergovernmental Panel on Climate Change (IPCC) set up three working groups to explore the major subjects related to global warming and climate change. Working Group I dealt with the scientific assessment of climate change and estimated the extent of warming that is likely to take place up to the year 2100. Under the IPCC business-as-usual scenario of greenhouse gas (GHG) emissions, the average global mean temperature increase during the next century is likely to be about 0.3°C per decade, with an uncertainty range of 0.2°C to 0.5°C (IPCC, 1990). Consequently, the likely increase in global mean temperature would be 1°C above present levels by 2025, about 2°C above temperatures during the pre-industrial period, and then rising to about 3°C above present levels before the end of the next century.

With respect to South Asia, the IPCC business as usual scenario indicates warming of 1° to 2°C in 2030 over the preindustrial period. Precipitation is predicted to increase 5 percent to 15 percent in the summer, but would change very little in the winter throughout the region. The moisture in the soil during summer is projected to increase by 5 percent to 10 percent. There are, of course, various uncertainties associated with these projections, and scientific doubts regarding their validity are widespread, but for the purpose of assessing likely impacts of global warming these magnitudes should be quite acceptable. The general circulation models (GCMs) on which predictions of climate change are based, unfortunately, have not attained a high degree of reliability with respect to regional climatic effects. However, there is a general convergence in their results and some conclusions are applicable to South Asia:

- in general, a warmer globe will also be wetter and more humid, particularly in tropical regions;
- sea level is likely to rise, along with an increase in the temperature of the oceans; and
- tropical regions are likely to suffer from increased risk of storm tides because of the already high frequency of tropical cyclones.

We will evaluate the implications of these possible changes later in these pages, but it would be useful first to assess the human dimensions of likely impacts that may occur in the region as a result of global warming. We will then quantify the number of human beings who would be affected by climate change. Population changes in the South Asian subcontinent are projected on the basis of World Bank estimates, as shown in Table 6.3.

In essence, we would be dealing with over one fifth of humanity when we assess

TABLE 6.3. *Projections of Population in Countries of South Asia* (thousands)

	1985 (Estimated)	1990	2000	2025
Bangladesh	101,147	114,783	146,603	212,672
Bhutan	1,362	1,516	1,891	2,831
India	769,183	847,925	1,010,954	1,292,640
Maldives	183	212	268	359
Nepal	16,915	19,037	23,223	31,126
Pakistan	103,241	120,701	153,307	230,257
Sri Lanka	16,108	17,162	19,011	21,859
Total Regional Population	1,008,139			1,791,744
Total World Population	4,853,848			8,466,516
South Asia (% of World Total)	22.77			21.16

Source: World Bank estimates.

the impacts of global warming in South Asia. Some of the impacts that we will discuss may also have implications for movement of people across international boundaries in the region, and the dimensions of this problem can be assessed on the basis of the populations projected for the countries of the region. In particular, population increases in Bangladesh, Nepal and Pakistan are not only projected to be higher than the average for the region, but constraints in the availability of good land for cultivation and human habitats would make these societies particularly vulnerable to climate change and the associated effects on agriculture and forestry.

Effects of Global Warming

Agriculture

The nations of South Asia have a high dependence on agricultural activities. Table 6.4 shows the value added in agricultural production, as well as GDP, for the countries of South Asia in 1988.

Because of the importance of agriculture, the economic well-being of the countries of South Asia is highly vulnerable to changes in climate. A variety of impacts can be anticipated from each of the potential consequences of global climate change. For instance, some work on assessing the effect of temperature on rice yields indicates that increases in temperature result in a reduced growing season and a decline in productivity. There appears to be an increasing consensus

among scientists that temperature increases would lead to lower yields in wheat and rice.

However, increased levels of CO_2 could be favorable. Several studies, as yet inconclusive, suggest that the responses of agricultural plants to higher levels of CO_2 concentration in the atmosphere depend on whether they belong in the C3 group—i.e., wheat, rice, legumes, oilseeds, cotton—or the C4 group consisting of sugarcane, sorghum or maize. The C3 grouping shows higher CO_2 assimilation, growth and yield in response to higher CO_2 concentration than do the C4 crops. One preliminary survey indicates that CO_2 enrichment caused a 26 percent increase in economic yield of mature crops and 40 percent by dry weight of immature shoot. In general, C3 crops showed a mean increase of 36 percent. At the same time, both C3 and C4 crops showed a decrease of 34 percent or more in transpiration with doubling of CO_2 concentration. This could lead to a net decrease in water requirements.

Some simulations using computer models for crops have been undertaken recently to study the effects of changes in all three sets of variables: temperature, water availability and CO_2 concentration. The indications from these studies are that increased temperatures would have a much greater impact in the higher latitudes, particularly under conditions of dryland farming, but under irrigated conditions the variation would be minor. It also appears that increased CO_2 concentration would compensate for any yield decreases caused by higher temperature, particularly in the case of maize.

TABLE 6.4. *Agricultural Production as a Share of Gross Domestic Product (1988)*

| | Production (millions of dollars) | | Share of |
	GDP	Agriculture	GDP (%)
Bangladesh	19,320	8,882	46
Bhutan	300	130	44
Nepal	2,860	1,601	56
India	237,930	76,618	32
Maldives Pakistan	34,050	8,935	26
Sri Lanka	6,400	1,685	26

Source: World Development Report 1990.

Forestry

While research on the impacts of global warming on agriculture in South Asia has been meager, assessment of the impacts on forestry has been more or less nonexistent. Yet, forestry activity is perhaps far more critical to the ecosystems and human activities of South Asia than any other sector. Since region-specific analysis of this problem has not been carried out, all we can do is fall back on research carried out elsewhere and then draw analogies between forests in other parts of the world and those in South Asia. The report of the United States EPA to Congress (Smith and Tirpak 1989) is perhaps the most comprehensive assessment of the impact of global warming on forests, and some of the assessments and conclusions contained in that report are of relevance to the situation in South Asia. Figure 6.1 provides an indication of the spread and distribution of different types of forests that exist in South Asia. In essence, the variability of climate in different parts of the subcontinent is such that we find in this region forests of every variety that occur in this figure, with a large predominance of deciduous forests. Hence, in

assessing the impacts of global warming we would necessarily have to research the specific impacts on each of the different agro-climatic regions in South Asia and derive implications for forestry activity accordingly.

In general, forests in South Asia, as in other parts of the world, are sensitive to variations in temperature and precipitation. Within a certain range, an increase in temperature would promote rapid growth of trees, but once the range of tolerance is exceeded, excessive warming can result in a reduction of growth and the destruction of plants. Similarly with rainfall, too much or too little precipitation can limit forest production and survival. An excess of rainfall can result in flooding or an increase in the water table, resulting in the submergence of roots and a reduction in oxygen intake. A drop in rainfall can also reduce growth and increase vulnerability to fire or pests, thereby leading to increased mortality. Higher CO_2 concentration levels could spur the growth rate of tree species with increased photosynthesis and higher efficiency of water use. This is particularly true of hardwood species. But most of the experimentation on this subject has been carried out in growth chambers, and

FIGURE 6.1. Approximate distribution of the major groups of world biomass based upon mean annual temperatures and precipitation. (Source: Hammond, 1972.)

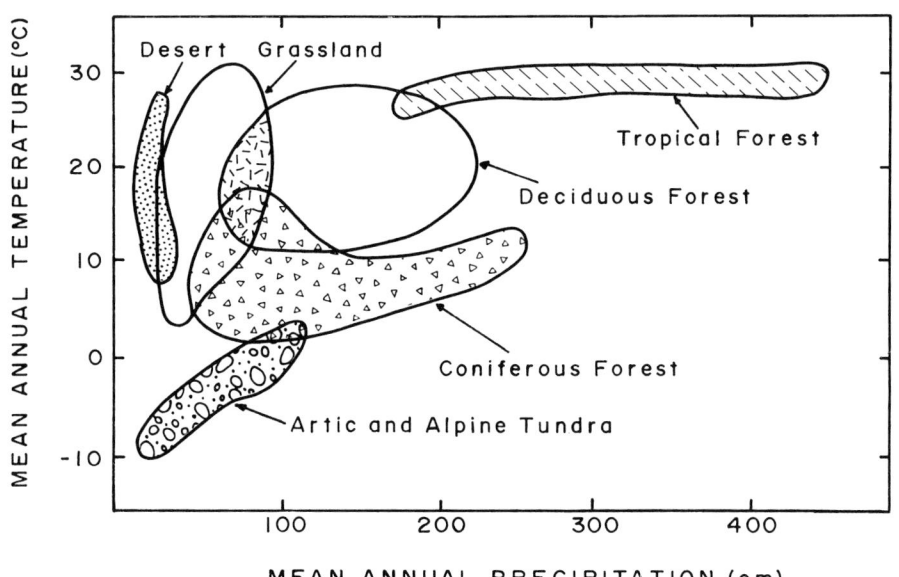

therefore extrapolation to real-world conditions is not always valid. A higher rate of CO_2 intake and forest growth could, to some extent, act as a balancing force, enhancing the ability of the forests to act as a CO_2 sink. Changes in the level of light, which could accompany climate change in the future, would also have a direct effect on forests in South Asia. Greater cloud cover would reduce the flow of sunlight and could, therefore, reduce the growth of trees.

Some models have provided outputs indicating changes in forest cover resulting from climate change in North America. But these results are tentative, not only in their applicability to North America, but far more so in their relevance to other regions. In essence, of course, a change in climate would bring about migration of certain species from regions where they grow currently to other regions that may become climatically more favorable. For instance, in the United States it has been suggested that, in the eastern part of the country, species such as spruce, northern pine and northern hardwoods could move north by about 600 to 700 km. Consequently, coniferous forests in New England could be replaced by hardwood forests, particularly oaks. In the southeastern part of the United States it was indicated that forestlands might get replaced by scrub, savanna or very sparse forest cover. In one of the scenarios that was developed, it was found that forest areas in South Carolina would become marginal, with biomass volumes roughly half of those existing currently. Given the fact that a large part of South Asia has very poor forest cover, it is not unlikely that extensive areas might get converted to grasslands and shrub-covered marginal terrains. Also, several species common in central India could migrate to the sub-Himalayan region and to the Ganga Basin of northern India. The exact process of migration is difficult to identify at this stage, but it could take place through a variety of possible movements, which could include changes in reproductive processes, such as flowering, pollination, seed setting and germination, as well as through an increase in droughts and changes in distribution of rainfall.

The ecological implications of changes in forestry include impacts on animals, soils, water and so forth. There would be consider-able changes in biomass production, and it is not certain whether, even with massive reforestation efforts, species might be as productive as existing forests. A change in the type and extent of forest cover would impact the survival of certain types of animals. Animals exert a considerable effect on forest structure, as a result of selective browsing of seedlings, insect attack, seed dispersal and other factors. There would be significant changes in soils resulting from changes in forests. The existence of bacteria, fungi and animals has an important effect on the decomposition of litter and, therefore, the availability of nutrients essential for forest growth. Hence, in effect, the delicate balance that exists currently could be disturbed, whereby further changes in the type and pattern of forestry and forest resources could come about.

An important dimension of changes in forestry in South Asia is the effect that these would have on human lives and patterns of livelihood. For instance, large numbers of people in South Asia are dependent on fuelwood for meeting their cooking needs. An estimate of dependence on biomass forms of energy within four countries of South Asia is shown in Table 6.5. It is evident from this that the already critical problem of fuelwood supply and constrained availability would be accentuated by a decline in forests. The availability of vegetation could reach crisis proportions as a result of projected population increases discussed earlier.

Biological Diversity

What is applicable in a limited sense to the balance between forests and the ecology of a region is relevant to a much greater degree in relation to biological diversity as a whole. It is, of course, difficult to arrive at specific predictions about how biological diversity would be affected by changes in climate, but some general directions of change and likely effects can be assessed to provide pointers to the future. Biological diversity would be influenced not only by direct impacts of climatic change, but also as a result of indirect influences. For instance, changes in population that may bring about changes in habitat, food availability and predator/prey relationships could prove far more important than

TABLE 6.5. *Total Household Biofuels and Total Modern Fuels: (ci. 1980)*

	Population (millions)	Household biofuels (Mtoe)	(Mtwe)	Modern fuels (Mtoe)	Biofuels as % of total
Bangladesh	88.5	10.3	(27.3)	3.0	77
India	673.2	71.9	(191.4)	97.3	43
Pakistan	82.2	8.5	(22.6)	12.7	40
Sri Lanka	14.7	2.8	(7.4)	2.0	58

Source: Leach, 1987.
Mtoe: 1 million tons oil equivalent (42.6 GJ per toe); Mtwe: 1 million tons wood equivalent (16 GJt-1)

direct effects of climate change. Barriers of various kinds, including roads, townships, mountains, bodies of water, land under agriculture and other elements of habitat, could block migration of species, thereby multiplying the extent of losses due to climate change. Rapid climate change will, of course, accelerate the destruction of bio-diversity which is currently taking place as a result of manmade activities such as deforestation and fragmentation of habitats. Freshwater fish in large bodies of water could increase in productivity as a result of warmer climates in some parts of South Asia, but it is equally likely that some species would actually die. The survival of migratory birds would also vary, with some birds actually benefitting from warming of climates, particularly those that migrate from Arctic regions. However, others could be threatened as a result of sea-level rise (SLR) inundating their wintering grounds or from increased temperature. There is already some indication that several types of butterflies in the Himalayas have become extinct, in the opinion of the famous collector F. Smetacek, as a result of warmer temperatures.

The importance of maintaining biological diversity can hardly be overemphasized. Bio-diversity has evolved gradually through millions of years in response to climate changes that have taken place slowly over time. The greatest concentration of biological diversity is seen in tropical forests, some of which exist in the hot and humid regions of South Asia. It is partly for this reason that the Silent Valley project, which would have resulted in large-scale inundation of rich tropical forests, was rejected by the government of India in the early 1980s.

Water Resources

A warming climate would have serious implications for the availability of water resources and their use. The South Asian region already has serious problems of water management, with growing scarcities during certain periods of the year in some parts of the subcontinent. Severe and recurring floods have tragic effects on the lives of many millions of inhabitants in the region, and a warmer and wetter climate could accentuate these effects.

In general, as the climate becomes warmer and drier, the demand for water will increase, particularly for irrigation and the production of electric power. Reduced flows of river water resulting from drier conditions could affect activities like production of hydropower, inland water transport and aquatic ecosystems. With reduced water availability, conflicts among users will increase. These disputes could involve the allocation of reservoir space for flood control, regulation of instream flows and water supply for agricultural, municipal and industrial users, both within and among nations.

Climate change is likely to affect both the supply and demand sides in the water cycle. Higher temperatures, in general, increase the demand for water, even for direct human consumption. With warming climates the extent of evaporation is likely to increase and, hence, there would be greater precipitation. At the same time, transpiration could be depressed as a result of increased carbon dioxide levels. The amount of snowfall and the timing of snowmelts in the northern region, along the Himalayan and Hind Kush mountains, could change.

Groundwater availability would also change as a result of altered recharge rates. All in all, the most unpredictable effect of climate change is its impact on rainfall patterns and precipitation rates. And yet, this is one set of effects that would have perhaps the most important impact on the lives of the people of South Asia. In particular, the population in the eastern part of India and all of Bangladesh may be affected most seriously, since not only would this region become highly prone to increased flooding and drainage problems, but also the timing and extent of monsoon rains could change substantially. This could have serious consequences for the agricultural practices and living patterns of people in the area. Much greater modeling and development of predictive capabilities on precipitation would be necessary before reliable forecasts can be made about the impacts on water resources in the subcontinent.

Human Health and Social Impacts

The South Asian region has several areas that experience very high temperatures for prolonged periods during the summer. Often a heat wave brings about a sharp rise in mortality, particularly among the aged and the very young. Table 6.6 shows temperatures for the month of June TABLE 6.6. 1989 as an example of the severe summer heat experienced by the large population in South Asia. The relationship between mortality and weather has been studied by several researchers, particularly Kutschenreuter (1959); Kalkstein and Valimont (1987). As would be expected, they found that mortality increases during cold winters and hot summers. Although a small but significant number of people (no more than 50 million) living above 4,000 feet could benefit from global warming, with a possible reduction in mortality, warmer temperatures in the year 2010 would adversely affect over a billion people living in the plains. On the basis of meteorological data, the hottest conditions would appear to affect a total of about 250 million people who live in the highest temperature regions of the subcontinent, where temperatures can go as high as 45°C.

An indirect health-related impact of global warming could be changes in employment to which people would have to adjust. For instance, various types of industrial plants, such as steel mills, foundries and forging units, generally experience much higher indoor temperatures than the outside environment. With global warming, the extent of sickness and productivity decline in some of these industries would adversely affect their competitive positions in the international market. Consequently, it may very well happen that some of these units would actually become nonviable. Future expansion, as well as relocation, may favor lower temperature regions in the subcontinent. Already there is a tendency to locate factories manufacturing electronic goods, watches and so forth in mountainous regions, largely because the transportation requirements of some of these units are very small, and relatively cold and dust-free conditions are an advantage. The growth of industrial centers like Bangalore, Pune (Poona) and Hyderabad in India is largely climate related.

Several diseases are directly related to higher temperatures, and it is likely that these would increase as temperatures in the region rise in the future. The indication is that outside the range of -5°C to +25°C mortality would increase as temperature goes up at the upper end and down below the lower end of the range. Hence, high summer temperatures are likely to increase mortality from cardiovascular, cerebrovascular and respiratory diseases in vast areas of the subcontinent, far in excess of reductions in mortality in those regions that experience cold temperatures.

There are also several vector-borne diseases that could increase as a result of likely changes in humidity and temperature. In particular, higher temperatures and larger

Mean Temperature on
Land Surface - June 1989 (°C)

Multan (Pakistan)	34.4
Jacobabad (Pakistan)	36.1
Bikaner (India)	34.8
New Dehli (India)	32.9
Trincomalee (Sri Lanka)	30.1
Kathmandu (Nepal)	23.8

Source: World Meteorological Organization.

volumes of standing water from higher precipitation would favor mosquito populations and mosquito-borne diseases are likely to increase. Perhaps the most dangerous of these is malaria, which, in any case, has increased in incidence in recent years. Unfortunately, mosquitoes have generally become resistant to several insecticides and sprays that were successful in earlier malaria-eradication programs. The most important social impact of higher temperatures and increased mortality and morbidity lies in the requirements for increased health care, particularly for infants and the elderly. Society would have to make provision in the future for a larger infrastructure and level of services to ensure the good health of the newborn and the aged.

Sea Level Rise (SLR)

A major cause for concern in South Asia relates to the impacts of global warming on the level of the sea. The danger of sea-level rise (SLR) is dramatized by the threat of total submergence of the Maldive Islands. But there are several other parts of the subcontinent that would also be at severe risk in the event of an appreciable rise in sea level. The primary effects of rising sea level will be increased coastal flooding, erosion, storm surges and wave activity. In order to assess the likely impacts of SLR, it would be useful to classify coastal activities in terms of (1) coast-dependent, such as ports, oil terminals and fish processing; (2) coast-preferring, such as tourism and coastal residential development; and (3) coast-independent, such as defense and other industries not directly linked with the sea. The extent of vulnerability of each of these has to be seen in relation to its linkages with coastal regions. The environmental effects of SLR is summarized in the matrix shown in Table 6.7.

While the island nation of Maldives, and other island settlements such as in the Andamans, Lakshadweep and Nicobar islands, are particularly vulnerable, the large coastline of the main landmass of South Asia makes coastal settlements and activities a far more serious problem in the event of SLR. For instance, India itself has a coastline of about 6,000 km. with about 55 percent of Indian shores being occupied by beaches. The eastern shore of India is rich in deltas formed by the rivers Ganga, Godavari, Krishna,

TABLE 6.7. *Relation Between Environmental Effects from SLR and Coastal Uses*

Marine Environ. Effects	1	2	3	4	5	6	7	8	9	10	11
Salinity			X	X	X	X	(X)	X	X		
Turbidity		X		X	X	X		X	X		
Temperature		X	X		X	X		X			
Dissolved O$_2$					X	X					
Nutrients		X			X	X		X	(X)		
Flora & Fauna	(X)	X	X	X	X	X		X	X		X
Primary Productivity		X			X	X		X	X		
Erosion	X	X	X		X	X					X
Deposition & Accretion		X	X		X	X					X
Submergence of Wetland		X			X	X	X			X	

Key: X: Major impacts; (X): Minor impacts
1: Urban development; 2: Tourism; 3: Ports and harbors; 4: Power stations;
5: Commercial fishing; 6: Mariculture; 7: Agriculture; 8: Desalination;
9: Sea-salt ponds; 10: Paper and pulp; 11: Navigation

Mahanadi and Kaveri. These delta regions generally have large deposits of clay and mud and contain vast marshy areas. The eastern coast extending into Bangladesh is particularly vulnerable to tropical storms and coastal flooding. Sea level rise is likely to increase the vulnerability of this region to tropical storms, storm surges and greater inundation. Some of the mangrove forests of this region are likely to be completely decimated.

In terms of adaptive responses, the ability of the region is rather limited. Governments and societies face four possible responses:

- No protection;
- Retreat from the affected areas and relocation of economic and human activity;
- Construction of some form of structural protection; and
- Adaptation to the changed environment and climatic conditions.

Given the uncertainties attached to global warming and its impacts, together with the extreme poverty of the region, it is unlikely that governments would make significant commitments of resources to take any action that could provide a suitable response to the anticipated problem of SLR. The problem is complex, because the cost, for instance, of holding back the sea would be such a high share of total government expenditure that it would be enormously more burdensome than for the developed countries. Additionally, to initiate action in light of such uncertainty would require tying up scarce resources in projects of less immediate importance, as compared to the immediate, basic needs of the countries of this region. Consequently, SLR could cause the most serious human disasters in South Asia, unless response strategies are financed and organized through some international action.

The Vulnerability of the Ganga Basin

The South Asian subcontinent is densely populated in general, but the most densely populated region is within the basin of the Ganga River. It is within this region that political, ecological and demographic changes would be influenced most seriously by global warming in the future. Based on censuses carried out during 1981, the Ganga Basin populations for India and the nations of Nepal and Bangladesh have been compiled in Table 6.8. Existing data indicate that population densities in the range of 300 to more than 600 persons per km^2 extend throughout this region, and the bulk of the region shows population densities exceeding 400 persons per km^2. A substantial area in the extreme southeast of the basin extending over West Bengal and parts of Bangladesh has population densities in excess of 600 per km^2. The districts of Hoogly and Howrah, which constitute a large part of the Calcutta metropolitan region, had district densities of 1,129 and 2,022 persons per km^2, respectively, in 1981. The total area of the Calcutta metropolitan region, which is about 104 km^2, has an average population density of about 32,000 persons per km^2, one of the highest in the world. Based on current projections tabulated by the UN, the Ganga Basin is expected to have a population of 485 million by the year 2000 and 570 million by the year 2010, more than an 80 percent increase over the 1981 census estimate of 319 million.

The land that is categorized as forest cover is approximately 16 percent (175,000 km^2) of the total surface area, but a large part of this forest area is highly degraded and sparsely forested. Not only is the Ganga Basin vulnerable in terms of its agricultural potential, which is already overstretched in parts, but the growth of urban settlements and continuing degradation of land would affect agricultural output adversely. Intensive forms of agriculture would, therefore, have to replace traditional systems, accentuating the demand for water. To this must be added the growing demand for water for direct human consumption. If the demand for water per capita is to remain at 25 liters per day in rural areas and is to increase to 200 liters per capita per day in urban areas by the year 2010, the total urban and rural demand for water in the year 2010 would translate to a mean annual flow of about 520 cubic meters (m^3) per second for the basin as a whole, the equivalent of a major surface water irrigation scheme. If there is a further increase in temperature, with warmer and longer summers, not only would water supplies be depleted through increased evaporation, but human

TABLE 6.8. *Population Density in the Ganges Basin, 1981, by Countries and Indian States*

Country	Total pop. (,000)	Area (km^2)	Density (/km^2)	Rural pop. (,000)	Area (km^2)	Density (/km^2)	Urban pop. (,000)	Area (km^2)	Density (/km^2)
India	283,283	851,390	333	225,867	835,686	270	57,416	15,704	3,656
Nepal	15,023	147,000	102	14,066	ND	-	957	ND	-
Bangladesh	20,945	39,000	537	18,495	ND	-	2,450	ND	-
China (Tibet)	ND	26,000	ND	ND	26,000	NEG	0	0	-
Total Basin	319,251	1,063,390	300	258,428	-	-	-	-	-

Source: Computed by author from UN and World Bank data.

consumption itself would increase over the per capita requirements mentioned above. Hence, water, which is already a scarce resource and the subject of territorial disputes, would become a bigger bone of contention in the region.

In addition, if rainfall increases during the monsoon season, there would be an additional requirement for major river projects for flood control and regulation of flows. This could also provide opportunities for supplying water after the annual monsoon period. However, this would require a greater understanding of and im-proved institutional mechanisms for resolving riparian disputes and implementing mutually acceptable projects in one country with international implications. In the absence of a harmonious approach to solving each country's specific water problems, the 700 towns and cities that already exist in the Ganga Basin will be without an adequate supply of drinking water in the decades ahead. This would only add to the other disruptive impacts that will exacerbate the even more intractable problem of migration from one country to another. Cooperative efforts and agreements may be necessary to reduce potential conflicts created by ecological refugees from those areas adversely affected by the impacts of global warming.

International Cooperation within South Asia

The problems of the Ganga Basin and likely disputes on the control and use of water are only one area of concern that requires cooperation on an unprecedented scale among the countries of the region. The growth in demand for electricity, which is currently at very low per capita levels throughout this region, would require greater trade of energy across international boundaries. For instance, Nepal possesses substantial hydroelectric resources (World Bank, 1988), which, if tapped even to the extent of 25 percent of currently estimated potential, could provide at least 20,000 megawatts (MW) of power with a ready market in India. Unfortunately, several schemes that were identified for implementation through agreement between the two countries have not made much headway because of political suspicions and lack of understanding of pricing and transfer arrangements. Consequently, Nepal has foregone substantial revenues that could have helped alleviate poverty and promoted faster economic development, while India has foregone development opportunities on account of shortages in power supply and loss of economic output. The case of natural gas in Bangladesh is identical. Suitable trade agreements between Bangladesh and India involving the sale of natural gas from Bangladesh to India would prove mutually beneficial. On the western side of the subcontinent, there is the interesting possibility of supplying natural gas from the southern portion of Iran to Pakistan and India by construction of a natural gas pipeline. A preliminary estimate (Ardekani, 1990) indicates that a pipeline costing approximately $12 billion would supply 100 million m^3 of gas daily. This gas would be consumed partly in the southern portion of Iran, with the balance going to Pakistan and India.

While these possibilities would require political understanding and a framework for economic cooperation, which seems almost unthinkable at this stage, there is the potential, even at present, for cooperation in fields like the development of renewable energy technologies. There are substantial advantages in organizations and institutions in the countries of the region working together on joint research and development (R&D) projects, whereby technologies for harnessing renewable forms of energy could be developed jointly. Even more relevant would be joint R&D in areas such as forestry and the application of biotechnology to agriculture. The possibility exists for developing species of trees and agricultural crops that could withstand the drought and saline conditions that are likely to worsen as a result of global warming.

The leaders of South Asia, particularly the late Gen. Zia-ur-Rahman of Bangladesh, established the South Asian Association for Regional Cooperation (SAARC), which has started functioning as the main instrument for promoting cooperation among the nations of the region. One of the subjects included under the present charter of SAARC is the environment, and some minor exchanges have already taken place in sharing of experiences and knowledge on environmental protection among the different countries of the area. It would be useful for SAARC to set up a small multination group to explore the possibilities of cooperation in the context of possible global warming. Such a group could mobilize experts within and outside the region to devise a blueprint for cooperation. Despite the political problems that hamper effective economic cooperation within the countries of South Asia, one should not underestimate the power of this type of analysis, which would clearly and unambiguously illustrate the benefits of cooperation within the region. In essence, the poverty, overpopulation, and stagnant economic growth of this region require greater cooperation, not only for some of the specific reasons discussed above, but also to ensure that the region emerges economically stronger in a short period of time. In the ultimate analysis, the ability of the countries and the people of this region to withstand the impacts of global warming will depend largely on their economic strength. If there is a peace dividend from removal of tensions and the threat of war in other parts of the world, such a dividend in South Asia is a prerequisite for survival and stability.

References

Ardekani, A.S., 1990. Asian gas pipeline: A pipeline for energy, environment, development and peace. *Proceedings of the Twelfth International Conference of the International Association for Energy Economics.* R.K. Pachauri and L. Srivastava (Eds.). Vikas Publishers, New Delhi India.

Gupta, S., and R.K.Pachauri (Eds.), 1990. *Global Warming and Climate Change: Perspectives from Developing Countries.* Tata Energy Research Institute, New Delhi.

Intergovernmental Panel on Climate Change, June 1990. Policymakers Summary of the Scientific Assessment of Climate Change. *Report to IPCC from Working Group I.* Meteorological Office, Bracknell, U.K.

Kalkstein, L.S., and K.M. Valimont, 1987. Effect on human health. In: *Potential Effects of Future Climate Changes on Forest and Vegetation, Agriculture, Water Resources, and Human Health.* D. Tirpak (Ed.). USEPA, Washington, D.C., pp. 122-52.

Kutschenreuter, P.H., 1959. *A Study of the Effect of Weather on Mortality.* New York Academy of Sciences, New York, pp. 126-38,

Leach, G., 1987. Household energy in South Asia. *Biomass.* Elsevier, U.K., pp. 155-84.

Smith, J.B., and D. Tirpak, 1989. *The Potential Effects of Global Climate Change on the United States.* USEPA, Washington D.C.

United Nations Development Programme, 1990. *Human Development Report, 1990.* Oxford University Press, New York.

World Bank, 1988. *Nepal: Power Subsector Review.* The World Bank, Washington, D.C.

———, 1990. *World Development Report, 1990.* Washington D.C.

Part II

**Analyzing the Regional Impacts
of Climate Change**

Computer models provide a unique opportunity for understanding a multifaceted problem like the impacts of global climate change. Global climate models have been used to simulate the effect of greenhouse gas concentrations on global temperature and precipitation. However, to date they have some severe limitations. For instance, they give valuable information on average global temperature changes, but as more specific information is required, as for example on regional or seasonal effects, the differences among the results obtained from different models reveal their limitations. Higher resolution models will need massive additional computing power and, at a conceptual level, a much improved understanding of cloud formation, the air-sea interface and the carbon cycle.

Nowhere are the limitations of the models more apparent than in their ability to predict changes in precipitation and soil moisture. The first paper in this section describes soil moisture data for the former USSR (now Russia), analyzed by Vinnikov. His data from paleoclimatic reconstructions of the soil moisture regimes for warm epochs of the past do not support the conclusions of climate models, namely progressive aridization of the continents. The models appear to strongly underestimate summer moisture values for the middle latitudes of the Northern Hemisphere.

Although it will be many years before we can adequately model all of the factors that affect global climate so that regional predictions are possible, by looking at specific aspects of climate at the regional level, some insights are possible. This approach also makes use of historical weather patterns known to have occurred in warmer times. In this way, the specific vulnerabilities of an area can be highlighted and adaptive responses developed.

This theme is the focus of the chapter by Glantz, who assesses the value of historical analogies, both physical and societal, for determining likely impacts and responses to climate change. He points out the pitfalls of extrapolating too literally from past climatic experiences, but explains their value for helping policymakers and the public gain an understanding of what conditions could be like in the future. In this context, historical analogies can provide useful insight into how well prepared a society is to cope with environmental change, allowing them to capitalize on their strengths and minimize their shortcomings.

Rosenberg and co-workers describe their study of four states in the United States (Missouri, Iowa, Nebraska and Kansas), referred to as MINK, in which they simulate the conditions of the 1930s, a time of warmer and drier conditions. By looking at historical conditions, they were able to distinguish among the more severe conditions likely to affect Nebraska and Kansas, as compared with Missouri and Iowa. Thus, generalizations about the area as a whole could be misleading, particularly since competition for groundwater in parts of Nebraska and Kansas is already a problem and will limit the potential for irrigation. The model assesses these impacts on agricultural productivity and how changes in this sector of the economy could affect other sectors, such as meatpacking. By incorporating potential technological advances into the model, they show how stable, wealthy economies have the capacity to adapt to climate change.

Significant climate change is expected in the Great Lakes region. A model developed by Botkin is used to simulate the impact of climate change on forest ecosystems. Because different tree species will have different levels of vulnerability, Botkin proposes that this information is a valuable tool for developing management practices over the next few decades. He demonstrates how the model can be used to maximize timber harvests and set priorities for land-use allocation.

The chapter by Panturat describes how the model RICE can predict rice yields using the following inputs: cultural practices, soils, climate and genetics of rice variety. The methodology has been used to modify the genetic coefficients of the rice varieties currently grown at seven selected locations across Thailand, in order to obtain a set of "improved" varieties that would produce higher yields at these locations under a "new" climate. The model can be used to optimize the sowing dates and fertilizer applications for the new varieties.

In the final chapter of Part II Ayres discusses the methodological problems of quantification and monetization of unpriced goods and services with respect to climate change. He explains how an assessment of the welfare implications of climate warming are complicated by the fact that most of the important consequences for humans are likely to be indirect rather than direct. He suggests that the major fault of traditional economic analyses is the implicit zero valuation of unpriced environmental assets and services, which may be of much greater magnitude than the included items. A practical research plan for developing a coherent, defensible valuation methodology for assessing the regional costs of climate warming in monetary terms is proposed.

These examples reveal the potential for studying climate change at the regional level and the types of response strategies that could be developed. Amassing this kind of information at the present time would involve a small investment of resources with potentially very large payoffs. With the new genetic engineering techniques currently being developed, significant opportunities could arise for developing new crop varieties that would be better adapted to new climatic conditions.

7.   ASSESSING PHYSICAL AND SOCIETAL RESPONSES TO GLOBAL WARMING

Michael H. Glantz

Because current atmospheric general circulation models (GCMs) are not yet able to produce credible, reliable regional detail of the consequences of global climate change, research into environmental and societal impacts must seek other ways to gain a glimpse of possible consequences of future regional climatic changes. The use of analogies (and analogical reasoning) has been at the center of attempts to understand what future climate change might be like, and to provide an insight into some aspects of societal responses to climate change at the regional level. Such glimpses provide only a first approximation of regional impacts and responses, but they can aid societies at all levels of economic development in their attempts to identify their strengths and weaknesses, opportunities and limitations, in coping with climate change, as well as with interannual climate variability. Proper use of analogies can take some degree of uncertainty out of the societal side of the global climate change question.

Webster's defines an analogy as an "inference that if two or more things agree with one another in some respects they will probably agree in others." Bruce Mazlish (1965, p. 5) suggested that "[a]nalogy is the most primitive and, at the same time, one of the most important forms of reasoning."

Analogies are composed of a base and a target; "known facts about the base are then used to make predictions about yet unobserved relationships in the target problem" (Hunt, 1989, p. 622). The appropriateness of a specific analogy in a specific situation has been difficult to assess. Should appropriateness depend solely on the basis of the number of similarities in structure and function or on the importance of those similarities? "Good analogical reasoning," to quote Dale Jamieson (1988, p. 81), "does not concern the number of similarities two objects share, but rather the significance of the similarities." This issue remains a central controversy in assessing the utility of analogies as a scientific methodological tool.

Some atmospheric scientists, expressing doubts about the utility of historical analogies, argue that climates of the future will not be like those of the past. History, they argue, can provide little reliable guidance or insights to policymakers seeking to develop climate-change-related policies. Yet, societies (institutional structures and functions) often change slowly and, therefore, societal reactions to environmental change (even surprises) in the near future would most likely be similar to those of the recent past.

Uses of Analogies

Analogies fulfill several functions. Thus, before one can properly judge the value of a particular analogy, the purpose of the analogy should be made explicit (e.g., Kedar-Cabelli, 1988, p. 72). While someone may understand why he or she resorted to a particular analogy, others with whom the person communicates will not necessarily be able to identify that purpose. Analogies related to global warming have been used (1) for general education, (2) to educate researchers about relationships and sensitivities and to generate hypotheses, (3) to give parameters to complex processes, (4) to forecast future states of systems, such as the atmosphere or society, (5) to generate policy options or responses, and (6) to fulfill a psychological need. These categories are suggestive, not exhaustive and they are examined more closely in the sections that follow.

General Education

Analogies can be used to educate nonspecialists about aspects of a complex situation by making reference to a different situation, about which they already have information. These have been referred to as "expository analogies" (quoted in Keane, 1988, p. 13). The greenhouse effect is such an example. People know about greenhouses. They are protected, glass-enclosed, often carbon-dioxide-enriched environments. Plants in a greenhouse are kept in a controlled environment, protected from harsh climatic conditions. The idea that carbon dioxide, an atmospheric trace gas, traps heat in the atmosphere in an analogous way to a greenhouse was apparently first suggested in the early 1800s.

Human activities such as the burning of fossil fuels, deforestation and food production processes, among other activities, are producing large amounts of atmospheric CO_2 and other greenhouse gases such as chlorofluorocarbons, nitrous oxides, tropospheric ozone, and methane. About half of the anthropogenically produced CO_2 remains airborne, acting to enhance the natural greenhouse effect (80 percent of which is due to water vapor). The more greenhouse gases that are injected into the atmosphere, the more heat will be trapped. The greenhouse analogy suggests that carbon dioxide in the atmosphere acts like the panes of glass in a greenhouse, trapping the heat generated by the penetrating sunlight. It also suggests that, if our production of CO_2 continues unabated, perhaps we could create somewhat of a runaway greenhouse effect, producing on earth a much warmer atmosphere than is presently projected (see Schneider, 1990, p. 103).

There is little scientific doubt about the theory behind the greenhouse effect. This, however, is where the scientific consensus begins to break down. What a warmer atmosphere would mean, for example, for cloud formation that, in turn, affects global albedo (reflected light) at the top of the atmosphere that, in turn, affects global and regional temperatures and precipitation patterns is still not known with confidence. Other consequences of a greenhouse warming, such as its implications for regional climates, environments, and societies of enhanced atmospheric greenhouse gas concentrations, remain controversial.

Although scientists continue to use the greenhouse analogy, they are well aware of the dissimilarities between the analogy's base (a greenhouse) and target (a CO_2-enriched atmosphere). The glass enclosure, for example, prevents the warmed air from mixing with the colder air outside the greenhouse. This analogy holds when used as a heuristic device to educate the uninitiated reader on global warming. It is also extremely useful for communication. By citing the greenhouse analogy, one can easily convey the idea behind the end result of an atmospheric greenhouse gas buildup, without having to go into detail about complex atmospheric physics, atmospheric chemistry, or even the industrial, social, or economic processes that lead to increased greenhouse gas emissions. The analogy, however, fails to educate us about the processes involved in the atmospheric greenhouse effect—a purpose for which it was not really intended.

Education of Researchers

More sophisticated global warming analogies enable researchers to understand changes in processes, interrelationships and sensitivities. These analogies might be based on warmer periods, such as the Altithermal (4,000 to 8,000 years ago) or the interglacial warming (125,000 years ago). Some scientists have used a set of the warmest Arctic summers this century as an analog to identify the possible regional climatic consequences of a warmer atmosphere. Others have targeted summertime as a warmer earth analogy. Still others have used regional analogs, by identifying existing regional climates that might be analogous to possible future climatic regimes in different regions (for example, a few degrees warmer global temperature might, decades in the future, create in Iceland the present-day climate of Scotland; see Parry et al. 1988).

Some scientists contend that there are no appropriate analogs to a global warming. Reasons usually cited include the fact that the boundary conditions (extent of sea ice, sea surface temperature distribution, vegetative cover, etc.) have changed so much and that factors that are forcing climate change (changes in albedo, amounts of atmospheric greenhouse gases, and changes in the solar constant or in the earth's orbital parameters) are so poorly understood for these earlier epochs that comparisons (e.g., analogs) are extremely risky. Others suggest that temperature increases induced by anthropogenically produced greenhouse gases will be unprecedented with respect to the level and rate of warming that could ensue. Yet, many prominent scientists have searched the climate record of Earth (and other planets) in order to identify possible climate change analogies.

These conflicting views underscore the importance of being both specific and explicit about the reasons why an analogy (or analog) was put forward in the first place. For the most part, historical (including geological) analogs have been presented as heuristic devices to enable researchers to test the sensitivities of a variety of components of the biosphere to changing atmospheric temperatures and not as forecasts of future states of the atmosphere.

To Give Parameters to Complex Processes

Analogical reasoning is a prominent part of general circulation modeling of the atmosphere. GCMs are "numerical models of weather and climate systems based on mathematical equations describing complex physical processes. . . . Many meteorological processes are simplified or omitted entirely from these models" (Pittock and Salinger, 1982, p. 26). These computer models can be expensive and time-consuming to run. Modelers must work with a fixed number of grid points, which must represent average conditions for a specified area. One GCM grid point, therefore, must represent a grid cell several hundred square kilometers in area. Thus, these models are simplifications of reality. Because of this simplification, complex atmospheric processes occurring on all spatial scales, including subgrid ones, must be set within parameters. The World Metereorlogical Organization (WMO, 1975, p. 19) has defined parameterization as "the expression of the statistical effects of various small-scale transport and transfer processes in terms of the large-scale variables explicitly resolved by the model." Where a lack of information exists about those processes, or

where there is a need to reduce the complexity of the atmospheric processes to meet the computational as well as economic constraints of computer modeling, parameterization must be undertaken.

Analogs are used in numerical modeling where there is a need to include important processes related to atmospheric circulation. Some of those processes may be well known but too complex for the needs or limited technical capabilities of the computers or the models. They must, therefore, be simplified. In other instances, the processes are not well known, yet they must be represented in the model in some form. As a result, there are simple base analogies that can, at the least, serve as adequate place holders in the models until those processes become better understood. For example, one reads about the "bucket" analogy for soil moisture in National Oceanographic and Atmospheric Administration's (NOAA's) Geophysical Fluid Dynamics Lab's (GFDL's) GCM (Manabe and Wetherald, 1986; Meehl and Washington, 1988). As another example, in the absence of reliable output from a variety of emerging interactive air—sea models, simpler, analogous statements have been used. GCM modelers have developed swamp models (simulates a fixed wet surface) and slab oceans (upper ocean moves uniformly). Thus, parameterizing atmospheric and oceanic phenomena is to a large extent dependent on the use of analogies to address spatial resolution problems, representing in the models such atmospheric aspects as clouds, cumulus convection, turbulence, boundary layer and diffusion processes, and rainfall. Such analogies can provide a first approximation of the desired parameters.

Curt Covey (1989, p. 23) noted that "all climate models. . . contain a number of adjustable parameters. These come from averaging over spatial and time scales which the model cannot resolve, and also from leaving components of the climate system out of the model . . . Consequently there always will be a need for parameterizations, and inevitably some of these parameterizations will be of doubtful validity." As these parameters are redefined and adjusted, as new knowledge of atmospheric processes is acquired, their role in the modeling process becomes increasingly valuable.

Forecasting

Heated debate, confusion, and skepticism are often generated by the use of climate analogs (such as those based on warmer periods, historical analogies based on societal responses to previous environmental changes, and scenarios based on the output of atmospheric GCMs) to develop forecasts of future regional climate. While an analogy may be used for any one of a variety of purposes, a troublesome purpose is to use it to forecast a state of the atmosphere or of society several decades into the future. It can, however, be used to make other kinds of projections about the nature and ability of different types of societal responses to cope with a variety of plausible (but not necessarily probable) future regional climatic changes.

Many examples abound where such analogies have originally been suggested to generate research hypotheses, only to be used later by others as forecasts. T.M.L. Wigley (quoted in Jager and Kellogg, 1983) has suggested that there is no analog to global warming in the twentieth-century climate record. Thomas Crowley (1989, p. i) suggested that there is no analog on prehistoric to geologic time scales that would suffice for insights into a future CO_2-induced global warming. Crowley was probably correct when he noted that "there may be no warm time period that is a satisfactory past analog for future climate," if he was talking about the lack of satisfactory geological analogs for the purpose of forecasting the regional impacts of a global warming. If he was referring to analogs as heuristic or hypothesis-generating devices when he noted that "paleoclimate studies may provide important insights into climate processes" (1989, p. i), then he was also correct. The same can be said for Wigley's comment about no good analogs in the twentieth century for global warming; a valid statement for forecasting, but an invalid one if the twentieth-century analog is used for other research-oriented purposes. Thus, analogies are useful for generating scientific hypotheses or for improving scientific understanding of the processes that may result from global warming. They are not very useful as forecasts of future states of the atmosphere or

society on which one might base specific policy responses to hypothetical regional climate changes.

Some scientists do explicitly provide reminders that their analogs are not predictions. For example, Jill Jager and W.W. Kellogg (1983) explicitly noted that "[n]one of these approaches can be used to make reliable predictions of the potential climate changes . . . They can be useful, however, in showing the nature of climatic responses and in increasing understanding of the complex climate system." Yet, many scientists at one time or another have suggested policy actions based on the output of their modeling efforts (e.g., U.S. Senate, 1988, passim).

Generation of Policy Options

"Analogies do not necessarily prove anything. They merely suggest a possibility" (Mazlish, 1965, p. 6). Plausibility of a physical or societal analogy is not a sufficient condition for use by policymakers, because several plausible but contradictory policies could be formulated based on different analogs drawn from the same pool of objective scientific information (see Martin, 1979). But, then, on which historical analog should a decision-maker rely?

The use of an analogy to generate specific policies related directly to global warming is a risky business. Unfortunately, some policymakers are not reluctant to use these analogies as guides to or support for the formulation of specific policy options to cope with the hypothesized regional impacts of global warming. Often, specific analogies are chosen by politicians to support specific policy choices they favor. For example, a senator who seeks to enact a national energy plan might favor a specific global warming regional analog to underscore the need for such a plan. Making matters more difficult, there are a few scientists who portray their models' output as being more reliable for use in the policy-making process than the models deserve. For example, the specter of the return of the Dust Bowl days is often cited as a probable regional impact in the North American midcontinent of a global warming (e.g., Bernard, 1980).

Although it is clearly within the realm of possibility that, given the normal variability of the region's climate, such a situation could temporarily return in the future, it is a misapplication of the analogy to suggest that permanent Dust Bowl conditions may be in the region's future. Such analogies, however, can be used to identify policy needs in order to eliminate shortcomings in societal responses to environmental change.[1] For example, the analogy of the Dust Bowl is a reminder that the region is drought-prone. It is a reminder that inappropriate land use practices can set the stage for desertification processes, if drought conditions were to return to the region. It is also a reminder that, despite the region's favorable level of economic development, farmers have not yet been able to buffer completely their activities from the vagaries of weather.

Recently, Abrahamson (1989, p. 3) highlighted an official statement of the 1988 Toronto Conference on "Our Changing Atmosphere" that drew an analogy between the consequences of nuclear war and global pollution leading to global warming: "Humanity is conducting an unintended, uncontrolled, globally pervasive experiment whose ultimate consequences could be second only to nuclear war." He noted that this analogy was not made by "idealistic, scientifically innocent environmentalists" (1989, p. 3). The use of this analogy by the Toronto Conference organizers was intended to compare an unknown situation, the consequence of global warming, with a known worst-case scenario, a nuclear holocaust, in order to capture the attention of policymakers, the public, and especially the media, and to urge prompt policy action to address the global warming issue. This explicit analogy was apparently meant to evoke an implicit analogy. It was an overt attempt to transfer all of the fears associated with the consequences of a nuclear war to the climate-change issue and to adopt policies stimulated by those fears.

The nuclear war analogy could be cited as an example to identify a "dread factor," with regard to societal impacts, much as physical scientists have suggested cataclysmic dread factors among plausible impacts of climatic change on the physical setting—for example, the disintegration of the West Antarctic ice sheet or the rapid switch in ocean currents (e.g., Glantz, 1988, pp. 57-81). Such changes in the physical system would greatly challenge societies to develop adequate response (coping) mechanisms based on dread factors.

Recent interviews with Toronto Conference organizers provide support for the assertion that the nuclear war analogy was invoked to "grab" the attention of the media. The consequences of a nuclear war are a worst-case scenario of societal impacts of human actions. For argument's sake, what if the organizers had favored a different, less severe, scenario such as the summertime analogy to global warming. Replacing the phrase "nuclear war" with "perennial summertime" would modify the Toronto Conference statement to read as follows: "The consequences of global warming are second only to the consequences of a perennial summertime." What effect, as a call to policy action, would this scenario (analogy) have on policymakers and the public?

Psychological Aspects of Analogies

Still another function that analogies can perform is the fulfillment of a psychological need. When confronted by unknown situations, analogies can provide us with a feeling of understanding. They offer a first step toward knowing or at least considering the unknown. As Bruce Mazlish (1965, p. 7) noted, "[a]nalogy provides an 'original,' an archetype, offering us the secure feeling of a familiar experience." Zashin and Chapman (1974, p. 313) suggested that "it is not hard to see why analogies can be persuasive; they promise order and intellectual control." In this regard analogies comfort us by providing a bridge between the past (the known) and the future (the unknown).

---

[1] Ever since the 1930s, soil conservation practices have been undertaken in the Great Plains to avoid the desertification processes that were catalyzed in the thirties by prolonged drought combined with inappropriate regional land-use practices and economic depression.

The Search for Regional Analogs to Global Warming: The Physical Sciences

This section briefly discusses attempts to identify the regional impacts of global warming by using climate analogs. Such analogs include the following periods, each of which has been suggested as a global warming analogy: (1) the Medieval Optimum, (2) the Altithermal, and (3) 2.5 million to 3.5 million B.P. Other attempts to identify global warming analogs include (4) summertime and (5) the 10 warmest Arctic summers.

The Medieval Optimum

Flohn (1981, p. 37) suggested that between the years 2000 and 2010 the global average temperature would increase by about 1°C, given continuance of the existing trends for carbon dioxide emissions to the atmosphere. He then proposed that the warmest period in the past 2,000 years, the Medieval Optimum, which occurred between about A.D. 900 and 1200, could serve as an analog for climate impacts with a 1°C warming. The period was characterized by warm conditions in the mid-latitudes, an absence of sea ice in the East Greenland Current, and a retreat of Atlantic sea ice north of latitude 80°N. In addition, Icelandic and Norwegian farmers were able to cultivate cereals up to latitude 65°N. The Caspian Sea was several meters lower than present levels. Higher tracking cyclones in this period meant extended summer droughts and winter blizzards in Europe.

Altithermal

The Altithermal (also referred to as the Holocene warming) occurred 4,000 to 8,000 years ago. It provides, perhaps, the most popular analog that has been suggested for about 1.5°C global warming. It has also been the most challenged analog. In the late 1970s, Kellogg used paleoecological information to construct a map depicting regional climates during this period, which was considerably warmer than at present (Figure 7.1). Flohn (1981) also cited the

Altithermal as an analog for identifying the possible regional implications of a 1.5°C global warming. He expected such a warming to occur between the years 2005 and 2050. Kellogg and Schware (1981, 1982) pointed out that Kellogg's Altithermal map "should not be taken as a prediction of what will happen, but rather a scenario of future climate describing what could happen."

Wigley *et al.* (1980) critiqued these reconstructions, noting that the data used are indirect and poorly dated, and Pittock and Salinger (1982) suggested that "what appears to be a synchronous spatial pattern may be a superpositioning of a number of episodes which all occurred at slightly different times." Wigley (1982, p. 2) also noted that the spatial coverage is generally poor with substantial data gaps over the land and the oceans. Another weakness of this analog is that the warming occurred only in summer (the winters were colder), whereas GCMs suggest that with a global warming, both summers and winters will be warmer, with winters warming to a greater degree than summers. According to Crowley (1989, p. 1), "This can be explained by variations in the seasonal cycle of insolation at the top of the atmosphere—the Milankovitch effect." Also, the cause of warming during the Altithermal was probably not due to elevated levels of CO_2 (Pittock and Salinger 1982, p. 36; Schneider and Londer 1984, p. 325).

2.5 Million B.P.

Just before the last large-scale glaciation of the northern continents began, Earth was warmer and the Arctic was "substantially ice free." This period, the late Tertiary, had different boundary conditions (such as changes in vegetation, sea level, sea ice extent, mountain formation). Flohn (1981) has used this period as an analog for a 4°C temperature increase, creating an ice-free Arctic region. This analog has been challenged by Crowley (1989) with respect to potential regional climatic impacts resulting from an ice-free Arctic, because the boundary conditions have greatly changed. Crowley (1989, p. i) also believed this analog would be misleading "because these warm periods had reduced polar ice cover, whereas future temperatures will be very warm but ice sheets

FIGURE 7.1. A reconstruction of the Altithermal (or Hypersithermal) Period of about 4,500 to 8,000 B.P., showing where the conditions were wetter or drier than now. The blank areas are not necessarily regions where no change occurred; the information is incomplete (from Kellogg and Schware 1981).

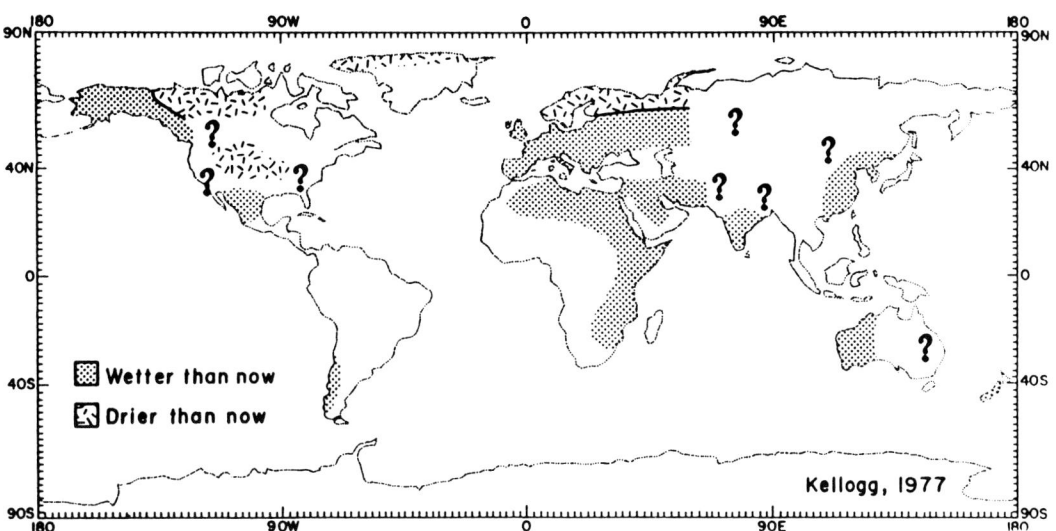

will persist because of their large thermal inertia. Due to the different time scales for the atmosphere, deep ocean, and ice sheets, there may also be a significant non-equilibrium component to the climate response that may not apply to past warm periods."

Summertime

With global warming, will climate variability increase, decrease, or stay as it is today? Kellogg and Schneider (1978) noted that "[w]e have a few hints concerning the variability of the patterns shown in a scenario, and arguments can be made for a decrease in variability. For example, one might say that it could be more 'summerlike' on a warmer earth and that therefore there would be a weaker general circulation and fewer extreme events. Indeed, variability of the current weather is known to be less in summer than in winter." Reduced variability would have major implications for strategies and tactics to cope with the regional impacts of future climate change. Some scientists have suggested that an increase in extreme meteorological events would likely occur

with a global warming (see U.S. GAO, 1990, p. 10). The United States EPA has recently funded research addressing the variability issue. Preliminary results suggest that the variability-global warming issue is complex and as yet unresolved, with contradictory results having been generated by different models' runs (Mearns, 1989; Mearns *et al.*, 1990).

As Katz and Brown (1992, p. 1) have suggested, "the frequency of [extreme] events is relatively more dependent on any changes in the variability than in the mean of climate." This has major implications for policy analyses that "rely on scenarios of future climate involving only changes in means" (1990, p. 1).

Ten Warmest Arctic Summers

Jager and Kellogg (1983; see also Williams, 1980) identified the 10 warmest Arctic summers in the 20th century and from this composite produced a picture of possible future regional climate anomalies in the northern and mid-latitudes under global warming. This approach was based both on a general belief and on GCM output suggesting that the Arctic region would be much more sensitive than the mid-latitudes to an increase in average global temperatures. This approach has been challenged for a variety of reasons. For example, Pittock and Salinger (1982) have argued that "the use of a set of individual warm years in a 'scenario' is open to question because the warmth was presumably a result of internal dynamical fluctuations, not a systematic change in the global radiation balance." In addition, they noted that using a set of years, as opposed to a run of years, does not allow for the cryospheric or the oceanic components of the climate system to come into equilibrium (Pittock and Salinger, 1982). Problems with this approach notwithstanding, the study did generate other studies based on the concept of a higher warming in the Arctic region than in the mid-latitudes, such as research that examined consecutive years above and below the long-term regional temperature mean.

Summary

These examples are not exhaustive of the list of analogs used to identify regional as well as global climatic changes that might be associated with a global warming. Others include a search of the recent climate record for warmer fluctuations, such as the 1920s and 1930s (e.g., Aspen Institute, 1978), comparing composites of the five warmest and five coldest years (Wigley *et al.*, 1980), a search of the global geological record (e.g., the Eemian, 125,000 years ago) or of regional and local records (e.g., Pittock and Salinger 1982, and Kates *et al.* 1984, respectively). As an example of a scenario of possible soil moisture patterns on a warmer Earth, Kellogg and Schware (1982) combined the results of three separate climate warming analogs—paleoclimatic reconstructions for the Altithermal, comparison of recent warm and cold years in the Northern Hemisphere, and a climate model experiment. The end result was yet another plausible climate analog, as shown in the following map (Figure 7.2).

Each of these analogs (or combinations of analogs) possesses its own particular set of strengths and weaknesses. Each has been challenged as imperfect, if not misleading. As noted earlier, Crowley has suggested that

there are no good paleoclimatic analogs, and Wigley has suggested that there may be no 20th-century analogs. On the other hand, scientists such as Flohn and Budyko strongly believe in the value of paleoclimatic analogs.[2] Flohn and Budyko have independently established a sequence of analogs that track proposed increases in CO_2 levels and in global temperatures, matching them to previous warm periods. Flohn's perception of analogs is shown in the Table 7.1.

One general argument usually cited against the reliability of historical analogs is that the forcing function for contemporary global warming is increased carbon dioxide and other anthropogenically produced greenhouse gases, whereas the forcing functions of earlier times have not all been identified. For those situations in which the forcing functions have been reliably determined, the value of the analogs has been challenged outright. Kutzbach and Webb (1980) have challenged Kellogg's use of the Altithermal as a possible analog because of differences in Earth's orbital characteristics. The position of Earth with respect to the Sun was such that there would have been warmer summers and colder winters, as opposed to warmer temperatures in all seasons, which is expected to occur in a CO_2-induced warmer Earth. An additional argument used against the search for historical climate analogs in the physical sciences is that the boundary conditions in the earlier periods were not the same as they are at present. Crowley (1989, p. 6) has recently suggested that analogies from the geologic past cannot provide useful analogs for global warming because "When all of these geographic effects are considered together, it is evident that, even if global temperatures were warmer, the regional climate patterns may have been significantly different than what we might experience in the future."

---

[2] During the early 1970s, when there was considerable speculation about the return to an Ice Age, several analogs were also put forward—but for a cooling of the global atmosphere. These include the Younger-Dryas (10,800 to 10,000 B.P.) cold interval, a cooling which was most intense in Europe, especially northwest Europe and the British Isles (Schneider et al., 1987, 403), and the Little Ice Age, which lasted for several centuries, ending about 1850.

FIGURE 7.2. Example of a scenario of possible soil moisture patterns on a warmer earth. It is based on paleoclimatic reconstructions of the Altithermal Period (4,500 to 8,000 B.P.), comparisons of recent warm and cold years in the Northern Hemisphere and a climate model experiment. Where two or more of these sources agree on the direction of change, the area of agreement is indicated with a dashed line and a label (from Kellogg and Schware 1982).

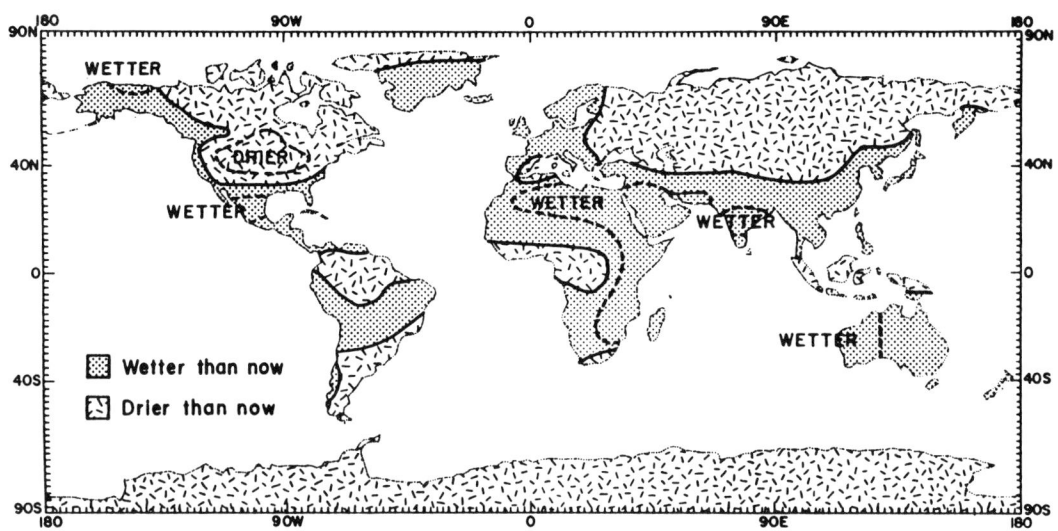

TABLE 7.1. *A Global Warming Timetable*

If by about this time . . . (calendar years)	2000-2010	2005-2050	2020-2050	2040-2080
Atmospheric carbon dioxide and other trace gases reach . . . (parts per million)	430	492	610	880
Global warming could reach . . . (degrees Centigrade)	1.0	1.5	2.5	4.0
Which corresponds with . . . (years ago)	1,000 (Early Middle Ages)	6,000 (Peak Holocene Period)	120,000 (Last Inter-glacial Period)	2.5-12 million (Late Tertiary Period)

Source: Flohn, 1981.

This section underscores the widespread discussion about the value to the physical science research community of plausible analogs to climate change impacts, and it exposes the controversy within scientific circles over the search for appropriate regional analogs to climate change. While several of these analogs might seem farfetched to the unsophisticated observer, they do play an important role in the search for improved understanding of global warming and its impacts at the regional and local levels. Nevertheless, readers should be warned that specific regional policies based on any specific analog would be risky at best.

Societal Analogies and Climate Change

It is widely acknowledged that GCMs of the atmosphere are not yet very reliable for identifying regional climate changes that might result from global warming. While the various GCMs are in moderately good agreement on projections of temperature changes, they agree much less on projections of precipitation and soil moisture. Yet policymakers require this regional and local information in order to make specific policies related to preventing, mitigating, or adapting to the impacts of global warming. While the modelers continue to seek to improve the reliability of their models for identifying regional climate change impacts, researchers interested in the societal (including policy) aspects of climate change have searched for other ways to undertake reliable, plausible regional assessments. Such approaches include (1) regional analogs, (2) CLIMPAX, and (3) historical analogies. Each is discussed below.

Regional Analogs

Parry *et al.* (1988, pp. 38-99) discussed the use of regional analogs as indicators of climate impacts. These researchers identified analogous regions based on temperature and precipitation projections for a CO_2 doubling (according to the Goddard Institute for Space Studies [GISS] model) for each of their case study areas, as shown in Figure 7.3. They suggest that such analogs can be useful for

interpreting the GISS $2xCO_2$ (i.e., CO_2 doubling) scenario, because: (1) they provide effective illustrations of regional impacts of climate change by serving "to highlight the magnitude of the future change in climate within regions in terms of the present-day differences in climate between regions." Parry and colleagues (1988) suggest, for example, that Iceland's climate under the GISS $2xCO_2$ scenario will be similar to the climate of northeast Scotland today. (2) They can be used as indicators of likely agricultural adaptations. Agricultural practices in analog regions "are a useful indicator of the adaptive strategies to retune agriculture to altered climatic resources . . ." (3) They are useful as indicators of potential productivity, based on the assumption that current agricultural productivity in the analog region may be of similar magnitude to that obtainable under a changed climate in the study area.

Parry *et al.* (1988) note that there are many difficulties associated with the use of regional analogs along the lines they have proposed. The main problems stem from the existence of regional variations in environmental, management and technological factors. The severity of climate impacts will be affected by cultural practices and these are most likely to be different between the case study and analog regions. As Firey (1960, p. 145) noted, "[t]he resource composition of a habitat varies as much with the activities of

FIGURE 7.3. Present-day regional analogs of the GISS $2xCO_2$ climate change estimated for the case study reions: Saskatchewan, Iceland, Finland, Leningrad and Cherdyn regions (USSR) and Hokkaido and Tohokyu districts (Japan).

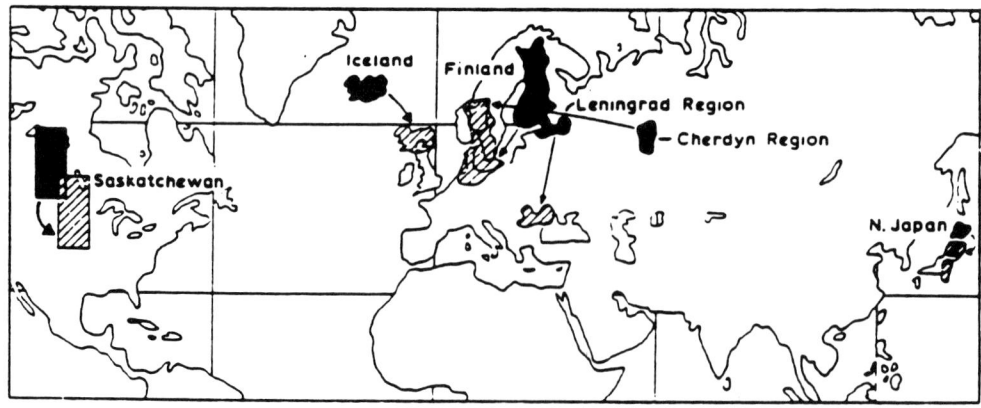

the people who occupy that habitat—their techniques, beliefs, knowledge, and social organization—as it does with the physical properties of the habitat itself." Parry *et al.* (1988) also point out that climate changes will affect other factors that can favorably or adversely alter agricultural production, such as latitudinal variations of daylight, water resources, soil fertility, soil erosion, plant diseases, and pests.

CLIMPAX

The Climate Impacts, Perception, and Adjustment Experiment (CLIMPAX) was undertaken in the mid-1980s, based on the view that "empirical evidence can be incorporated into arguments about the impacts of future fluctuations" (Kates *et al.*, 1984, p. 1). The experiment was an empirically based search for different types of past climatic fluctuations in the United States, some of which were then "chosen to emulate a CO_2 change or other scenarios of future climate" (p. 1).

Study team members sought to identify what they defined as "persistent periods (decades or greater) of climate variation or slow cumulative climate change." They referred to this as "the consecutive epoch analysis" (Kates *et al.*, 1984, p. 5). CLIMPAX focused on the meteorological variables of rainfall and temperature. Kates and colleagues (1984) suggested that "CLIMPAX offers a complementary approach to [simulations of expected climate change based on guesstimates of what might be expected to occur with a global warming, e.g., 1°C and 10 percent decline in pre-cipitation] . . . by identifying regions where changes similar to those projected actually took place over a sustained period" (p. 7).

The authors of the study pointed out the possible weaknesses of their approach by noting that "these empirical analogs differ from projected CO_2-induced climate change in very significant ways. The changes are almost instantaneous, are not sustained and cumulative as predicted in the CO_2 cases, and are smaller than changes projected by some models at high levels of CO_2 doubling" (Kates *et al.*, 1984). Another problem with this approach is that the relatively modest

quantitative changes in meteorological variables over short time periods often prompt subtle changes in human behavior. These societal changes may not be identifiable until the fluctuation has passed, if at all. In addition, other intervening factors could make it almost impossible to attribute the changed behavior to relatively minor changes in climate statistics (on the attribution problem, see Katz, 1988).

Historical Analogies

Historical analogies, like several other analog methods, provide an alternative to sole reliance on GCM-generated regional scenarios and can provide a first approach in attempts to ascertain the level of societal preparedness for the impacts of a global warming. As noted earlier, the various GCMs do not agree on the possible regional distribution of precipitation, although they are in somewhat better agreement on possible regional temperature changes (Hansen, 1988). A recent study by Adams and co-workers (1990), based on a comparison of the regional output of two GCMs, shows how different conclusions could be derived from a reliance on different GCMs.

Another approach, based on the use of historical analogies, has been referred to as "forecasting by analogy" (Glantz, 1988, 1990). This approach was proposed in the early 1980s in an attempt to identify potential responses to regional climate change by assessing societal responses to changes in the water balance of the American Great Plains. This approach does not attempt to forecast the future state of the atmosphere or of society. It suggests that we can learn how societies have dealt with recent environmental changes (regardless of degree or rate of change) and apply that knowledge about societal responses to improve the way that societies might cope with such changes in the future. Historical responses can be used to identify societal strengths and weaknesses with the hope that the weaknesses can be addressed.

This approach is based on the premise that societies in the near to midterm future will likely conduct activities as they have done in the recent past, barring unforeseen shocks of a magnitude that would fundamentally alter their traditional patterns of interaction.[3] Decision-makers can usually relate to real-world predicaments with which they or their predecessors most probably have had to cope (e.g., Neustadt and May, 1986). Thus, such studies have a reality about them that gives their findings a level of credibility that are lacking in computer-generated scenarios that project future trends.

The Dust Bowl era of the American Great Plains in the 1930s has been proposed as an analogy to what might happen in that region with a global warming. This analogy has existed for at least a decade and has been cited as a worst-case scenario (Bernard, 1980; Warrick, 1984). It catalyzed studies that looked at changes in the water balance within the Great Plains region as a possible analog to societal responses to changes in the regional water balance (e.g., Glantz and Ausubel, 1984). Models now suggest that, with a higher global temperature, mid-continental areas in North America, southern Europe, and Siberia would tend to become warmer and drier (e.g., Manabe, 1988).

The specter of a return of the Dust Bowl days is frequently used to spark concern about the possible long-term adverse effects of global warming. In a recent article on the possible impacts of global warming in eastern Colorado, the Dust Bowl image was once again invoked by suggesting that "In a worst-case scenario, large-scale activity [such as the drying of vegetation and the exposure of fixed dunes to prevailing winds] on the high plains would create severe droughts that would make the Dust Bowl look mild" (Rovin, 1990). It is a scenario to which many policymakers as well as farmers in North America can relate, having grown up with a recollection as well as a fear of a return of thirties-like droughts and their socioeconomic and environmental consequences for the region. The "fore-casting by analogy" approach has since been applied to other environmental changes in North America (Glantz, 1988) and to marine fisheries around the world (Glantz and Feingold, 1990). These cases have also served to stimulate researchers in other climate-sensitive regions to undertake

sensitivity analyses using their own climatic and environmental records. Similar studies have also been undertaken in Hungary (Antal and Glantz 1988), Brazil (Magalhaes and Neto, 1990), and Vietnam (Ninh and Hien, 1989).

Summary

The need for information on the regional aspects of global warming is very great, as the desire for policy action at all levels of social and political organization to prevent, mitigate or adapt to climate change continues to accelerate. Analogies, when properly used, can provide some guidance to the search for policy responses to yet unknown future changes in regional climatic conditions.

Texas High Plains Case Study

Scientists have suggested that a global warming would increase dryness in the mid-continental regions of the mid-latitudes around the world (e.g., Budyko and Sedunov, 1988; Meehl and Washington, 1988). They have come to this conclusion based on GCM model output, the 1920s-1930s warming, and paleoclimatic reconstructions for earlier warm epochs in Earth's history. The American Great Plains is one area that would be affected by such a situation.

The Texas High Plains region deserves special attention because it is on the climatic dry margin of the southern Great Plains. It serves to illustrate what could (not will) happen in other parts of the Great Plains. It can also serve as an analog to itself at different points in time (as opposed to forming an historical analogy between two different places at the same point in time). This example is used only to generate understanding of regional responses to change.

For over a century, the High Plains of Texas have developed economically as the direct result of judicious societal responses to changes in regional climatic (and, later, groundwater) conditions. High Plains settlers began as cattle ranchers and as dryland farmers seeking to find, through trial and error, sustainable economic activities that would mesh with the region's climatic and other

---

[3]Recall that even the energy crises of the 1970s have not radically altered the way American society consumes energy, temporary response adjustments notwithstanding.

environmental and economic conditions.[4] Both ranchers and farmers were dependent for the most part on the vagaries of the weather. Because of recurrent, extended, severe drought, among other factors, inhabitants of the region who did not abandon their land have been forced to devise ways to cope with anomalous water supply conditions.

The waves of migration into the Texas High Plains were often sparked by a belief that rainfall could support agricultural activities, at least at the local farm level. During wet periods, many new settlers believed, as a result of overzealous if not unscrupulous advertising by land speculators and railroad companies, that local rainfall conditions would be enhanced by farming the land (e.g., Fite, 1966). Whenever drought returned to the region, however, many settlers were forced to abandon their homesteads.

In specific response to droughts and to the realization that the region overlies portions of an underground reservoir called the High Plains Aquifer, farmers began to seek ways to tap the groundwater. Initially, in the latter years of the nineteenth century, parts of the aquifer closest to the land's surface were tapped with the use of windmill-driven pumps, whose rates of flow, however, were too low for the irrigation needs of large agricultural activities.

Thus, awareness of the underground water resource preceded, by decades, either the farmers' ability or their desire to draw on the groundwater in quantities necessary for extensive irrigation. In addition, there was also a prevailing distrust of, if not an aversion to, irrigation. In the 1890s and the first decade of the twentieth century, land speculators and town boosters sought to "sell" the value of irrigation in hope of enticing migrants to the region. In this pe-

---

[4]Some of the changes with which farmers and local governments have had to cope include the following: alternating short, as well as prolonged, wet and dry periods; severe decade-long drought that coincided with a worldwide economic depression; fluxes of weather-induced immigration and emigration; alternating periods of per-acre profits and losses over which they had little or no control; changes in the types and prices of energy used in irrigation; and groundwater exploitation from discovery to depletion.

riod (1896 to 1907), however, rainfall was adequate for dryland farming, while the costs of installing and operating irrigation facilities remained quite high.

Most of the 1930s, however, with the exception of two years (1932 and 1937), were drought-plagued (e.g., Worster, 1979). In the early part of the decade, there was a sharp drop in prices for agricultural goods grown in the region because of the nationwide Great Depression. The federal government, through its Agricultural Adjustment Act of 1933, aided distressed farmers with subsidies, used by many to mechanize their farms (Green, 1973, p. 123). These factors prompted a major shift in regional attitudes toward the exploitation of groundwater. New wells started to appear after 1934 as a reaction by farmers to the harsh environmental conditions in the region during the Dust Bowl era. By 1940 the High Plains of Texas had become the most important irrigated region in the country.

Severe droughts in the southern Great Plains in the 1950s reminded inhabitants of the vulnerability of their fields and their livelihood to drought and dust storms. In addition to the implementation of conservation measures for water and land, these droughts also prompted another sharp increase in the construction of wells for irrigation. Today more than 70,000 deep wells tap the Texas High Plains Aquifer. Expansion of irrigated acreage paralleled the increase in the number of wells, going from 80,000 acres in 1936 to a high of 5.2 million acres in 1981. In 1989 about 4 million acres were under irrigation.

Several reasons have been given for the receptivity to irrigation in the 1930s and afterwards, including technological improvements in irrigation facilities, the emergence of preferred cash crops (cotton, grain sorghum, and winter wheat), and the availability of credit to farmers for developing irrigation plants. Besieged by drought and economic depression, High Plains farmers turned to irrigation as a form of crop insurance (Green, 1973, p. 139). They became more adept at irrigating their fields and more efficient at irrigation. They began to look at irrigation in a different light: It was for use not only in drought emergencies, but also to maximize crop production. Groundwater assessments at least as early as the 1950s were already alluding to the depletion of the aquifer, declining water tables, and the possible long-term consequences for the region of the mining of the underground reservoir (e.g., Firey, 1960; Hendrick, 1958, 1959). The predominant mind-set of Texans on the High Plains had shifted from an earlier perception of an unlimited supply of groundwater (the region was once referred to as the "land of underground rain") to one of the mining of the groundwater resource. Today the longe-vity of this resource is no longer measured in terms of centuries but of generations.

Inhabitants of the region are at present in the midst of a transition into an era of scarce, if not depleted, groundwater resources, prompting different responses. For example, some people in the region abandoned their land as their wells dried up or because the water table dropped, thereby increasing the energy requirements to pump ever-deeper water onto their fields. Resource managers in the region have been developing viable options to land abandonment. Tactical responses to the declining water tables have included a switch to crops with lower water requirements, a shift to dryland farming, and the development of such new irrigation methods as surge flow, furrow diking, and low-energy precision application (LEPA). The controversial scientific practice of cloud seeding has also been advocated (e.g., High Plains, 1979; see also Reisner and Bates, 1990, pp. 115-122). Another tactical response of recent interest has been the possible use of plant-growth-regulating chemicals that can inhibit the moisture loss of plants by as much as 20 percent. There is also interest in the possibility of using precipitation captured by some of the 17,000 playa basins on the Texas High Plains as a potentially important means of recharging the High Plains aquifer (see U.S. Department of Interior, 1987, pp. 3-4). These options have all been designed to stretch out the remaining groundwater supply for a few more decades.

The region has undergone a variety of changes, some of which have been environmental, others technological, still others social and economic. Societal responses to these and other socioeconomic changes suggest that the region will adapt well to the potential regional consequences of a climate change. With or without a CO_2-induced global climate change, climate in this semi-arid region will continue to vary from one year to the next, as well as on other time scales.

Focusing on one region at different points in time provides us with a relatively "clean" analogy with respect to societal resilience in the face of a variety of environmental, societal, and technological changes. It is important to note, however, that even the use of a region at one point in time as an analog to itself at another point in time will not enable the researcher to develop the perfect analogy. Times have changed. Cultural factors have evolved. Perceptions and attitudes about resources may also have changed.

Conclusions

What Analogies Can Do

Despite the ongoing philosophical as well as practical debate on the value of analogies in physical and social scientific inquiry, it appears that the careful use of analogies can provide some insights into the possible regional implications of a global warming.

As noted at the outset, analogies have been used as heuristic devices both to educate the lay public and generate scientific hypotheses worthy of additional research. They have been used by modelers in attempts to improve their understanding of atmospheric processes under conditions of climatic changes involving both warmer and colder global average conditions and also to parameterize atmospheric processes, an essential part of general circulation modeling activities.

Analogical reasoning and analogies are implicitly, if not explicitly, an integral part of human thought and communication processes. Analogies can be used to generate a first approximation of the regional impacts of a global climate change. They can also provide reasonable insights into possible societal responses to those impacts. We must state why such reasoning is used so that others will better know how to judge the appropriateness of the use of analogies in a particular context. We must accept them as important approaches to scientific inquiry and we must be explicit about their strengths and shortcomings.

Physical scientists often cite their use of global models as providing the research community with sensitivity analyses of atmospheric processes and, therefore, climate under a variety of conditions. Similarly, analogies can provide researchers with a tool for undertaking sensitivity analyses of societal responses to changing environmental conditions.

What Analogies Cannot Do

Bruce Mazlish, like Ludwig von Bertalanffy (1960), warned that, when used carelessly, historical analogy can be a misleading guide to action. Mazlish identified two special problems with the use of historical analogies in the social sciences; fair sampling and self-fulfilling prophecy. With respect to fair sampling, there is a greater possibility of achieving an adequate sample of cases in the physical than in the social sciences. Thus, analogies in the social sciences are often based on only a few instances, so that their findings are somewhat more tenuous, and their usage in need of greater care. With respect to self-fulfilling prophecies, societies respond to signals they perceive in their environment. Thus, projections could turn out to be false or true depending on perceptions of, or responses to, environmental change.

Analogies have also been proposed to suggest possible alternative futures and to stimulate specific policy formulation to cope with the adverse aspects of those futures. The latter, of course, is one of the aspects of the use of historical analogies in the global warming issue most fraught with risk. While all the other functions of analogies are directed toward improving our knowledge of a marginally understood climate system in terms of the impacts of feedback mechanisms in global warming projections, the forecasting function for purposes of specific policy action based on a specific analogy remains one of the riskiest applications of analogies (see, for example, Jager and Barry, 1991).

The Search for the Perfect Analogy

There is no perfect regional analogy to the global warming expected to occur during the next several decades. Important boundary conditions that affect atmospheric processes, such as the spatial and latitudinal extent of sea ice, the extent of vegetative cover including tropical forests, solar parameters, Earth's orbital characteristics and so forth, are constantly changing. It is necessary to recognize that analogies have different strengths and shortcomings for inquiry about societal capabilities to cope with environmental change. From the standpoint of societal impacts research, one can distinguish between first-, second- , and third-order analogies, sometimes called "downstream analogies." First-order analogies provide direct insights into what we wish to know, such as, for example, a glimpse of possible societal responses to regional climatic changes. The analogy should provide that glimpse directly. Second- and third-order analogies are based on the output of a preceding analogy. Analogies derived in this fashion are subjected to cascading uncertainties.

Analogies can provide clues, generate ideas, and spark reactions that lead to different searches for new analogies. Each analogy provides us with additional information about the target problem. Even if one does not accept particular analogies to future global warming, critical assessments of them can improve our understanding of the potential regional impacts of global warming.

Does History Have a Future?

The use of analogies and analogical reasoning has been central to our understanding of many facets of the physical, social and humanistic aspects of the regional impacts of global warming. Analogies must be used with care and the purpose(s) for their construction must be made explicit. Historical analogies can provide useful insights into how well prepared societies are to cope with environmental change, allowing societies the opportunity to capitalize on their strengths and to minimize their shortcomings. One could conclude, then, that the use of historical analogies has a future in researching most aspects of the global warming issue.

References

Abrahamson, D.E., 1989. *The Challenges of*

Global Warming. Island Press, Washington, D.C., pp. 3-4.

Adams, R.M., C. Rosenzweig, R.M. Peart, *et al.*, 1990. Global climate change and U.S. agriculture. *Nature,* 345:219-24.

Antal E., and M.H. Glantz (Eds.), 1988. *Identifying and Coping with Extreme Meteorological Events.* Hungarian Meteorological Service, Budapest.

Aspen Institute, 1978. *The Consequences of a Hypothetical World Climate Scenario Based on an Assumed Global Warming Due to Increased Carbon Dioxide.* Report of Symposium and Workshop. Aspen Institute for Humanistic Studies, Boulder, Colo.

Bernard, Jr., H.W., 1980. *The Greenhouse Effect.* Ballinger Press, Cambridge, Mass.

Budyko, M.I., and Yu. S. Sedunov, 1988. Anthropogenic climate change. Paper presented for the World Congress on Climate and Development, November 7-10.

Covey, C., 1989. Mechanisms of climate change. In: *Global Climate Change.* S. Fred Singer (Ed.). Paragon House, New York, pp. 11-65.

Crowley, T.J., 1989. Are there any satisfactory geological analogs for a future greenhouse warming? Applied Research Corp., Texas A&M University, College Station.

Firey, W., 1960. *Man, Mind and Land: A Theory of Resource Use.* Free Press. Glencoe, Ill

Fite, G.C., 1966. *The Farmers' Frontier: 1865-1900.* Holt, Rinehart and Winston, New York.

Flohn, H., 1981. Life on a Warmer Earth: Possible Climatic Consequences of Man-Made Global Warming. Executive Report No. 3. Laxenburg, Austria: IIASA.

Glantz, M.H. (Ed.), 1988. *Societal Responses to Regional Climatic Change: Forecasting by Analogy.* Westview Press, Boulder, Colo.

———, 1990. Does history have a future? Forecasting climate change effects on fisheries by analogy. *Fisheries,* 15:39-44.

———, and J.H. Ausubel, 1984. The Ogallala Aquifer and carbon dioxide: Comparison and convergence. *Environmental Conservation,* 11:123-31.

———, and L. Feingold (Eds.), 1990. *Climate Variability, Climate Change, and Fisheries.* ESIG/NMFS/EPA Study. National Center for Atmospheric Research, Boulder, Colo.

Green, D.E., 1973. *Land of the Underground Rain: Irrigation on the Texas High Plains, 1910-1970.* University of Texas Press, Austin.

Hansen, J.E., 1988. The Greenhouse Effect: Impacts on current global temperature and regional heat waves. Testimony presented at Hearing on the Greenhouse Effect and Global Climate Change, Committee on Energy and Natural Resources, U.S. Senate, One Hundredth Congress, First Session. U.S. Government Printing Office, Washington D.C.

Hendrick, R.L., 1958. *A Study of Ground-Water Supplies and Their Relation to Irrigation Farming on the South High Plains of Texas.* Travelers Insurance Company, Hartford, Conn.

———, 1959. *Ground Water Supplies and Irrigation Farming on the North High Plains.* Travelers Insurance Company, Hartford, CT.

High Plains Underground Water Conservation District No. 1, 1979. *A Summary of Techniques and Management Practices for Profitable Water Conservation of the Texas High Plains.* Report 79-01. High Plains Underground Water Conservation District No. 1, Lubbock, Texas.

Hunt, E., 1989. Cognitive science: Definition, status, and questions. *Annual Review of Psychology,* 40:603-29.

Jager, J., and R. Barry, 1991. Climate. In: *The Earth as Transformed by Human Action.* B.L. Turner, (Ed.). Cambridge University Press., Cambridge, U.K.

———, and W.W. Kellogg, 1983. Anomalies in temperature and rainfall during warm Arctic seasons. *Climate Change,* 5:39-60.

Jamieson, D., 1988. Grappling for a glimpse of the future. In: *Societal Responses to Regional Climate Change.* M.H. Glantz (Ed.), Westview Press, Boulder, Colo, pp. 73-93.

Kates, R.W., S.A. Changnon, Jr., T.R. Karl, *et al.*, 1984. *The Climate Impact, Perception, and Adjustment Experiment (CLIMPAX): A Proposal for Colla-*

borative Research. Clark University, Worcester, Mass.

Katz, R.W., 1988. Statistics of climate change: Implications for scenario development. In: *Societal Responses to Regional Climate Change.* M.H. Glantz (Ed.) Westview Press, Boulder, Colo. pp. 95-112.

————, and B.G. Brown, 1992. Extreme events in a changing climate: Variability is more important than averages. *Climate Change,* in press.

Keane, M.T., 1988. *Analogical Problem Solving.* Ellis Howard Ltd., Chichester, U.K.

Kedar-Cabelli, S., 1988. Analogy—From a unified perspective. In: *Analogical Reasoning.* D.H. Helman (Ed.) Kluwer Academic Publishers, Dordrecht, Netherlands, pp. 65-103.

Kellogg, W.W., and S.H. Schneider, 1978. Global air pollution and climate change. *IEEE Transactions on Geoscience Electronics,* 16:44--50.

Kellogg, W.W., and R. Schware, 1981. *Climate Change and Society: Consequences of Increasing Atmospheric Carbon Dioxide.* Westview Press, Boulder, Colo.

————, and R. Schware, 1982. Society, science and climate change. *Foreign Affairs,* 60:1088.

Kutzbach, J.E., and T.Webb III, 1980. The use of paleoclimatic data in understanding climate change. In: *Proceedings of the Carbon Dioxide and Climate Research Program Conference.* L. Schmitt (Ed.) pp. 163-71.

Magalhaes, A.R., and E.B. Neto, 1990. Socioeconomic impacts of climate variations and policy responses in Brazil. A report to UNEP (United Nations Environment Programme) and the State of Ceara (Brazil). Secretariat for Planning and Coordination, Fortaleza, Brazil.

Manabe, S., 1988. Changes in soil moisture. In: *The Challenge of Global Warming.* D.E. Abrahamson (Ed.) Island Press, Washington D.C., pp. 146-50.

————, and R.T. Wetherald, 1986. Reduction in summer soil wetness, induced by an increase in atmospheric carbon dioxide. *Science* 232:626-28.

Martin, B., 1979. The Bias of Science. O'Connor, ACT, Australia: Society for Social Responsibility in Science.

Mazlish, B., 1965. Historical analogy: The railroad and the space program and their impact on society. In: *The Railroad and the Space Program: An Exploration in Historical Analogy.* B. Mazlish (Ed.) MIT Press, Cambridge, Mass, pp.1-52.

Mearns, L.O., 1989. Variability. In: *1989 Potential Effects of Global Climate Change on the United States.* J. Smith and D. Tirpak (Eds.). USEPA, Washington D.C., pp. 29-54.

————, S.H. Schneider, S.L. Thompson, *et al.,* 1990. Analysis of climate variability in GCMs: Comparison with observations and changes in variability in $2xCO_2$ experiments. *Journal of Geophysical Research—Atmosphere,* 95: 20469-90.

Meehl, G.A., and W.M. Washington, 1988. A comparison of soil-moisture sensitivity in two global climate models. *Journal of Atmospheric Sciences,* 45:1476-92.

Neustadt, R.E., and E.R. May, 1986. *Thinking in Time: The Uses of History for Decisionmakers.* Free Press, New York.

Ninh, N.H., and H.M. Hien, 1989. Preliminary studies on climate-related impact in Vietnam. *Report of the Center for Natural Resources Management and Environmental Studies.* Hanoi University, Vietnam.

Parry, M.L., T.R. Carter, and N.T, Konijn (Eds.), 1988. The impact of climatic variations on agriculture. *Vol. 1: Assessment in Cool Temperate and Cold Regions.* AIR, Dordrecht, Netherlands.

Pittock, A.B., and M.J. Salinger, 1982. Toward regional scenarios for a CO_2-warmed earth. *Climate Change,* 4:23-40.

Reisner, M., and S. Bates, 1990. *Overtapped Oasis: Reform or Revolution for Western Water?* Island Press, Washington, D.C.

Rovin, A., 1990. Megadunes could create another Dust Bowl. *High Country News.* August 27, p. 4.

Schneider, S.H., 1990. *Global Warming: Are We Entering the Greenhouse*

Century? Sierra Club, San Francisco.

————, and R. Londer, 1984. *The Coevolution of Climate and Life.* Sierra Club, San Francisco.

————, D.M. Peteet and G.R. North, 1987. A Climate model intercomparison for the Younger-Dryas and its implications for paleoclimatic data collection. In: *Abrupt Climate Change.* W.H. Berger and L.D. Labeyrie (Eds.). D. Reidel Publishing Company, Dordrecht, Netherlands, pp. 399-417.

U.S. Department of Interior, 1987. *High Plains States Groundwater Demonstration Program: Phase I Report,* December. U.S. Department of Interior, Washington D.C.

U.S. GAO (General Accounting Office), 1990. *Global Warming: Adminstration Approach Cautious Pending Validation of Threat* (GAO/NSLAD-90-63). U.S. Government Printing Office, Washington D.C.

U.S. Senate, 1988. Testimony presented at Hearing on the Greenhouse Effect and Global Climate Change, Committee on Energy and Natural Resources, U.S. Senate, 100th Congress, First Session. U.S. Government Printing Office, Washington D.C.

von Bertalanffy, L., 1960. *General Systems Theory* (rev. ed.). George Baziller, New York.

Warrick, R.A., 1984. The possible impacts on wheat production of a recurrence of the 1930s drought in the U.S. Great Plains. *Climatic Change,* **61:** 5-27.

Wigley, T.M.L., 1982. *Geographical patterns of climatic change: Analogues of possible CO_2-induced global warming.* Report to the U.S. Department of Energy, Contract No. DE-AC02-79EV10098, March 3, 1982.

————, P.D. Jones and P.M. Kelly, 1980. Scenario for a warm, high-CO_2 world. *Nature,* **283:**17-21.

Williams, J., 1980. Anomalies in temperature and rainfall during warm Arctic seasons as a guide to the formulation of climate scenarios. *Climatic Change,* **2:**249-66.

WMO (World Meteorological Organization), 1975. *The Physical Basis of Climate and Climate Modelling.* Report of the International Study Conference in Stockholm, July 29-August 10, 1974. GARP Publication Series No. 16. WMO, Geneva, Switzerland.

Worster, D. 1979. *Dust Bowl: The Southern Plains in the 1930s.* Oxford University Press, New York.

Zashin, E., and P.C. Chapman, 1974. The use of metaphor and analogy: Toward a renewal of political language. *The Journal of Politics,* **36:**290-326.

8. THE PROBLEM OF SOIL MOISTURE WITH GLOBAL WARMING

Konstantin Ya. Vinnikov

No single conclusion about possible changes in climate conditions as a result of anthropogenic global warming has produced such a strong impression on the public as the one about summer desiccation of the continents. Probably no international statement on climate variation, calling for coordinated international efforts to mitigate or prevent further increases in atmospheric pollution by greenhouse gases, could avoid dramatizing the possibility of aridization of the continents. Thus, the problem of soil moisture changes has often assumed a significant place in the minds of politicians and public representatives.

However, most specialists concerned with the problem of climate change are being faced with the subject of soil moisture for the first time, and they know hardly more about it than the politicians. Nevertheless, modern science has a significant amount of information on soil moisture at its disposal, though this information has seldom been made public and has probably never been studied by a wide community.

The information referred to is based on experimental data, including soil moisture measurements at an extensive network of agrometeorological stations of the former USSR, that has been summarized in a number of reference books and monographs. The use of such empirical data allows one to shed new light on the conclusions about possible changes in the soil moisture regime, such as those obtained by means of climate models. Another source of information of soil moisture regimes in the future, independent from the models, appears to be paleoreconstructions of soil moisture distributions for relatively warmer epochs in the past. These data also allow us to estimate to what extent the predictions based on climate models are true.

This chapter, then, is aimed at investigating the problem of possible future changes in soil moisture, using as a basis both available theoretical estimates and empirical data.

Climate Models Predict Soil Moisture Changes Better Than Do Changes in Precipitation

Initial attempts to use climate models to estimate the likely changes in regional climates under conditions of increased atmospheric carbon dioxide were essentially focused on the analysis of changes in surface air temperature and precipitation. Quite soon it was found that the effect of doubling or quadrupling atmospheric carbon dioxide on the model climate system was not strong enough for the estimates of regional changes in precipitation to be statistically significant and stable, even by sign. As a result, maps of average precipitation were, as a rule, not given in the original papers by the authors. However, it turned out that in each model changes of quite a different moisture characteristic—i.e., the moisture content of the active soil layer (usually 1 m deep)—are well pronounced relative to the background noise. It seems that, at the present level of knowledge, we cannot predict changes in atmospheric precipitation as a result of a global warming process, but we can forecast definite changes in the soil moisture regime.

Climatologists are well aware of the fact that atmospheric precipitation is characterized by a large spatial and temporal variability. That is why their climatic publications usually include procedures for temporal and spatial averaging. Even at the outset of meteorological station networks, it was understood that the sites for a network of precipitation measurements should be several times more densely spaced than those for measuring other meteorological variables. In any case, the possibility of identifying long-term trends in precipitation from the data of existing precipitation measurements remains unclear.

The main reason for doubt appears to be the fact that the instruments for precipitation measurements have been continually improved so as to reduce water loss from its adherence to the rain gauge interior and from evaporation between observation times. As a result, alterations in the instruments and the methods of measurement result in an apparent increase in the amount of precipitation measured—even if there are no actual changes in the natural precipitation regime. Therefore, a positive trend in mean annual precipitation of about 6 percent per 100 years for the continents of the Northern Hemisphere in the 35-70°N zone (excluding northern Canada and China) for the period 1981-1988 does not seem to be sufficiently reliable, even though the sign of the trend has been independently

estimated by Bradley *et al.* (1987) (see also Vinnikov *et al.,* 1990; Groisman, 1990).

If one were to assume that this trend is governed by the process of mean air temperature increase in the Northern Hemisphere, then we obtain a simple estimate. Over the same 100 year time period, the temperature increased 0.4°K. Thus, with a 1°K increase in mean annual surface temperature in the Northern Hemisphere, the mean annual amount of precipitation over the continents in the 35-70°N zone (without northern Canada and China) would increase by 15 percent, as compared with its mean annual amount averaged over the entire period (see Figure 8.1).

This empirical estimate (Vinnikov *et al.,* 1990) can be compared with those from two climate models. For the same territory a numerical experiment on the effects of carbon dioxide doubling with the GFDL model

FIGURE 8.1. (a) Annual surface air temperature of the Northern Hemisphere (0°-90°N) and (b) mean annual amount of atmospheric precipitation at the continents in 35°-70° zone (without northern Canada and China). Observation period from 1891 to 1988. The underlying trends, 0.4°C/100 years for temperature and 6 percent/100 years for precipitation, have a 95 percent statistical confidence level.

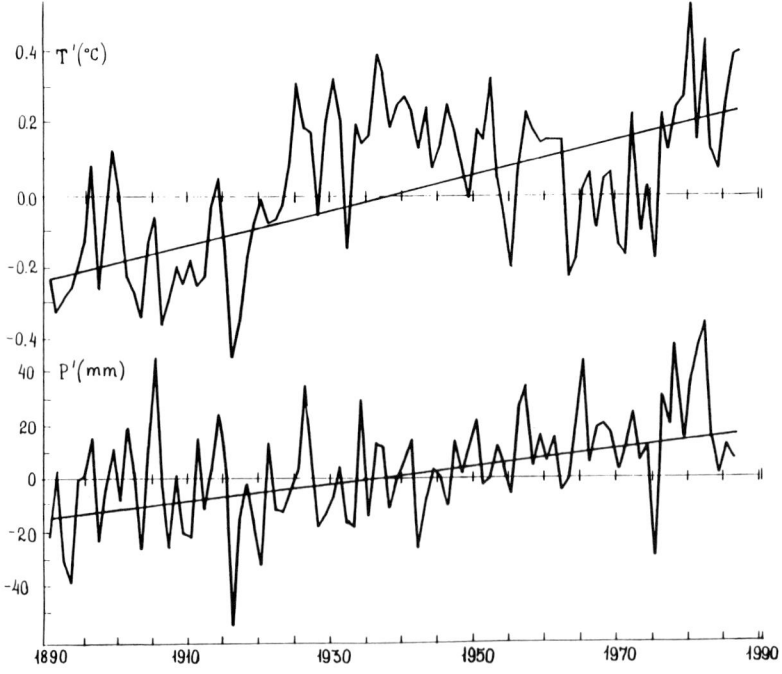

(Manabe and Wetherald, 1987) indicates a sensitivity of mean annual precipitation equal to 2 percent with a 1°K hemispheric warming, while the OSU model (Schlesinger and Zhao, 1989) shows it to be equal to about 4 percent per 1°K warming.

Even without digitized results from the model experiments, using the models with high spatial resolution one can, based on the estimates mentioned above, already draw a conclusion about a significant discrepancy between the theoretical (model) and empirical estimates of precipitation sensitivity to warming. These estimates, though coinciding as to sign, differ by half an order of magnitude, and in no way can they be considered as satisfactorily in agreement. Moreover, both the empirical and model estimates seem to be uncertain.

Then a further question arises concerning the suitability of the existing observation network that has been established to detect large anomalies in a monthly or even seasonal time scale, and also to monitor trends in precipitation changes over a time scale from several decades to centuries. The answer is probably affirmative, but before this affirmation is established it is necessary to carry out some studies to collect a larger number of the observations made during the period of meteorological measurements. (At present the analyses use, at best, only several percent of the available data.) When necessary, the homogeneity of precipitation time series should also be retrieved.

For these reasons, large scale and long term changes in precipitation of the kind being predicted by climate models in global warming experiments are unlikely to be found and identified in the observations at present. This does not imply that, in some sufficiently sensitive regions such as, for example, the Sudan-Sahel zone of Africa, the existing observational system will not record significant trends in precipitation.

Then what can be said about changes in soil moisture? If climatologists inform politicians and the general public about possible changes in the soil moisture regime, would they then be able, on the basis of this information, to make the right decision? Let us consider further the analyses of empirical data.

Soil Moisture Measurements

In the former USSR water content measurements in the active soil layer have been carried out at a network of agrometeorological stations since the 1930s. At present the number of such stations conducting measurements in agricultural fields is about 3,000. These measurements are summarized in the form of reference books in which the statistical characteristics of soil moisture in fields with different agricultural crops, averaged by districts, areas, republics, and economical regions of the country, are published (Kelchevskaya, 1979, 1989; Zhukov, 1986). The monographs of Verigo and Razumova (1973), Meshcherskaya et al., (1982), Kelchevskaya (1983), and other publications present a detailed analysis of the results of measuring soil moisture in agricultural fields. As soil water content in the fields depends both on the type of crop and agricultural practices, these measurements are not always representative of the territories surrounding these fields.

Measurements of soil moisture over natural surfaces (usually meadows) were started much later, in 1967, in support of existing water balance (runoff) stations, and also at some agrometeorological and hydrometeorological stations. The network consists of approximately 250 stations, but both the period and the quantity of the observations vary.

Usually soil moisture measurements are made at permanent sites with an area of at least 0.10 to 0.15 hectare (ha) all year round. In warmer seasons measurements are made on the 8th, 18th, and 28th day of each month; in cold seasons measurements are made only at the end of each month. Measurements are made by a thermostat weight technique in the following way: 10-cm diameter soil cores are extracted from different parts of the site down to a depth of 1 or 1.5 m. Samples from each of these cores are weighed before and after drying in the thermostat. The results of measurements are expressed in units of millimeters of plant-available moisture. Unavailable moisture is determined experimentally for each site as the moisture of steady wilting of plants. If soil moisture decreases to this level, the plants die and never return to life. The agrohydrological parameters of the soil are also determined experimentally for each site, the most important of them being:

W_f = field capacity, the maximum water content that can be kept in the soil by the forces counteracting gravitation when there is no contact with groundwater, and W_o = total water-holding capacity, the maximum water content that can be contained in the soil if all its pores are filled with water.

Field capacity is usually used in climatic models to parameterize evaporation and runoff processes. In most of the models field capacity is assumed to be 150 mm for all land regions of the globe. Data analysis shows that, for the former USSR territory alone, this "constant" changes in the range of 70 to 210 mm, depending on the type and mechanical composition of the soil (Vinnikov and Yeserkepova, 1991). Also, the use of field capacity in the parameterization of evaporation is correct only if the groundwater depth exceeds 3 m. If the groundwater table is closer to the surface, then the characteristics of the maximum capillary water-holding capacity of the soil should be used in the parameterization of runoff. Since the groundwater table has quite significant geographical variability and seasonal variations, the parameterizations used in the models are not always and everywhere sufficiently correct.

Long-term soil moisture measurements carried out in the former USSR at a sufficiently extensive station network serve as an empirical basis to investigate the soil moisture regime and its current changes, and to develop experimentally based parameterizations of the water exchange processes in the active soil layer.

Modern Regimes of Soil Moisture

It appears that one of the best known attempts to construct global maps of mean monthly soil moisture was undertaken by Mintz and Serafini (1989), although earlier calculations of mean soil moisture content were often a by-product of computations of the evaporation regime over land surfaces

(Zubenok, 1976). However, the importance of such calculations was not understood by the authors, and the results were usually left unpublished.

Mintz and Serafini (1989) used an algorithm for calculating components of the land water balance, employed in a corresponding climatic model, to construct maps of relative moisture content in the active soil layer for the present climatic conditions. The computational algorithm is based upon the water balance equation and the method of calculating evaporation and evapotranspiration developed by Thornthwait. It is well known, however, that the Thornthwait method used in evaporation calculations does not provide for fulfillment of the equation of heat balance of Earth's surface. In addition, this calculation uses one and the same value of field capacity, equal to 15 cm, for all land regions. This seems to explain why the authors could not obtain a realistic geographical pattern of soil moisture distribution. This is found by a comparison of the maps, mentioned above, with those constructed from data of the former USSR station network (Vinnikov and Yeserkepova, 1991). This comparison indicates that the maps by Mintz and Serafini are in a more or less satisfactory agreement with the empirical data only for the winter and spring, and for the summer and fall the computed values are only half as large. In addition, in that part of the former USSR for which the measurements are available, the minimum of annual soil moisture occurs in the summertime, while in the calculations of Mintz and Serafini they occur in the fall. It should no doubt be noted that annual runoff estimates presented by these authors appear not to be comparable with the maps of the annual runoff (Korzoun, *et al.*, 1974), which were constructed to a large extent from empirical data. In the regions of excessive moistening the calculated runoff turned out to be strongly underestimated, whereas in the arid zones it is significantly higher than the observed values.

Figure 8.2 presents schematic maps of soil moisture for the winter and summer seasons constructed by Vinnikov and Yeserkepova (1989, 1991) using observational data from the sites with a natural vegetation cover. (Let us note that the model calculations did not result in any

FIGURE 8.2. Moisture content of the upper 1 m soil layer from measurements over natural vegetation—solid isolines (a) winter; (b) summer. The shaded area (1) shows mountains and areas for which the data are not available; dashed line (2) shows isolines of mean soil moisture in the fields of winter wheat; and dotted line (3) shows isolines of mean soil moisture in the fields of spring wheat.

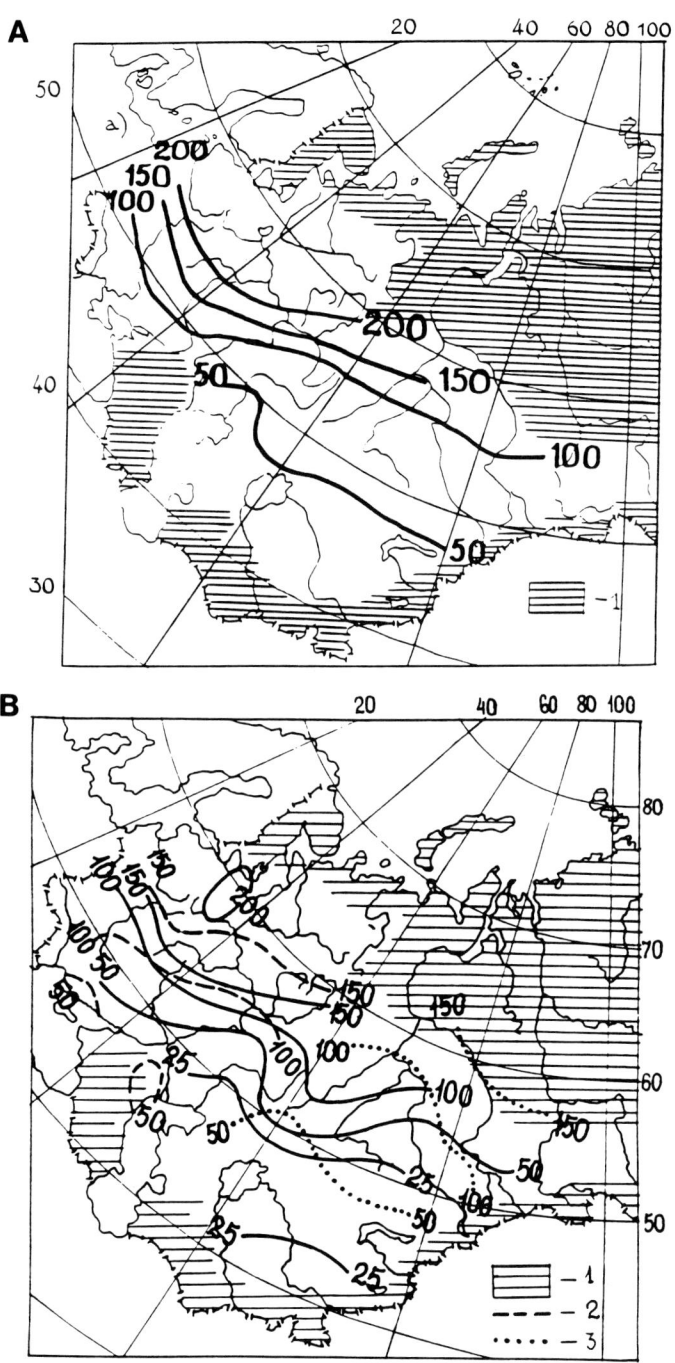

values of soil moisture content exceeding 15 cm, because of the assumption that this value corresponds to the field capacity.) In nature the groundwater table strongly affects the soil moisture regime. In northern latitudes, where the groundwater table is near the surface, mean soil moisture content values can exceed those of field capacity and be close to the total water-holding capacity values. This is clearly defined in Figure 8.2, particularly for the winter season. A dashed line on the maps of mean soil moisture content under a natural vegetation cover gives moisture values for agricultural fields containing winter and spring wheat crops in the European and Asian territories of the former USSR.

Here we are faced with a dilemma. What data can be considered to be representative of such places as the Ukraine or Kazakhstan, where the largest part of these areas is occupied by agricultural fields? Is it worthwhile to choose sites here with a natural vegetation cover? One should note a pronounced characteristic of such lands. In more northern areas with a sufficient or excessive precipitation the soil moisture over agricultural fields is less that over areas with natural surfaces. Southward, in the areas of insufficient precipitation, it is vice versa, and the summer soil moisture over the agricultural fields is more than that of natural surfaces.

These data suggest that those routine agricultural practices employed over vast territories in what was once the USSR contribute to creating the soil moisture regime, and this is more favorable for plants than the regime where the natural vegetation cover has been preserved. A feedback appears to exist here, in which agricultural use of a territory will mitigate adverse changes in the soil moisture regime connected with global climate change.

While the estimates of patterns of mean soil moisture content obtained from the data on current temperature and precipitation (Mintz and Serafini, 1989) are in poor agreement with the empirical estimates, the consistency of the latter with soil moisture patterns obtained by the current climate models is even worse. In this case, both an inadequate parameterization of hydrological processes and an insufficiently accurate simulation of major components of the current climatic state will contribute to the poor agreement between the calculated values and the measurements.

A comparison was made for four large regions of the former Soviet Union in the latitude band 45°-60°N, as indicated in Figure 8.3, between the measured and calculated values of the current soil moisture distribution by the models of GFDL (Manabe and Wetherald, 1987), OSU (Schlesinger and Zhao, 1989), and UKMO (Mitchell, 1989). The results are given in Figure 8.4. All these models employ similar parameterizations to calculate evaporation and runoff. The same value of field capacity, 15 cm, is used in all three models. All models demonstrate the effect of summer desiccation of the continents with a carbon dioxide increase in the atmosphere. Analysis shows that all three models simulate more or less successfully only the shape of annual variation in soil moisture, but the quantitative agreement is quite unsatisfactory.

The largest inconsistencies between the measurements and the model calculations have been observed in Region 1 (55°-60°N, 22.5°-45°E). Here the measured values are two times or more larger than the values calculated by the models for all months of the year. The reason for such a discrepancy is that the actual location of the groundwater table is not taken into account in the models, and instead the underestimated value of water-holding capacity is used.

The second significant discrepancy is well evident in all four regions for the GFDL and OSU models: The models yield unrealistically low (close to zero) soil moisture values for the summer months. In most of the cases the differences between the various models, and also the differences between models and observed data, considerably exceed those soil moisture changes that are predicted by these very models with a doubling of carbon dioxide concentration.

Seasonal Structure of the Trend: Summer Desiccation

The seasonal structure of the soil moisture changes due to a CO_2-induced global warming, as predicted by climate model experiments, appears to be their most consistent feature (i.e., soil moisture

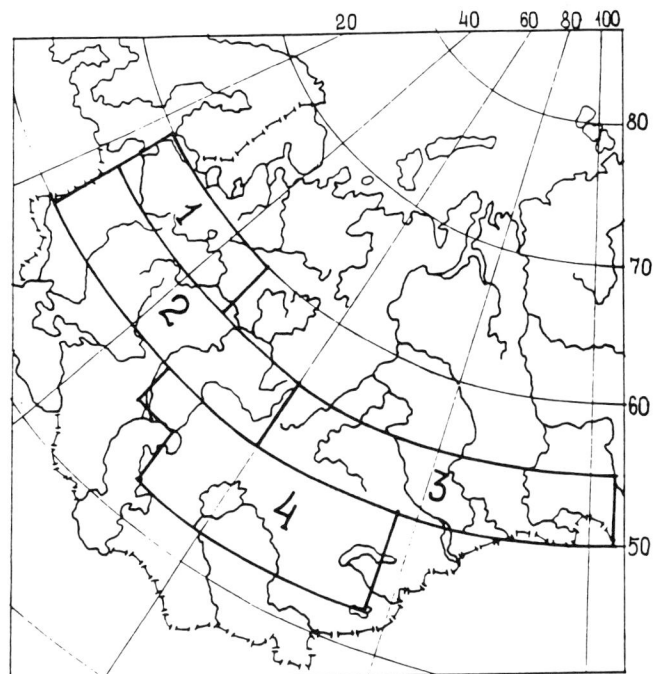

FIGURE 8.3. Four regions in the former Soviet Union for which soil moisture regimes have been studied.

FIGURE 8.4. Comparison of seasonal variations of soil moisture simulated by the GFDL (thin line), the OSU (dashed line), and the UKMO (line with crosses) models with soil moisture measurements (solid line) for the regions shown in Figure 8.3; (a) region 1; (b) region 2; (c) region 3; (d) region 4.

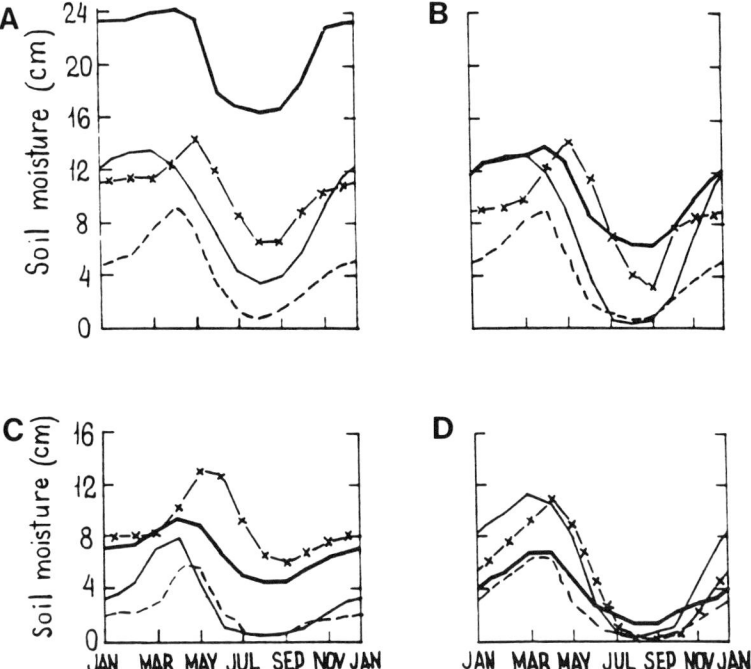

increases during the cold part of the year, decreases during the warm part). These changes are particularly well pronounced in the northern parts of the continents in the Northern Hemisphere. Figure 8.5 shows these changes calculated by two model equilibrium experiments for a doubling of CO_2 concentrations in the atmosphere (Manabe and Wetherald, 1987; Schlesinger and Zhao, 1989).

This author, dealing with the regional climate change problem, found repeatedly that many characteristic features of the geographical distribution of climatic parameter changes were not specific with regard to the cause of the global climate change. This hypothesis was suggested by

Vinnikov and Groisman (1979, 1982) and Vinnikov (1986) as a postulate in developing an empirical model of climate change to relate regional climate changes with those of mean annual surface air temperature in the Northern Hemisphere. This hypothesis was confirmed by Manabe and Wetherald (1980) and Hansen *et al.* (1984) who, using numerical experiments with GCMs, showed that response of the climate system to atmospheric carbon dioxide and solar constant changes is not specific to the cause.

It is interesting from this standpoint to consider the seasonal structure of long-term trends in soil moisture if, of course, such trends do occur. Vinnikov and Yeserkepova (1989, 1991) obtained estimates of seasonal

FIGURE 8.5. Model estimates for differences in the soil moisture for the climates with double the present day concentrations of carbon dioxide: Solid line—Manabe and Wetherald (1987); thin line—Schlesinger and Zhao (1989). Estimates are given for the regions shown in Figure 8.3: (a) region 1; (b) region 2; (c) region 3; (d) region 4.

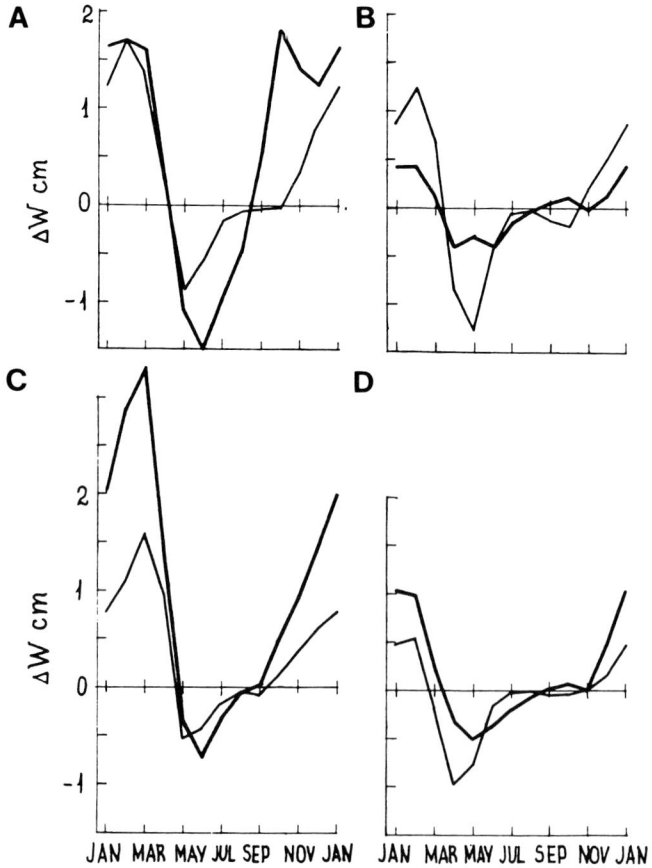

variation in the parameters of the linear trend of mean monthly values of soil moisture content and monthly sums of atmospheric precipitation for the period of 1972 to 1985 (see Figure 8.6) for the same four regions as indicated in Figure 8.3. The trend appeared to be of the same positive sign for regions 1, 2 and 3 in the latitude band 50°-60°N for all months of the year, with the soil moisture content increasing in all seasons. Owing to the limited amount of observational data in region 4, the trends can only be estimated for the warm month of the year. The trends in this zone were found to be close to zero, and thus no seasonal variations can be discerned.

Calculations showed that in quite a vast intercontinental area of Eurasia in the latitude band of 50°-60°N, where one might expect such a phenomenon as "summer desiccation," nothing occurs, despite a mean rate of increase of 0.2°K/10 yrs in mean annual surface air temperature in the Northern Hemisphere (Vinnikov *et al.* 1990). One can attempt to attribute the increase in soil moisture to the comparatively short observation period under consideration (14 years) and a nonstationary process of global warming. However, the absence of a negative correlation between the trends of the warmer and colder parts of the year, as found in the empirical data, is difficult to explain, other than to attribute it to the erroneous concept of summer desiccation of the continents.

Another significant conclusion, suggested by an analysis of the estimates given in Figure 8.6, is that the observed trends in soil moisture resulted mainly from the increased amount of precipitation, the trends of which happened to parallel that of soil moisture. Let us also note the scale of the

FIGURE 8.6. Seasonal variations of linear trend (cm/decade) of soil moisture (thick line) and precipitation (thin line) for the measurement period 1972 to 1985. a-d represent the data for the regions 1, 2, 3 and 4 shown in Figure 8.3.

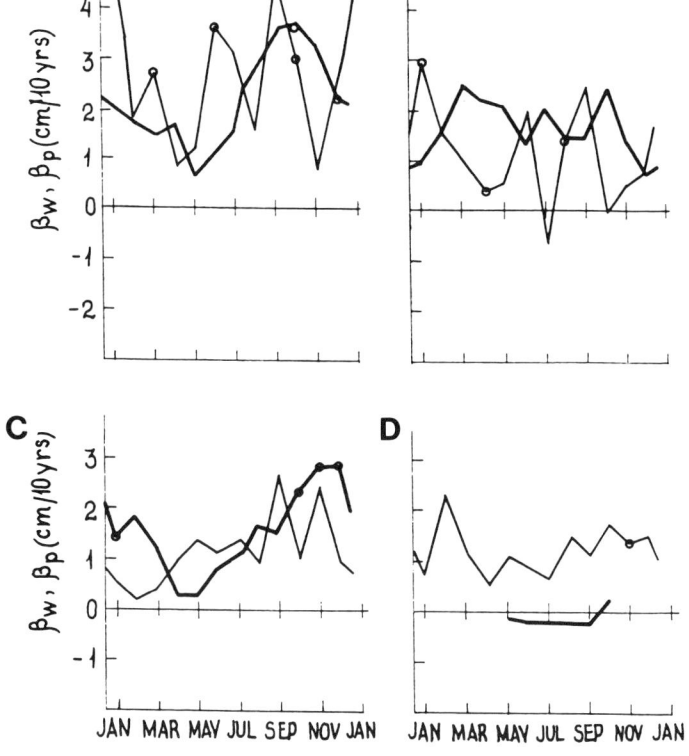

observed trends in soil moisture, being approximately 1-2 cm/decade.

The schematic maps in Figure 8.7 provide a better understanding of summer and winter trends in soil moisture content. In winter, in the entire band 50°-60°N, from the western boundary of the former Soviet Union to about 90°E, soil moisture for the 1972-1985 period increased with a mean rate of 2-3 cm/decade. In summer soil moisture in the same band also mainly increased, although the magnitudes of the trend are a little less than in winter. The area of negative trends (close to zero) in the southern part of the Asian sector of the territory under consideration is a feature that should be considered more carefully.

In the fall (the map of estimates is not given here) the positive trends have the largest value in the northwest region of the European portion of the country, reaching 4 cm/decade and more. Weak negative (close to zero) trends are observed only in the southern part of Central Asia.

It is curious that all these changes in soil moisture during the 1972-1985 period actually remained undetected. However, during this same time period, experts in agriculture in the USSR noticed exceptional weather problems, including the increasing frequency of occurrence of very severe droughts, which resulted in a decreased grain crop yield in the major grain-production regions of the country.

Therefore, a very significant increase in soil moisture content, objectively recorded in middle latitudes of the USSR, has not been detected in a timely manner, and its influence on the economy, including grain production, has not been identified either. One may, of course, suggest that soil moisture reduction would have been quite quickly revealed. However, there are no data yet available necessary to answer the question of whether the enhanced soil moisture in the territories with a natural vegetation cover was accompanied by a similar soil moisture increase over agricultural fields.

Again, referring to publications by Menzhulin, Nikolaev, and Savvateyev (1983) and private communications of experts who analyzed long-term series of the grain crops of medium yield in the U.S.S.R. territory, one can say that it was not only the changes of weather (climatic) conditions that can account for the yield variations. This series of events clearly shows certain features connected with the "historical decisions of the Plenum of the Central Committee of the Communist Party of the Soviet Union on agricultural issues." It also follows from the analyses that in some historical periods the availability of combines for harvesting was a limiting factor in the increase of grain crop yield. This phenomenon was been particularly evident during the exploration of the virgin land in Kazakhstan and western Siberia, which significantly extended the land area under grain crops, but with the same amount of agricultural equipment.

Thus, one can suggest that the changes in soil moisture regime in middle latitudes of the former Soviet Union by themselves, even over vast territories, possibly do not have a noticeable effect on agricultural production.

We have considered the empirical data on the current soil moisture regime and its variations. Now let us consider the epochs of the distant past, the climate of which was warmer than the current one. What was the soil moisture regime like? How much did it differ from the current one?

Soil Moisture Regimes in the Warmer Epochs of the Past

With the prospect of an enhanced global warming, attempts to find ways to use the information about past climatic conditions become more frequent. The first among them were probably the publications of Kellogg (1977), Budyko et al. (1978), and Budyko (1980). Later on, the idea was further developed by Budyko and Izrael (1987) and the IPCC's Scientific Assessment of Climate Change (1990). Although in the IPCC report this idea is presented in quite a critical way, a critical analysis of its advantages and disadvantages will contribute to its further development. The referenced works consider two past warm epochs as the climates most nearly analogous to the future with a 1°K and 2°K global warming relative to the present. They are the Holocene Optimum (5,000-6,000 B.P.) and the Last Interglacial (125,000 B.P.), mentioned in some other sources as the Mikulino, the Sangamon, the Riss-Wurm, or the Eem. A.A. Velichko and I.I. Borzenkova and colleagues in their works

FIGURE 8.7. Winter (a) and summer (b) trends of the soil moisture content (cm/decade) for the measurement period 1972 to 1985. Shading shows: 1—mountains and areas for which the data are not available; 2—regions of negative trends.

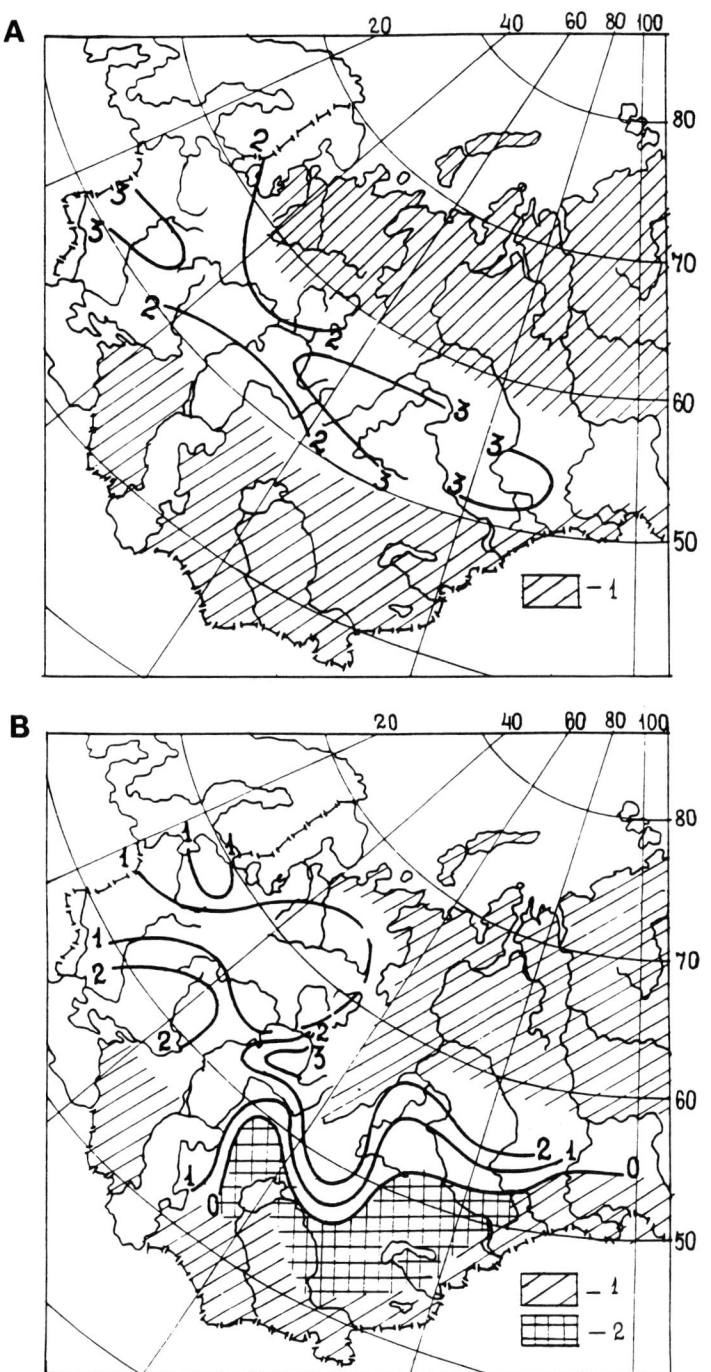

have constructed schematic maps of winter and summer surface air temperature and mean annual precipitation. Data of these paleoreconstructions for the extratropical part of the Northern Hemisphere were published in a preliminary form in the monograph of Budyko and Izrael (1987). Later on they were improved, and this author had an opportunity to use the improved temperature and precipitation maps, expressed in terms of the deviations of the corresponding variables from the mean values in the current epoch.

Vinnikov and Lemeshko (1987) and Vinnikov, Lemeshko, and Speranskaya (1990) developed a method for reconstructing soil moisture regimes for those same epochs of the past. It is based on a complex procedure for calculating evaporation from the land surface, suggested by Budyko (1971) and widely used to investigate the current water balance of Earth's continents (Korzoun, *et al.*, 1974; Zubenok, 1976). This method provides for fulfillment of two main physical laws, written in the form of an equation for Earth's surface energy balance and an equation for the soil active layer water balance. In addition, two parameterizations are employed, one to calculate evaporation and the other to calculate runoff. Both parameterizations were suggested by M.I. Budyko, based upon the analysis of observations from hydrometeorological and agrometeorological stations, and are in agreement with the empirical data.

To reconstruct the moisture regime of the two past warm epochs mentioned above from the paleoreconstruction of winter and summer air temperature and annual precipitation levels, some less unquestionable additional hypotheses were introduced. These hypotheses addressed the unchanging cloud regime and effective relative humidity, and the character of the change in seasonal precipitation variations. Though creating certain errors, they do not strongly influence the calculated results.

Figure 8.8 presents differences in the summer moisture content (in percent from the present) for the Holocene Optimum and the Last Interglacial, based on the data mentioned above. The authors are sure that the possible errors in these reconstructions result, first of all, from the erroneous paleoreconstructions of temperature and precipitation, rather than by the method used. Regretfully, the calculation for the

Interglacial could be made only for Eurasia, as no data on temperature and precipitation in this epoch are available for North America.

Both reconstructions for the Holocene Optimum and for the Last Interglacial for Eurasia indicated a similar pattern, with the summer soil moisture content being lower in the north than in the current epoch, and higher in the south. A zero isoline of change during the Holocene Optimum coincided, on average, with the 50°N latitude circle, but it was more to the north during the Last Interglacial (in a warmer epoch) and closer to 60°N latitude. Summer soil moisture in the northern regions is reduced by about 10 percent, as compared with the current norm (i.e., by a value not exceeding 2 cm). It is unlikely to be significant for the areas not already suffering moisture deficiency. One can suppose that such moisture reduction may have a favorable influence on the economic exploitation of the northern territories. At the same time to the south, where the moisture is currently insufficient, both warm epochs of the past were characterized by an enhanced summer soil moisture content. This increase was larger for the warmer of the two past epochs. During the Last Interglacial the summer soil moisture in the forest steppe and steppe zones of Eurasia exceeded the current one by up to 2-2.5 cm. These changes appear to be quite favorable for both natural vegetation and agricultural crops.

During the Holocene Optimum there is a similar picture in the territory of North America. However, the zone between the areas of a decreased summer soil moisture content in the north and an increased one is situated approximately 10 degrees more to the south than in Eurasia.

It seems to be difficult in the present study to establish to what extent the pattern of changes in summer moisture regime in the active soil layer, deduced for the two warm periods of the past, can be considered as a prediction of the future with a global warming of approximately 1° and 2°K, respectively. However, one would think that such changes should be considered possible. In any case, they have actually taken place. On the contrary, if during climate modeling a parameter is used that is not quite correct, then the output can be something that has never existed in nature and it will not do so

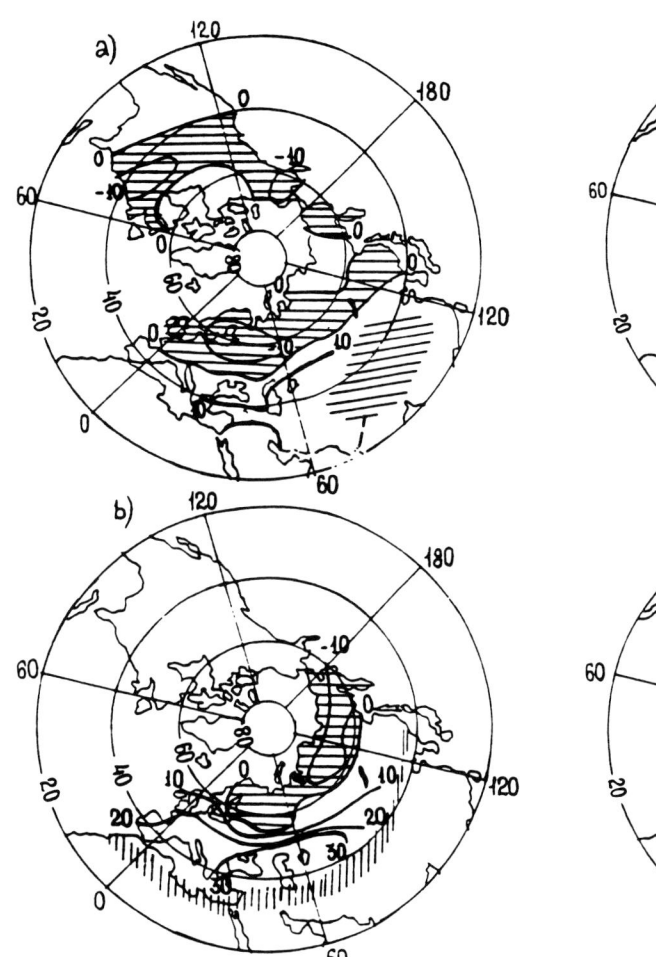

FIGURE 8.8. Changes in summer water content in the active 1 m soil layer for (a) the Holocene Optimum (5,000-6,000 B.P.) and (b) the Last Interglacial (125,000 B.P.). Data presented as a percent of present-day values.

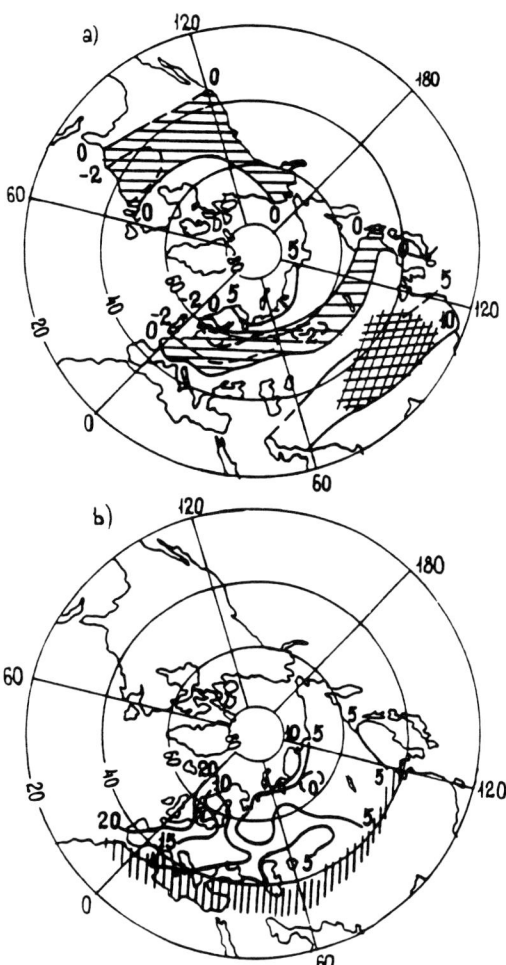

FIGURE 8.9. Changes in annual runoff for (a) the Holocene Optimum (5,000-6,000 B.P.) and (b) the Last Interglacial (125,000 B.P.).

under any circumstances.

Let us compare the paleoreconstructions considered for the territory of the former Soviet Union with the current trends according to the scale of the changes and their sign. The scale of the currently observed changes in summer soil moisture content is about 1-2 cm/decade. As a result, the total soil moisture change over the 1972-1985 period is of the same order of magnitude as

that for each of the two past warm epochs, a time interval extending from several thousand to more than 100,000 years ago. But mankind did not even notice these changes. It is possible that the changes relating to the development of an anthropogenic global warming will not be noticed, just because they will not induce adverse side effects.

The sign of the changes in summer soil moisture in 50-60°N observed over the 1972-1985 period is positive. During the Holocene

Optimum the summer soil moisture content in this northern zone was lower than during the present epoch. In Central Asia summer soil moisture during this past epoch was higher to some extent than the present one, and the map of the trend for 1972-1985 indicates an insignificant but slight decrease.

Let us make a comparison with the best model results, as given in the Scientific Assessment (IPCC) report (1990). The decrease in the summer soil moisture in the north of Eurasia and North America in paleoreconstructions is quite consistent with model results, having the same magnitude and sign. However, this is not the case for the areas of summer moisture content increase in the southern part, which are well defined in the paleoreconstructions but are missing in the model maps of change in summer soil moisture with a doubling of carbon dioxide. Some indications of such an area of positive change (but small in value) are found only in the results obtained by the GFDL model, but they are absent in the calculations by the CCC (Canadian Climate Center) and UKMO (United Kingdom Meteorological Office) models. Model estimates are also opposite in sign to the observed trends of enhanced soil moisture content in the 50-60°N zone. This means that the observed trend is not connected in any way with the current process of global warming, or else that the model estimates in this region are incorrect (an alternative that unfortunately cannot be eliminated either).

Let us consider one more aspect of the soil moisture problem, one related to water resources. Even if summer soil moisture content decreases, with a general increase of available renewable water resources it would be difficult to consider this as a disaster.

Soil Moisture Problems and Water Resources

The annual runoff, determined as the difference between precipitation and evaporation, is usually considered to be a measure of the renewable water resources in a region. Naturally, only part of this runoff can be actually used by mankind. Owing to insufficient reliability of model estimates of the change of the precipitation regime with global warming, any conclusions on possible

runoff changes are not very reliable either. The task is further complicated as the dependence of regional precipitation on mean global air temperature appears to be nonlinear, in any case in the middle latitudes of the continents in the Northern Hemisphere (Budyko, 1980; Budyko, *et al.* 1978; Budyko and Izrael, 1987; Drosdov, 1981). That is why the use of model estimates of an equilibrium response of the climate system to a doubling of carbon dioxide for forecasting future climate is not reliable for the intermediate stages of a warming.

Analysis of paleoclimatic analogs used in forecasting of climatic conditions with global warming, in particular the climates of the Holocene Optimum and the Last Interglacial, confirms the hypothesis of a nonlinear dependence of regional precipitation on the magnitude of the mean global air temperature increase (Budyko and Izrael, 1987; MacCracken *et al.*, 1990). It is fortunate that the calculation of the annual runoff change was a by-product of the paleoreconstructions of the soil moisture regime made by Vinnikov and Lemeshko (1987) and by Vinnikov, Lemeshko, and Speranskaya (1990). Results of these calculations for the same two warm epochs of the past are given in Figure 8.9. If we are to have some confidence in these epochs as paleoanalogs, then the conclusion of a nonlinear relation between global temperature and precipitation should also be applied to the relation between global temperature and runoff. In fact, in middle latitudes, mainly in the 50-60°N zone of Eurasia and in the 35-50°N zone of North America in the Holocene Optimum epoch (about 1°K warmer than the present), the calculations show a band of small but distinct runoff decrease. The southern boundary of this band coincides with the southern boundary of the area of summer soil moisture decrease.

A simultaneous decrease in soil moisture and water resources (annual runoff) can be a cause for concern, as it can undoubtedly influence plant available moisture in agricultural operations. In any event, a more detailed analysis is required here. The situation is quite different if we consider the runoff estimates for the Last Interglacial epoch (about 2°K warmer than the present). Actually, in that case the difference between the annual precipitation and evaporation (the runoff) increases everywhere. This increase is especially significant in Western Europe, being 10-20 cm/year. With a noticeable runoff increase the river flow will probably be greater and the groundwater table will increase.

It is probable that the runoff will continue to increase with a further increase of mean global temperature. If there are some additional water resources, even if in some regions the problem of summer soil desiccation occurs, it can be solved if necessary by well known ways.

Figure 8.10 shows areas where, according to equilibrium model experiments with doubling the carbon dioxide concentration, the summer soil moisture decreases while the annual runoff does not increase. Both GFDL and OSU models demonstrate such areas in North America. These regions will possibly have significant difficulties from global warming.

Discussion and Conclusions

Let us now consider the main statements and conclusions of each section of this chapter and briefly comment on them.

1. The problem of soil moisture changes, which has arisen because the modeling of it has generally been made without reference to any empirical or experimental data, appears to be part of a larger problem, namely to evaluate the influence of anthropogenic global climate changes on the biosphere.

2. Conclusions obtained by different modeling groups vary on the possible changes in soil moisture regime with an increase of carbon dioxide concentration (Kellogg and Zhao, 1988; Schlesinger and Mitchell, 1987; Zhao and Kellogg, 1988). Three research groups, GFDL (e.g., Manabe and Wetherald, 1987),

FIGURE 8.10. The area in which, after doubling of the carbon dioxide concentration, summer desiccation will not be accompanied by an increase in mean annual runoff: (a) data from the GFDL model (Manabe and Wetherald, 1987); (b) data from the OSU model (Schlesinger and Zhao, 1989).

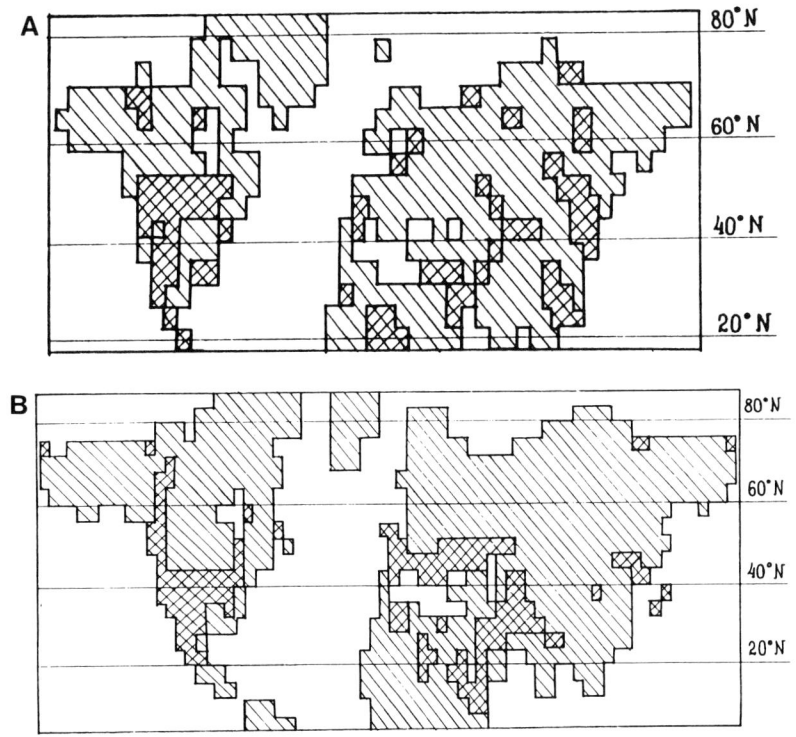

OSU (Schlesinger and Zhao, 1989), and UKMO (Wilson and Mitchell, 1987) concluded that there would be a summer aridization of the mid-latitude continents of the Northern Hemisphere. It is these and other works by these groups that have created "the problem of soil moisture." At least two other research groups, NCAR (Washington and Meehl, 1984) and GISS (Hansen *et al.*, 1984), came to the conclusion that there would be a general absence of summer desic-cation. However, these latter results are no more convincing than the first ones, and we prefer not to consider them at all. The NCAR and GISS groups did not arrive at conclusions of soil moisture changes that might be unfavorable, and that would have required developing a response strategy.

Model estimates of possible precipitation changes with global warming are not large enough to be distinguishable by means of the current observational meteorological network. These measurements indicate some trends, and they are significantly larger than the predicted ones. However, there is no confidence that these trends are real. Soil moisture in future climate predictions appears to be a suitable alternative to precipitation.

3. Measurements of soil moisture have been carried out at some thousands of stations in the former USSR for several decades. These data serve as a comprehensive empirical basis to verify the parameterizations used in models, and model results.

4. The current soil moisture regime, as simulated by the climate models, does not seem to be realistic. The models strongly underestimate summer soil moisture values for the middle latitudes of the Northern Hemisphere. In high latitudes the groundwater, located close to the surface and influencing considerably the soil moisture regime, is very important, but it is not taken into account.

Measurements indicate that hydrophysical soil characteristics, such as field capacity, differ quite signifi-cantly, exhibiting a dependence on the type of soil and its mechanical composition. A comparison of the observational data over agricultural fields with those over natural surfaces shows that the agricultural practices employed at present in the former Soviet Union effectively mitigate unfavorable changes in the soil moisture regime. As a result, in agricultural fields summer moisture content in the areas of excessive moistening is less than in adjacent meadows. And in the regions with insufficient moistening, the summer soil moisture in agricultural fields is larger than in adjacent areas with natural meadow vegetation. Therefore, an anthropogenic effect of a feedback type occurs in the areas of intensive agriculture, preventing the development of soil moisture changes that are unfavorable for agricultural plants.

5. Analysis of observational data exhibits quite significant soil moisture trends, its scale being 1-2 cm/decade or more for the observational period of about 15 years. These trends, however, have one and the same sign for all months of the year. Nothing similar to a specific seasonal trend was observed, as is occurring in climate model experiments in which soil moisture decreases in the warm season are accompanied by an increase in the colder season. That is, situations of the "summer desiccation" type were not detected in these observations.

6. Comparisons of the present soil moisture regime with its paleoreconstructions for past warmer epochs, assumed to be climatic analogs of the future, indicate a significant summer soil moisture decrease taking place only in the northern areas of the continents, usually currently excessively moistened. To the south, in the areas of insufficient moistening in summer, soil moisture was increasing. During the warm epochs of the past there was no general aridization of the continents.

The scale of moisture content changes in the active soil layer for these epochs is comparable to the scale of those changes in soil moisture that are found in the former USSR territory in the current 15-year data set.

On the whole, the analysis of paleoclimatic reconstructions of the soil moisture regime for the warm epochs of the past (the Holocene Optimum and the Last Interglacial) does not exhibit significantly unfavorable changes for the biosphere in terms of the moisture regime of the soil active layer in the extratropical zones of the Northern Hemisphere.

7. The annual runoff, which is a measure of annually renewable water resources, will probably depend nonlinearly on the global warming in some regions. However, with a sufficiently large warming, similar to the one that occurred in the Last Interglacial epoch or still larger, a generally significant annual runoff increase outside the equatorial part of the Northern Hemisphere is quite possible. During the late stages of a global warming, even the occurrence of a phenomenon like a summer desiccation of the continents will probably be accompanied by an increase of annually renewable water resources, and it will therefore not be harmful for the biosphere and economic activity. (Of course, this is a highly simplified conclusion. Summer desiccation is mostly very unpleasant even if there is enough water in the rivers.) However, during the earlier stages of a global warming some decrease in the annual runoff can be expected in some regions of the middle latitudes. A possible coincidence of these areas with those areas where summer desiccation will occur will require particular attention.

The conclusion about a coming climatic catastrophe, namely aridization of the continents with the advance of anthropogenic global warming of the earth, was drawn from the use of climate models. An analysis indicated that the various model estimates are in poor agreement, not only between themselves but also with the empirical data on the current soil moisture regime, and also with data from paleoclimatic reconstructions of soil moisture regimes during warmer epochs of the past.

The above estimates and the conclusions reached are not so accurate and definite that they can be recommended as a basis for making political decisions. However, some weak links have been highlighted in this problem that are responsible for the largest uncertainty in the estimates. It becomes clear that major difficulties stem from the fact that modeling of the soil moisture regime has been made in isolation from the empirical data, and without sufficient experimental verification. It is concluded that the data from long-term soil moisture measurements made at an extensive network in the former Soviet Union can be used to provide an experimental basis. That is why quick progress can be achieved on the basis of the already available data, and it will not require the development and implementation of long-term and expensive experimental programs.

The scientific data on possible changes of soil moisture regimes now under consideration do not allow one to draw the conclusion that these changes will have an adverse impact on the biosphere. However, the response of plant communities to the change in soil moisture regime can also depend on other factors, such as the carbon dioxide concentration in the atmosphere. The response can result in replacement of one plant community by others, and such replacement in agricultural fields can be considered as part of the biospheric adaptation process to changing external conditions.

Having clearly understood all this, we can conclude that only the results of a comprehensive analysis of all the consequences of global warming can serve as a basis for recommendations in terms of political decisions, both on a national and international scale.

References

Bradley R.S., J.F. Diaz, J.K. Eischeid, *et al.*, 1987. Precipitation fluctuations over Northern Hemisphere land areas since the mid-19th century. *Science*, **237**:171-75.

Budyko M.I., 1971. *Climate and Life.* Gidrometeoizdat, Leningrad (Russian edition). (English translation, D.H. Miller (Ed.). 1974. Academic Press.)

———, 1980. *The Earth's Climate: Past and Future.* Gidrometeoizdat, Leningrad (Russian edition) (English translation: Academic Press, 1982.).

———, and Yu. A. Izrael (Eds.), 1987. *Anthropogenic Climate Changes.* Gidrometeoizdat, Leningrad (Russian edition) (English translation: Arizona University Press, Tucson 1990.)

———, K. Ya. Vinnikov, O.A. Drozdov, *et al.*, 1978. Future climatic changes. *Proceedings of the USSR Academy of Sciences, Geographical Series*, **6**:5-20 (Russian).

Drozdov O.A., 1981. Formation of land moistening in climate fluctuations. *Meteorologia i Gidrologia*, **4**:17-23. (Russian).

Groisman P.Ya., 1990. Data on present-day precipitation changes in the extratropical part of the Northern Hemisphere. *Proceedings of the USSR Academy of Sciences, Geographical Series*, **3**:20-30. (Russian).

Hansen J., A. Lacis, D. Rind, *et al.*, 1984. Climate sensitivity analysis of feedback mechanisms. In: *Climate Processes and Climate Sensitivity*. J. Hansen and T. Takahashi (Eds.). *Geophysical Monograph*, **29**:130-163. American Geophysical Union, Washington D.C.

Kelchevskaya L.S. (Ed.), 1979. *Mean Long-Term and Probability Characteristics of Productive Water Stores Under Winter and Early Spring Cereals*. Vol. 1. Gidrometeoizdat, Leningrad (Russian).

———, 1983. *Soil Moisture in the European USSR*. Gidrometeoizdat, Leningrad (Russian).

——— (Ed.), 1989. *Mean Long-Term Stores of Productive Water Under Winter and Early Spring Cereals in Districts*. Vol. 2. Gidrometeoizdat, Leningrad (Russian).

Kellogg, W.W. 1977, 1978. Effects of human activities on global climate. Part 1: *WMO Bulletin 26*. pp. 2229-40 (1977); Part II:*WMO Bulletin 27*. pp. 3-10 (1978).

———, and Z.-Ci Zhao, 1988. Sensitivity of soil moisture to doubling of carbon dioxide in climate model experiments. Part 1: North America, *Journal of Climate*, **1**:348-66.

Manabe S., and R.T. Wetherald, 1980. On the distribution of climate change resulting from an increase in CO_2 content of the atmosphere. *Journal of Atmospheric Science* **337**:99-118.

———, and R.T. Wetherald, 1987. Large-scale changes in soil wetness induced by an increase in atmospheric carbon dioxide. *Journal of Atmospheric Science* **44**:1211-36.

Menzhulin G.V., M.V. Nikolaev and S.P. Savvateyev, 1983. Estimates of economical and weather component of cereal yield changes. *Proceedings of State Hydrological Institute*, **280**:111-19. (Russian).

Meshcherskaya A.V., N.A. Boldyreva and N.D. Shapayeva, 1982. *Mean Regional Stores of Soil Available Moisture and Snow Cover Thickness: Statistical Analysis and Examples of Usage*. Gidrometeoizdat, Leningrad (Russian).

Mintz Y., and Y.V. Serafini, 1989. Global monthly climatology of soil moisture and water balance. LMD Internal Report N 148, LMD, Paris.

Mitchell J.F.B., 1989. Personal communication.

MacCracken, M.C., M.I. Budyko, A.D. Hecht, and Y.A. Izrael (Eds.), 1990. *Prospects for Future Climate: A Special US/USSR Report on Climate and Climate Change*. Lewis Publishers, Inc.

Schlesinger M.E., and J.F.B. Mitchell, 1987. Climate model simulations of the equilibrium climatic responses to increased carbon dioxide. *Review of Geophysics*, **25**:760-98.

———, and Z.-Ci Zhao, 1989. Seasonal climatic changes induce by doubled CO_2 as simulated by the OSU atmospheric GCM/mixed–layer ocean model. *Journal of Climate*, **2**:459-95.

Scientific Assessment of Climate Change, 1990. *Report Prepared for IPCC by Working Group I*. J.T. Houghton, G.J. Jenkins, and J.J. Ephramus (Eds.) Cambridge University Press, Cambridge, U.K.

Verigo S.A., and L.A. Razumova, 1973. *Soil Moisture*. Gidrometeoizdat, Leningrad (Russian).

Vinnikov K.Ya, 1986. *Climate Sensitivity*. Gidrometeoizdat, Leningrad (Russian).

———, and P. Ya. Groisman, 1979. An empirical model of the present-day climatic changes. *Meteorologia i Gidrologia*. No. 3:25-36 (Russian).

———, and P.Ya. Groisman, 1982. Empirical study of climate sensitivity. Izvestiya of Academy of Sciences USSR. *Atmospheric and Oceanic Physics*, **26**:1159-69 (Russian).

————, and N.A. Lemeshko, 1987. Soil moisture content and runoff for the USSR territory with global warming. *Meteorologia i Gidrolgia*, No. 12:96-103. (Russian).

————, and I.B. Yeserkepova, 1989. Empirical data and results of modeling the soil moisture regime. *Meteorologia i Gidrologia*, No. 11:64-72. (Russian).

————, and I.B. Yeserkepova, 1991. Soil moisture: Empirical data and model results. *Journal of Climate*, **4**: 66-79.

————, P. Ya. Groisman and K.M. Lugina, 1990. Empirical data on contemporary global climate changes (temperature and precipitation). *Journal of Climate*. **3**:662-77.

————, N.A. Lemeshko and N.A. Speranskaya, 1990. Soil moisture content and runoff in the extratropical part of the Northern Hemisphere under global warming. *Meteorologia i Gidrologia*. No. 3:5-10 (Russian).

Washington W.M. and G.A. Meehl, 1984. Seasonal cycle experiment on the climate sensitivity due to doubling of CO_2 with an atmospheric general circulation model coupled to a simple mixed layer ocean model. *Journal of Geophysical Research*, **89**:9475-9503.

Wilson C.A., and J.F.B. Mitchell, 1987. A doubled CO_2 climate sensitivity experiment with a global climate model including a simple ocean. *Journal of Geophysical Research,* **92**:1315-43.

Korzoun, V.I., A.A. Sokolov, M.I. Budyko, et al. (Ed.), 1974. *World Water Balance and Water Resources of the Earth.* Gidrometeoizdat, Leningrad.

Zhao Z.-Ci., and W.W. Kellogg, 1988. Sensitivity of soil moisture to doubling of carbon dioxide in climate model experiments. Part II: The Asian Monsoon region. *Journal of Climate,* **1**:367-78.

Zhukov A. A. (Ed.), 1986. *Mean Long-Term Stores of Productive Water Under Winter and Early Spring Cereals in Districts, Regions, Republics and Economic Regions.* Vol. 1. Gidrometeoizdat, Leningrad (Russian).

Zubenok L.I., 1976. *Evaporation on the Continents.* Gidrometeoizdat, Leningrad (Russian).

9. METHODOLOGY FOR ASSESSING REGIONAL ECONOMIC IMPACTS OF AND RESPONSES TO CLIMATE CHANGE: THE MINK STUDY

Norman J. Rosenberg, Pierre R. Crosson, William E. Easterling III,
Mary S. McKenney, Kenneth D. Frederick and Michael Bowes

Scientists believe that a serious change in the climate of the earth could occur in the course of the next two to five decades as a result of warming caused by the rapid accumulation of radiatively active trace gases in the atmosphere (Bolin *et al.*, 1986; IPCC, 1990; Schneider and Rosenberg, 1989). There is concern that not only the amount of warming but the rate at which it occurs could be unprecedented, at least since the current interglacial period began.

Scientific uncertainties remain in our understanding of the climatic changes that may follow from greenhouse warming. Nevertheless, large and rapid changes in regional climates are conceivable. The impacts of such changes as the 8°C increase in mean summertime temperature in the central United States accompanied by a 1 mm/day decrease in mean precipitation (Manabe and Wetherald, 1986) would be extremely severe. This prediction is more radical than others that have been made (Schlesinger and Mitchell, 1985). Nonetheless, as long as the direction of change is credible, efforts are warranted to identify just what kinds of impacts to expect if society chooses to allow climate to change or cannot stop it from changing, and just what might be done to adjust to those impacts.

Additionally, it is known from the results of laboratory and growth chamber studies that when carbon dioxide, one of the primary greenhouse gases, increases in concentration in the atmosphere, plants respond with increased rates of photosynthesis and reduced rates of evapotranspiration. Hence, their water use efficiencies, measured as yield/unit of water consumed, improve (Cure and Acock 1986;

Kimball 1983, 1986; Rosenberg 1982). However, the strength of the CO_2-effect under field conditions still is uncertain.

Three broad policy strategies for dealing with prospective climate change are possible. One is to do nothing, accepting whatever change is in store and depending on induced individual and institutional responses to hold the consequences of change within acceptable limits. Another strategy is to adopt policies to facilitate adaptive responses; the third is a set of policies designed to reduce greenhouse gas emissions sufficiently to achieve acceptable warming and associated climatic change.

Each of the three policy strategies has its adherents. However, fruitful discussion of the merits of each strategy is inhibited by the great scientific uncertainty concerning the amount and regional incidence of prospective climate change, and its socioeconomic consequences. The discussion is impeded also by the absence of well developed methodologies for the study of the consequences.

The research we report here does not deal with the issue of scientific uncertainty. Our aim, instead, is to improve the methodological base for assessment of the regional impacts of climate change and of responses to it. In the following pages we explain the need for improved methodology and describe the analytical framework we have developed to accomplish this. We illustrate our approach by examining the impacts of climate change, CO_2 enrichment and adaptation on the productivity of agriculture, and the reliability of water resources in a particular region now and in the future. Finally, we analyze the implications of these impacts and responses on the current and future economies of the region chosen for study.

Methodological Limitations of Previous Studies

We identify four methodological limitations that typify most studies of the regional impacts of climate change.

1. *The climate of tomorrow is imposed abruptly on the world of today.* Significant climatic impacts of greenhouse warming are unlikely for at least the next few decades and possibly longer. Futurists assert that the economic structure of most regions of the world will change greatly in the coming decades (e.g., Singer, 1987). Regional vulnerabilities and capacities for adaptation must also change. Hence, impacts of climate change and responses will almost certainly be quite different in the future from what they would be today.

2. *The natural temporal and spatial variabilities of climate characteristic of large regions are ignored.* Scenarios of climate change employed in most prior studies, whether derived from general circulation model (GCM) runs or paleoclimatic evidence, obscure natural temporal and spatial variability in climate. The imposition of fixed temperature and precipitation changes on grid boxes hundreds of kilometers on a side does not allow the large natural intraregional variability in climate to be expressed and may, therefore, provide biased results.

3. *The complexities of the regional economy, its linkages with the rest of the world, and the place in it of the natural resource based industries and other economic sectors are not taken into full account.* Certain sectors—e.g., agriculture, forestry, and water—are apparently more sensitive to climate than are others. However, all sectors of the regional economy and the interactions among sectors must be considered if the real impact of climate change on a region is to be understood. Few if any studies of regional climate change impacts take this comprehensive approach.

4. *The full range of available technologies, management techniques, and policy tools that can lessen (or capitalize on) the impacts of climate change are not fully considered.* Many types of adjustments and adaptations within the limits of current technology might be applied to moderate the negative effects of climate change and/or capitalize on its positive impacts. Certainly the range of such technologies will increase in the future.

In an effort to overcome these limitations and strengthen the art of climate impacts assessment, four organizations have come together in a program supported by the U.S. Department of Energy, Office of Health and Environmental Research. Resources for the Future (RFF) has been responsible for the analytical aspects of the program, Oak Ridge National Laboratory (ORNL) for databases and geographical information systems, Sigma Xi for outreach, and Pacific Northwest Laboratories (PNL) for coordination. Sigma Xi and PNL are also involved in analysis of the uncertainties involved in the climate impact assessment methodology developed under this program.[1]

The Analytical Framework

To overcome the limitations described above, an analytical framework has been developed that aims to

- provide baseline information on current functioning of regional-scale economies and how they might develop in the future in the absence of climate change;
- analyze how various kinds of climatic change may alter baseline resource productivity (e.g., crop production, forest output, runoff to rivers, and water storage);
- study the ways in which the primary enterprises affected (e.g., farms, timber companies, water resource districts) may respond to these first-order effects.
- study how the responses of primary enterprises may affect the regional economy as a whole.

---

[1]This paper deals exclusively with the analytical aspects of the program. The multiorganizational team was coordinated by M.J. Scott of Pacific Northwest Laboratories. RFF participants included N.J. Rosenberg, P.R. Crosson, W.E. Easterling III, K.D. Frederick, M. Bowes, M. McKenney, R. Sedjo, J. Darmstadter, K. Lemon and L.A. Katz. Oak Ridge National Laboratory was represented by R. Cushman and Sigma Xi by T. Malone and G. Yohe.

TABLE 9.1. *The Mink Study: Analytical Tasks*

Task A.	Current baseline description of the region and its economy with emphasis on agriculture, water resources, forestry and energy.
Task B.	Assessment of qualitative and quantitative impact of climate change on various economic sectors and on the economy as a whole assuming today's resource base, technology, economy and institutional structure.

B_1 — Climate change only, no adaptive responses

B_2 — Climate change plus 100 ppm increase in atmospheric CO_2, no adaptive responses

$B_{3.1}$ — Scenario B_1 incorporating adaptions for which a technology or policy base currently exists

$B_{3.2}$ — Scenario B_2 incorporating adaptions for which a technology or policy base currently exists

Task C.	Baseline description of the economic, technical and institutional structure as it might be in 2030.
Task D.	Assessment of qualitative and quantitative impact of climate change on various economic sectors and on the economy as a whole assuming a resource base, technology, economy and institutional structure consistent with the year 2030 scenario.

D_1 — Climate change only, no adaptive responses

D_2 — Climate change plus 100 ppm increase in atmospheric CO_2, no adaptive responses

$D_{3.1}$ — Scenario D_1 incorporating adaptions for which a technology or policy base currently exists plus new policies and adaptation techniques developed in response to the fear, perception or observation of climate change

$D_{3.2}$ — Scenario D_2 incorporating adaptions for which a technology or policy base currently exists plus new policies and adaptation techniques developed in response to the fear, perception or observation of climate change

The full study deals with all of the important resource sectors identified above and with all of the orders of effects and linkages. However, because of space limitations, in this chapter we demonstrate our methodology by focusing on the agricultural and water resources sectors and their linkages with the rest of the economy.[2]

[2]Details of this research are to be found in a series of reports published by the Department of Energy under the title "Processes for Identifying Regional Influences of and Responses to Increasing Atmospheric CO_2 and Climate Change—The MINK Project," N.J. Rosenberg *et al.*, 1991. DOE/RL/01830T-H5-12. TR052 Parts A-H. Washington D.C.

The analysis consists of four specific tasks and a set of subtasks listed in Table 9.1. Our aim has been to develop a sound methodology that can be applied to the assessment of the regional impacts of climate change and of the possible responses to such change. The framework of this methodology must be theoretically sound. Its subcomponents need not be fixed, however. Each database, model or analytical tool used should be open to improvement or to replacement as better tools become available.

The Region of Study

The methodological design described above has been applied, in the first case, to the

FIGURE 9.1. The MINK region (above) and its Land Resource Regions (below): G = Western Great Plains Range and Irrigated Region; H = Central Great Plains Winter Wheat and Range Region; M = Central Feed Grains and Livestock Region; N = East and Central Farming and Forest Region. (Source: USDA, Soil Conservation Service Agricultural Handbook, p. 296.)

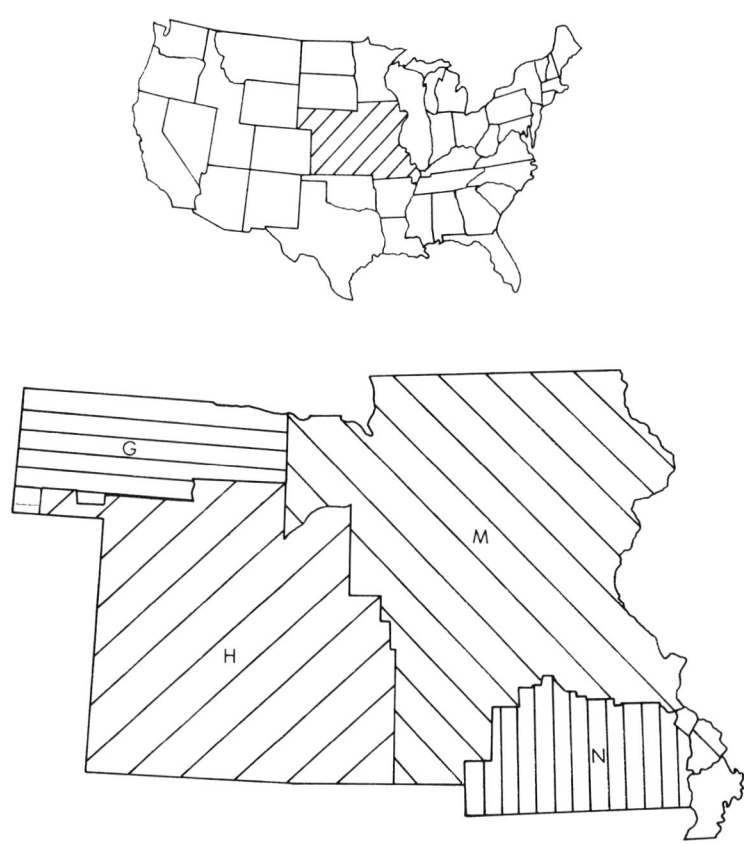

region composed of four states in the central United States—*Missouri, Iowa, Nebraska* and *Kansas* (hereafter, the MINK region, Figure 9.1). Compared with regions we might have chosen, MINK is simple and coherent with no deserts or maritime areas within or adjacent to it. Moreover, relative to the rest of the nation, the MINK economy is specialized in natural resource-based sectors most likely to be impacted by climate change. Finally, certain GCMs predict dire changes in climate for the region (e.g., Manabe and Wetherald, 1986).

The total area of the MINK region is about 734,000 km² (U.S. Department of Commerce, 1987). The region is mostly level to gently rolling except in the Ozark region of the southeast. Elevation increases gradually from 70 m in southeastern Missouri to 1,655 m in western Nebraska. Land use in the four states can be categorized in terms of Major Land Resource Areas (Figure 9.1). Iowa and the northern two thirds of Missouri are in the "Central Feed Grains and Livestock region." The southern third of Missouri is in the "East and Central Farming and Forest Region." Virtually all of Kansas and the southern half of Nebraska fall in the "Central Great Plains Winter Wheat and Range Region." The remainder of Nebraska is in the "Western Great Plains Range and Irrigated Region."

Current Climate and Scenario of Climate Change

The MINK region, far removed from moderating influences of large bodies of water, is typically continental and characterized by large seasonal swings in temperature and precipitation. Winters are cold and dry, summers hot with moisture and precipitation declining from east to west.

Summer temperatures greater than 37.8°C (100°F) occur everywhere in the region as do temperatures below -17.7°C (0°F). On average, northern Iowa experiences 6 and southern Kansas 70 such hot days annually, while Missouri experiences 1-2 and northern Iowa 30 such cold days. Length of the growing season, defined as the period between last frost in spring and first in fall, varies across the region from 200 days in the southeast to 120 days in the northwest.

Precipitation in the MINK region is controlled by two physiographic features: the Rocky Mountains, which remove moisture from maritime Pacific air from the west, and the Gulf of Mexico, which provides moisture-laden maritime-tropical air to the region. Distance from the Gulf and the presence of the Rocky Mountain rainshade explain the sharp decline in precipitation from east to west. Precipitation is least in winter and greatest in summer, except in Missouri where the peak occurs in spring.

Scenarios of climate change can be developed by simulation as with GCMs, paleoclimatic reconstructions, or through use of the historic climatic records (Lamb, 1987). We have chosen to use the latter approach and have drawn from the historic record a segment—the decade of the 1930s. We believe this is appropriate because the 1930s in MINK were hotter and dryer than the present climate. These characteristics approximate those predicted by GCMs for the MINK region sometime in the first half of the twenty-first century when the total radiative equivalent of a doubled preindustrial CO_2 concentration may occur.

For all of the analyses that follow, the climate of 1931-40 is compared with that of the period currently defined by WMO convention as the "normal" i.e., 1951-80. Temperatures were higher across the region in the 1930s, particularly in summer. Winters were unusually warm in Iowa and Missouri

as were autumns in Nebraska and Kansas. On average, mean monthly precipitation in the 1930s was lowest (compared to the control period) in spring and summer. Winters on average were a bit wetter. Although precipitation was lowest in the 1930s, the interannual variability in that decade was also less. Interannual variability in temperature was greater, however.[3]

It is incorrect to declare, simply, that the climate of the 1930s was hotter and drier throughout the MINK region than during the most current thirty-year normal. Climatology assures us that this could not have been the case everywhere in the region and all of the time. A better way to capture the richness and complexity that the analog approach provides is demonstrated in Figure 9.2. Here the Palmer Drought Severity Index (PDSI) (Palmer, 1965) is plotted for each month from January 1931 through December 1980 for each of the four MINK states. The PDSI considers drought severity as a function of rainfall deficit, antecedent soil moisture condition and potential evapotranspiration (calculated in this case as a function of temperature; Thornthwaite, 1948).

Drought experience since 1931 has been different in each of the four states. The western states descended rapidly into severe drought in the 1930s and did not emerge until the beginning of the 1940s. Although droughts did occur in the eastern states in the 1930s, they were less protracted and severe than in Nebraska and Kansas. In fact, drought was more severe in Missouri and Iowa in the 1950s than in the 1930s.

The decade of the 1940s was generally benign in the MINK region. The 1940s are excluded from consideration as part of the climatic normal by convention, while the droughty years of the 1950s are included. Hence the contrast between the climate of the 1930s and that of the climatological normal is less than it would have been had all years since 1940 been included.

---

[3]By definition, the last three decades for which there are complete records. Climatic means for the 1980s have not yet been computed. After that is done the climatic normal will be 1961-1990.

FIGURE 9.2.   Monthly values of the Palmer Drought Severity Index for the four MINK states, 1931 to 1980.

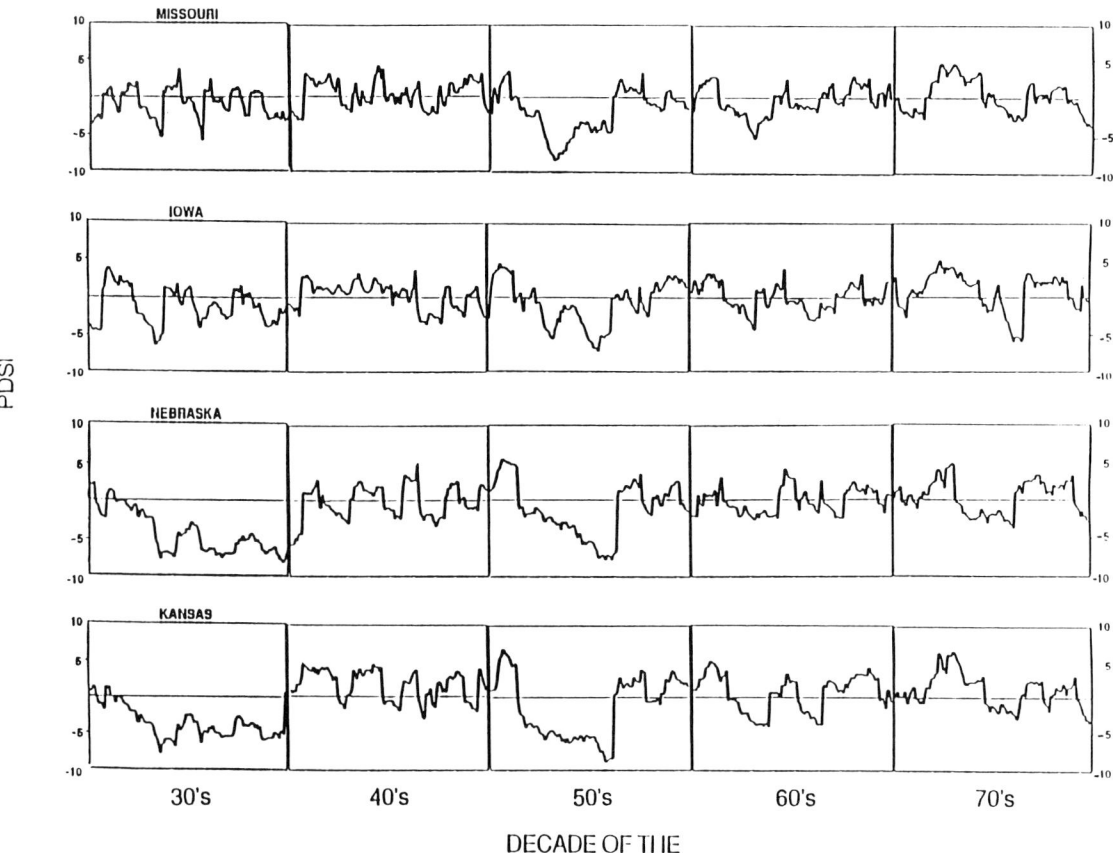

TASK A: Agriculture and Water Resources in MINK— Current Baseline

Agriculture in the MINK Region

The MINK study involved analysis of four natural resource—based sectors—agriculture, forestry, water resources and energy, and the interactions among them and interactions with other sectors. Space does not allow us to provide detail on all these sectors and linkages. We have chosen, therefore, to illustrate our approach by highlighting the agricultural sector and its linkages to the rest of the economy, and to provide information on water resources to the extent that it bears on the future of the agricultural sector. First, however, we describe agriculture and water

resources in the MINK region as they are today. Additionally, in the crop modeling work reported below we emphasize, for this chapter, those crops grown under both dryland and irrigated conditions—corn, sorghum and wheat. For analysis of the overall impact of climate change on the regional economy, we also include information on soybeans.

In the base period 1984-1987[4] MINK accounted for 34 percent of the value of the nation's production of corn for grain, 30 percent of its soybeans and winter wheat (the only wheat grown in the region), and 50 percent of its grain sorghum. Average yields

---

[4]This is the "base period" against which a future scenario of the MINK economy is developed. Statistics available when the MINK study began carried only through 1987.

of these four crops were higher and production costs per bushel were less in MINK than in the rest of the country.

Nebraska had 64 percent and Kansas 30 percent of the irrigated land of the region. Corn in Nebraska accounted for 53 percent of all the irrigated land in MINK.

MINK also accounted for 28 percent of national production of cattle and calves and 42 percent of its hogs. Nebraska produced 35 percent of MINK's cattle and calves and Iowa 63 percent of the region's hogs.

Value added by agriculturally related manufacturing industries was 77 percent of value added on the region's farms. Meatpacking alone accounted for 31 percent of value added in agriculturally related manufacturing during the baseline 1984-1987 period.

Analysis of linkages between agriculture and the rest of the regional economy showed that, in the baseline period, a change of $1.00 in on-farm production generated an additional change of $0.65 in production in the rest of the economy. The regional impact of activities tied to agriculture, such as meatpacking, was even greater (e.g. a $1.00 change in meatpacking output generated an additional change of $1.57 in the rest of the economy).

The MINK economy is much more specialized in agriculture than the rest of the nation, the percentage of farm income to total income in MINK being 3.4 times the percentage of farm income to total income in the nation as a whole. Measured by the share of farm income in total income, Nebraska is the most agricultural and least industrial of the four states, and Missouri is the least specialized in agriculture. Manufacturing is the most important single sector in Iowa, Kansas, and Missouri, followed closely by services. Finance, insurance, and real estate are most important in Nebraska, followed by services and manufacturing.

Although MINK clearly is specialized in agriculture relative to the rest of the country, that sector's share of regional income (3.7 percent) suggests that MINK is, nonetheless, not primarily an agricultural region. On the contrary, it could more accurately be called a manufacturing-services region since the shares of manufacturing and services in the economy are four times greater than that of agriculture. These interpretations, however,

seriously understate the importance of agriculture to the MINK economy and, hence, the region's vulnerability to climate change for two reasons: First, a significant amount of manufacturing in MINK is directed to processing farm outputs and manufacture of farm inputs (meatpacking, fertilizers, machinery); second, agriculture's percentage contribution to regional exports—the key element in the economic base of the region—is several multiples of its contribution to regional income.

Water Resources of the MINK region

The four MINK states do not comprise an independent hydrologic region. Because they include parts of four major water resource basins (the Missouri, Arkansas-White-Red, Upper Mississippi and Lower Mississippi), the water available to MINK states depends not only on events in MINK, but also on circumstances in states located up river. Moreover, water use in MINK may be restricted by obligations to downstream states. Instream flow needs for navigation, fish and wildlife habitat are large relative to withdrawal uses and mean streamflows.

Irrigation accounts for about half of the water withdrawal and 89 percent of the consumptive use of water within MINK. Nebraska and Kansas account for 97 percent of all irrigation water use within MINK. Groundwater provided 80 percent of all irrigation water and 48 percent of all water withdrawals in 1985. Intensive pumping and low recharge rates have resulted in the mining of water from the Ogallala formation underlying western Kansas and much of southwestern Nebraska. Conflicts over alternative uses of available water supplies have increased in recent years. Institutions for allocating water supplies within the region were stressed by drought during the 1980s and by a growing demand for more water for recreation and fish and wildlife habitat. Any climate change that tended to diminish runoff to rivers and reservoirs, that reduced recharge of groundwater where recharge is consequential, or that increased demand for water for irrigation or for competing uses could have profound effects on agriculture.

Modelling Impacts of the Analog Climate

Representative Farms

One important feature of our modelling approach has been to consider the impacts of climate change on representative farms, which are treated as economic entities, rather than on individual crops. A representative farm is a cohesive, functional enterprise that typifies most of the farms in its particular region. Forty-eight such farms distributed over 11 Major Land Resource Areas (MLRAs) throughout the MINK region were designed for this purpose.

Essentially, a representative farm is defined by its volume of crop output, soils, crop rotation, tillage, fertilization and other management practices and by the costs of production inputs it requires. The representative farms were designed in consultation with agricultural experts in each of the MINK states. Although alfalfa, soybeans and wheatgrass are included in the analysis, for our present purposes we offer results for only three crops—corn, sorghum and wheat—which are grown in the region under both rainfed and irrigated conditions.

Crops: The EPIC Model

A mechanistic model of crop growth is needed to allow us to predict the response of current crops to a climate change in the MINK region (Task B). We use it also for predicting the productivity of crops in about 2030 (Task C) and how crop productivity would be affected by climate change in 2030 (Task D). We have chosen to work with a family of simulation models known as EPIC (Erosion Productivity Impact Calculator). The EPIC models were developed at the Texas Agricultural Experiment Station by the Agricultural Research, Soil Conservation and Economic Research Services of U.S. Department of Agriculture (USDA) and are described in Williams *et al.* (1984).

Its ability to accommodate numerous parameters descriptive of weather, crop phenology and physiology, soil physical and chemical condition, farming practices—including tillage, irrigation, and pest control—and farm economics makes EPIC highly flexible. The model operates on a daily time step so that the daily weather is required as an input. Weather data are drawn from actual records or produced by stochastic simulation in EPIC. Moreover, EPIC simulates biomass production and yield, evapotranspiration (ET), irrigation requirement and number of days on which growth is constrained by temperature, moisture or nutrient stress, and it also calculates farm profitability. Maturity and harvest date are also simulated. Normally EPIC is run for a number of years so that interannual variability in yields and the other factors calculated can be assessed. A major reason for our choice of EPIC in the MINK analysis is its flexibility, which permits simulation of alternative crop maturities, rotations, tillage practices, planting dates, irrigation and fertilization strategies such as might be used by farmers attempting to adapt to climate change. Preparation of EPIC for the purpose of the MINK study required modification to account for the direct effects of increasing atmospheric CO_2 concentration, parameterization of a large number of representative farms that together characterize the great diversity of crop production of the region, and calibration or verification of the model's results.

Carbon dioxide enrichment of the atmosphere increases photosynthetic rate, especially in C3 species (the small grains, legumes, most trees and grasses) and reduces ET in both C3 and C4 plants—the tropical grasses (corn, sorghum, millet, etc.) (Rosenberg *et al.*, 1990). For us to evaluate the impact, not only of forecasted climate change (or its analog) but also the direct effects of CO_2 enrichment, it was necessary to modify EPIC to account for direct effects of changing CO_2 concentration on photosynthesis and ET. These modifications are described in Stockle co-workers (1992, in press).

We have no reason to expect EPIC to simulate yields, ET and other factors perfectly for each crop rotation—soil—weather combination encompassed in our representative farms. Nonetheless, little confidence is earned by a simulation model which, when realistically parameterized, fails to approximate real-world conditions.

We tested EPIC for realism in three ways: (1) assuming 1984-1987 technology and using the weather records for 1951-80, actual yields for the period 1984-1987[5] were compared with the means of a 60-year EPIC run of annual yields; (2) the simulated means were compared with expert judgments of what the yields should be on the representative farms with current technologies; and (3) the means, standard deviations and range of yields, and ET generated by EPIC were compared with results of pertinent experiments identified in an extensive review of the agronomic literature.

The results of these tests are discussed in detail in Report IIB: Farm Level Agriculture, of the Department of Energy Reports (1991). Suffice it to say here, we concluded that EPIC simulations of yields and ET are sufficiently realistic to permit use of the model to compare crop production in the baseline period with that predicted to occur under the analog climate.

Water Supply and Demand

Irrigation is the principal withdrawal use of streamflow in the Missouri Basin today and the principal consumptive use of water in the states of Nebraska and Kansas. For our present purposes, we use EPIC to estimate changes in irrigation requirements caused by the analog climate. The economic demand for water, therefore, would not necessarily equal requirements (i.e., the amount of water needed to avoid water stress).

To assess how the analog climate might alter the supply of water we require different methods. Ideally, to estimate the effects of the analog climate on streamflow, we would compare natural flows in the decade of the 1930s with those during the control period. However, flows in the basins today are affected by diversions, reservoirs and consumptive use that did not exist in the 1930s. For example, most of the 12,000 reservoirs within the MINK region and many more in the upstream states were built after the 1930s. Evaporation from the surface occupied by these reservoirs would certainly be greater than if the land surface they occupy were still vegetated or bare; and the impact of

an increased temperature and lower humidity on evaporation would be amplified because of their presence.

Consequently, only a very small number of the thousands of gauging stations in the region appear to have been unaffected since the 1930s by human impacts. Those can be used as a proxy for estimating natural streamflows for the regions of interest.

We estimated gross evaporation in the reservoirs under control and analog conditions by two well known methods—Harbeck (1962) and Penman (1948). Precipitation levels for each period were then subtracted from gross evaporation to yield net evaporation. Harbeck's method is highly empirical, but considers reservoir surface area, which would likely be lower under the analog climate. Penman's method has a firmer theoretical base but does not consider reservoir surface area. Results of the analyses concerning water supply in the MINK region are presented below.

Task B: Impacts of a Climate Change on the MINK Region

Crop Yields

Tasks B_1 and B_2 impose the analog climate without and with a 100 ppm increment of CO_2. In Task B_3 adaptations judged by expert opinion to be economically feasible were applied to the grain crops, again with and without the CO_2 effect. These adaptations included earlier planting, except for wheat, use of longer season varieties for the annual crops, and furrow diking for dryland row crops. Crop substitutions were also permitted in the B_3 runs, including shifts of irrigated corn land in Nebraska and Kansas to dryland production of wheat and sorghum. In all runs involving irrigation, water is applied in accordance with EPIC's estimates of requirements determined by climatic conditions. In essence, the crop is irrigated when soil available water in the root zone has been depleted by 10 percent.

The EPIC simulations of harvestable yields were averaged by crop across all representative farms in the MINK region. Results are shown for all B-tasks in Figure 9.3 for dryland and irrigated corn, sorghum and wheat. Bars in Figure 9.3 represent the

---

[5]U.S. Department of Agriculture, 1984-1987. County Level Estimates Tape ASB 111, National Agricultural Statistical Service, Washington, D.C.

FIGURE 9.3. EPIC-simulated yields for dryland and irrigated corn, sorghum and wheat under analog (1931-1940) climate conditions with and without additional atmospheric CO_2 and with and without adjustments to the climate change. Yields are normalized with respect to the control climate (1951-80). *I* indicates irrigation.

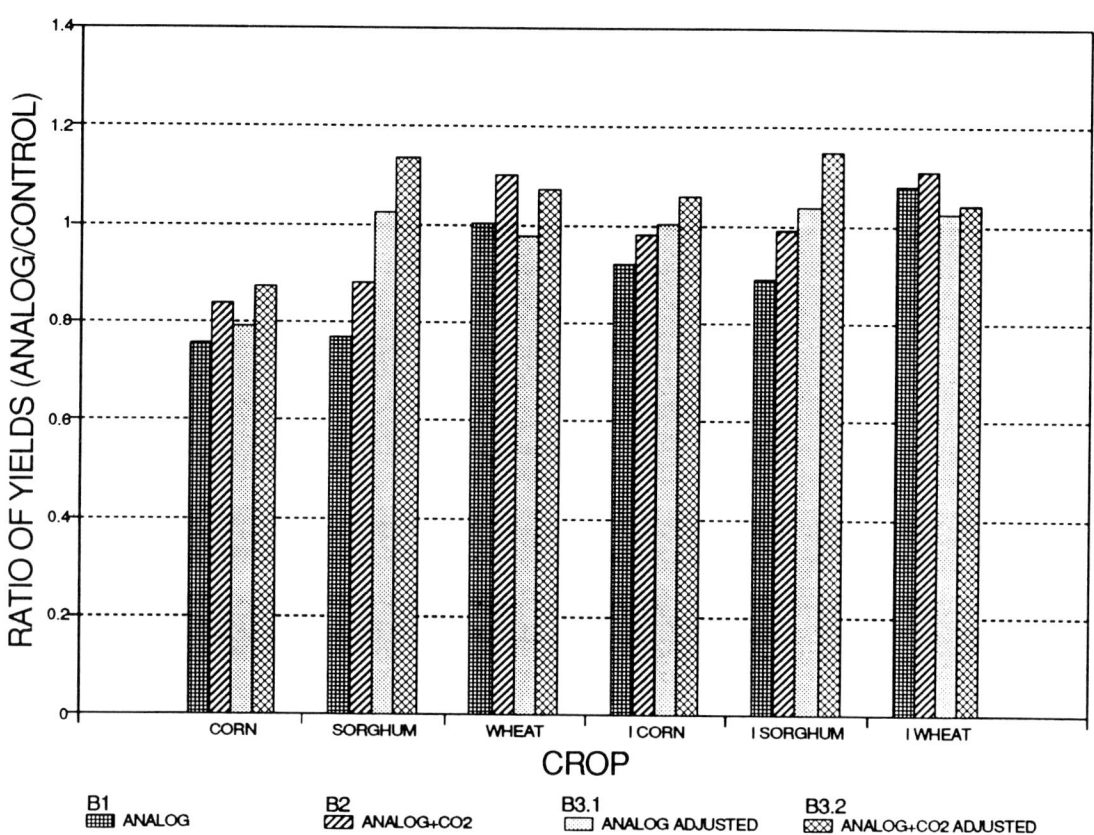

ratio of yield under analog climate conditions to those obtained under control (1951-1980) climate conditions. Both sets of yields assume existing technology and management practices. All inputs other than weather were identical (e.g., cultivars, tillage, etc.). Averaging, of course, obscures the regional differences due to soil, crop cultivar, rotation and local weather.

The greatest impacts of climate change are on corn and sorghum, particularly under dryland conditions. However, increasing the CO_2 concentration will partially mitigate these effects in the case of dryland corn and sorghum, and actually increases the yield of irrigated wheat. Applying adaptive responses also reduces the yield loss in the case of corn, and eliminates it altogether in the case of sorghum.

The loss of yield of dryland corn and sorghum under climate change is due primarily to a shortened growing season. However, increasing the CO_2 increment diminishes yield loss by reducing evapotranspiration and stimulating photosynthesis. The mean yield of dryland wheat is unaffected by the analog climate, and is actually increased with the CO_2 increment. Loss of yield in corn and sorghum due to the analog climate is considerably smaller under irrigation than on dryland, and the yield of irrigated wheat increases under the analog climate, with or without the addition of CO_2.

The use of currently available adjustment measures (e.g., earlier planting and longer season cultivars) is most effective in sorghum, both dryland and irrigated. Adjustments improve yields of dryland corn slightly in $B_{3.1}$ and eliminate losses in irrigated corn. Carbon dioxide and adjustments in $B_{3.2}$ raise corn yields further, an improvement that is more apparent under irrigated rather than dryland conditions. Earlier planting does not apply to winter wheat, and the longer season cultivars encounter temperatures too high for beneficial growth. Furrow diking is also not appropriate for wheat. Therefore, on both dryland and under irrigation, with or without the CO_2 increment, wheat yields are not improved by the simple adjustments made, which may say more about the selection of adjustments than about the ability of wheat to be adapted to the climate change.

Detailed explanations for the behavior of the various crops when exposed to climate change, increased CO_2 and the relatively simple and limited set of adaptations are to be found in the DOE Report on agriculture.[6] In these working papers we evaluate the extent to which limitations in EPIC may generate artifacts—for example, the relative insensitivity of wheat to the climate of the 1930s in which, records show, yields of that crop were reduced.

Water Demand and Supply

In the B_1 scenario the analog climate causes evapotranspiration (ET) to fall in dryland grains— from 5 percent in wheat to 12 percent in sorghum (Figure 9.4). Evapotranspiration is unaffected in irrigated sorghum and increased by about 5 percent in irrigated corn and wheat. Decreases in stomatal conductance caused by increased CO_2 in conjunction with the analog climate (B_2) futher reduce ET in the dryland crops despite increased leaf area. The increases in ET on irrigated corn and wheat are eliminated by CO_2, and ET declines about 5 percent in sorghum.

Evapotranspiration is indicative of but not identical with irrigation requirements. The latter considers that not all water applied actually penetrates the root zone or remains within it. Also, not all the water in the soil

is extracted prior to maturity and harvest. Changes in ET of the sort simulated by EPIC for the irrigated crops would have profound effects on the agriculture of the two western MINK states.

Average changes in irrigation requirement across all irrigated farms are shown in Table 9.2. With no adjustment, the analog climate increases irrigation requirement by 11 percent in wheat to 24 percent in corn. The increase is moderated by CO_2 for the corn and sorghum, and is eliminated in wheat. Adjustments increase irrigation requirements because of the longer growing season they impart. Carbon dioxide moderates this effect by reducing evapotranspiration, resulting in an overall reduction in total consumptive use for irrigation by about 6 percent.

Based on a scaling-up of the EPIC simulations in Table 9.2 for all of the irrigated farms considered, assuming all demand could be met by current irrigation systems and water supplies, water withdrawals under the analog climate (B_1) would increase by 39 percent in Nebraska and 14 percent in Kansas. Specifically, irrigation requirement increases 39 percent for corn in Nebraska and by 12 percent for corn, 11 percent for wheat and 20 percent for sorghum in Kansas.

On the supply side, our analysis indicates that during 1931-1940 actual natural streamflows were 72 percent of their 1951-1980 averages for both the Missouri and Upper Mississippi basins and 93 percent for the Arkansas River Basin. Increased evaporation from the surface of current reservoirs would decrease availability of streamflow only an additional percent. On the other hand, the CO_2 increment could also reduce evaporation from the watersheds, thus contributing to the streamflow. However, we did not assess this possibility.

It is unlikely that the increases in irrigation implied by the EPIC simulations under the analog climate would be realized. Competition for water in Nebraska and Kansas suggests that irrigators would not be able to acquire additional surface water rights. It is also unlikely that farmers would pump more water to produce the lower yields implied by the EPIC simulations.

---

[6]Report IIB: Farm Level Agriculture (see footnote 1).

FIGURE 9.4. EPIC-simulated evapotranspiration (ET) for dryland and irrigated corn, sorghum and wheat under analog (1931-1940) climate conditions with and without additional atmospheric CO_2 and with and without adjustments to the climate change. ET is normalized with respect to the control climate (1951-80). *I* indicates irrigation.

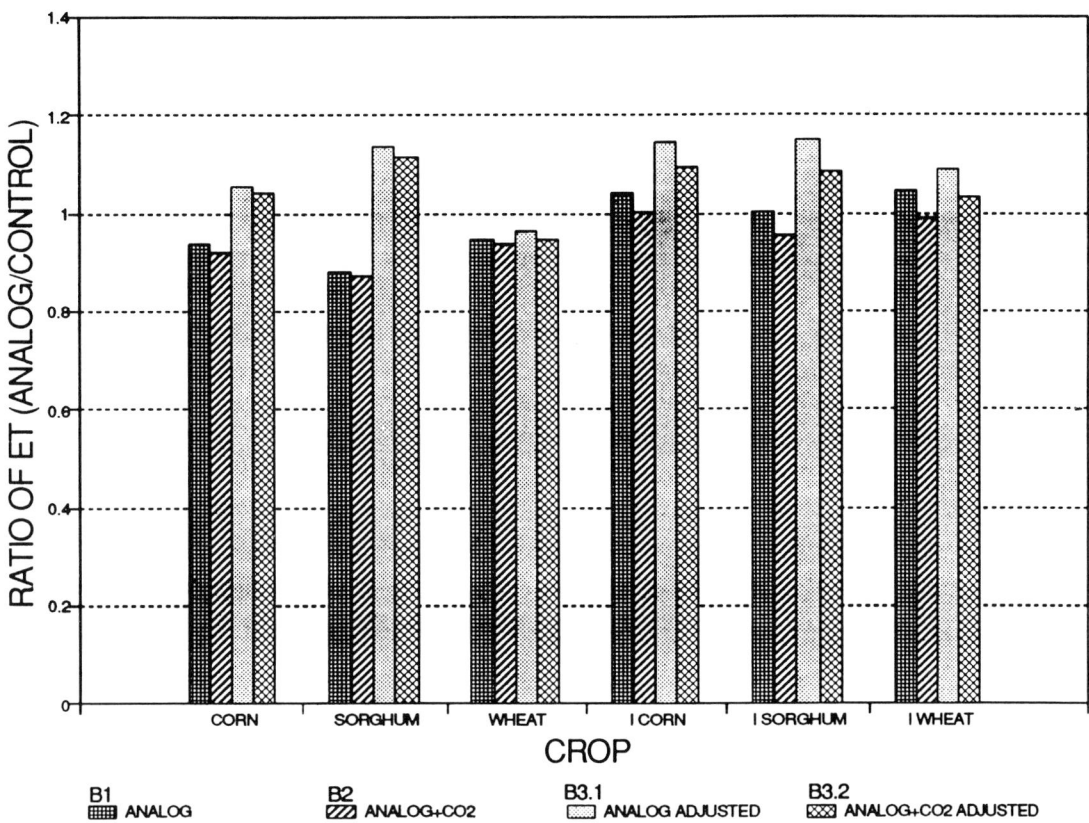

TABLE 9.2. *EPIC–Simulated Percentage Changes in Baseline (Task A) Irrigation Requirement Under the Analog Climate.*

	No adjustments		With adjustments	
	No CO_2 effect (B_1)	CO_2 effect (B_2)	No CO_2 effect ($B_{3.1}$)	CO_2 effect ($B_{3.2}$)
Corn	+24	+15	+43	+32
Sorghum	+20	+9	+39	+26
Wheat	+11	-1	+4	-3

Instead, irrigators are likely to adapt to scarce and more costly water by investing in more efficient irrigation techniques, switching to more drought-tolerant cultivars (if available) or to different crops or by abandoning irrigation. The analog climate change would likely accelerate the decline of irrigation in those portions of Kansas and Nebraska where farmers are already adversely impacted by rising energy costs, increasing pumping depths and declining well yields. In the B_3 scenario we assume that some presently irrigated land in corn in these states is shifted to dryland production of wheat and sorghum.

Other impacts of the analog climate on water resources might affect agriculture. Competing demands of recreation, hydropower, and fish and wildlife habitat would likely reduce water supplies for navigation on the Missouri River, with a consequent rise in the costs of shipping agricultural inputs and products.

Task C: Future Baseline for the MINK Region

Crop Yields in the Year 2030

To use EPIC to estimate the impact of the analog climate in the D scenario we must have projections of crop yields in the absence of climate change, that is, in the year 2030 baseline scenario. We constructed two sets of such yield projections. In one we assumed that the yield trends established in MINK in 1972-1988 would continue to 2030. The other set of yield projections was made by incorporating in EPIC various new technologies that the literature and expert opinion suggest could be economically available to farmers by 2030 (U.S. Congress, Office of Technology Assessment 1986; Ruttan 1989; personal communication with Alan Jones, one of the developers of EPIC). The following technological advances were assumed:

- increased photosynthetic efficiency by increasing the light-use efficiency term by 10 percent;
- an increase in the harvest index (ratio of harvestable organs to total biomass) of 110 percent;
- a 15 percent improvement in pest control;
- a shift in the timing of leaf development such that maximum leaf area is achieved 5 percent sooner in the growing season. This results in an increase in photosynthetic capacity when water stress is less limiting; and
- an improvement of 10 percent in harvest efficiency through reduced field losses during the harvest process.

The yields from EPIC incorporating these technologies are not very different from the yields obtained by simply projecting trends. The reasons for the differences that do exist are not obvious, and space precludes discussion of the issue. We note, however, that this dual approach to projecting future yields may have some important advantages relative to the traditional approach, which relies on extrapolations of historical trends tempered by the views of informed opinion. A crop growth model such as EPIC permits experimentation with specific technologies that are only implicit, or not represented at all, in trend projections. We believe this modelling approach, not as a substitute for but as a complement to the traditional procedure, merits further investigation.

In this study we use the EPIC yield projections for the year 2030 because this permits us to estimate the yield impacts of the analog climate under simulated conditions of 2030. The five technological improvements would increase yields of dryland and irrigated corn and sorghum 55 percent to 74 percent relative to the current baseline. In soybeans dryland yields would increase 60 percent, and total yields by 64 percent, reflecting a small increase in irrigated soybean production in the eastern part of MINK. In wheat the increases would be 58 percent for dryland and 93 percent for irrigation. The greater response for wheat may be due to the constrained supply of nitrogen in the "current baseline" (Task A) simulations of that crop. In all future (Task C) runs, nitrogen (N) is made to be nonlimiting, assuming that biological N-fixation might be possible in nonleguminous as well as leguminous plants and that N application and conservation technology will be much better than today.

Water Demand and Supply

Evapotranspiration increases, but only by less than 4 percent in the crops of 2030 as compared to those of the current baseline. Irrigation demand increases by 6 percent in corn and wheat and by 3 percent in sorghum. The primary cause for these increases in ET and irrigation demand is the earlier leafing and larger plant size caused by the adjustments in timing of leaf area maxima and light use efficiency (hence, photosynthesis).

Although the EPIC simulations show

that water requirements for irrigation in 2030 in the absence of climate change would be little different from the current baseline, we know that water supplies for irrigation could be considerably reduced. Specifically, changes in upstream water use and new reservoir development may reduce the quantities of surface water available to MINK in the year 2030. The greatest uncertainties and largest potential impacts concern the future of the Pick-Sloan Missouri Basin program, authorized as part of the Flood Control Act of 1944 (Campbell 1984). The program has already transformed the region with more than 400,000 hectares (ha) now flooded by reservoirs that provide downstream flood protection and more reliable water supplies for a number of in-stream and off-stream uses. The remaining finished or planned projects under Pick-Sloan would increase irrigated acreage substantially within the upper Missouri Basin. However, budgetary and environmental concerns have stalled most major new irrigation projects in recent years and their eventual outcome is in doubt. If these plans do proceed, surface water supplies to the MINK states will be severely curtailed.

During the mid-1970s net groundwater depletions from the 15 water resource subregions located at least in part within the MINK states, particularly from the Ogallala aquifer underlying the Nebraska and Kansas High Plains, averaged about 18 million cubic meters per day. Depletion continued into the 1980s at rates that, unless checked, will force major reductions in groundwater use in some parts of MINK well before 2030.

Unlike the situation in Nebraska and Kansas, current water use poses little or no strain on groundwater supplies in Missouri and Iowa. Groundwater use in these states could be maintained or increased in most areas with little impact on the quantity in storage.

The increasing scarcity and cost of groundwater in western Nebraska and Kansas imply substantial reductions in groundwater withdrawal for irrigation in those parts of MINK by 2030. Surface water for irrigation may also decline slightly in coming decades. These waters are already scarce, and new withdrawals for agriculture will have to overcome formidable economic and environmental barriers. Water supplies will not impose such constraints on irrigation in the eastern portions of Nebraska and Kansas and in Iowa and Missouri. If economic conditions are favorable, these areas might experience a substantial percentage increase in irrigation.

Task D: Impacts of the Analog Climate on MINK in 2030

Crop Yields

Table 9.3 shows the percentage impact of the analog climate on year 2030 baseline yields with and without the CO_2 and on-farm adjustment effects. The results are similar in amount and pattern to those found when we imposed the analog climate on the current baseline situation in the B_1, B_2, and B_3 cases: The impact is most severe when no allowance is made for the CO_2 effect and on-farm adjustments, and becomes progressively less severe when these conditions are relaxed. However, the change from the D_1 scenario to the D_3 scenario with the CO_2 and on-farm adjustments is more marked than the comparable change from the B_1 to the B_3 scenarios. The reason is that the D_3 scenarios incorporate two technological innovations not reflected in the B_3 scenarios: increased irrigation efficiency and increased resistance to drought (modeled in EPIC as increased stomatal resistance). The logic for introducing these innovations in the D scenarios is that, should the climate in MINK move in the direction of the analog, farmers, as well as scientists in both public and private research institutions, would have increased incentive to find ways to offset the increasingly hotter and dryer conditions of the analog climate. More research might be directed toward overcoming the technical and economic obstacles to widespread adoption of such methods as trickle irrigation, for example, and toward expanding knowledge of how to impart drought resistance to plants.

Our purpose here is not to emphasize these particular innovations, but to make the point that the study of climate change impacts should explicitly include research and technology development among the likely social responses to the impacts.

TABLE 9.3. *EPIC-simulated Percentage Changes in 2030 Baseline (Task C) Crop Yields Under the Analog Climate (Task D)*

		No Adjustments		With Adjustments	
		No CO_2 effect	CO_2 effect	No CO_2 effect	CO_2 effect
		D_1	D_2	$D_{3.1}$	$D_{3.2}$
Corn					
	Dryland	-28	-20	-16	-6
	Irrigated	-12	-7	+2	+8
Sorghum					
	Dryland	-21	-11	+11	+25
	Irrigated	-10	*	+7	+18
Wheat					
	Dryland	-12	+3	+1	+17
	Irrigated	-1	+9	+6	+18
Soybeans					
	Dryland	-25	-15	-11	+2
	Irrigated	–	–	+1	+12

Source: Calculated from EPIC.
*Less than 1 percent.

TABLE 9.4. *EPIC-simulated Percentage Changes in Future (Task C) Irrigation Requirement Under the Analog Climate (Task D).*

	No Adjustments		With Adjustments	
	No CO_2 effect	CO_2 effect	No CO_2 effect	CO_2 effect
	D_1	D_2	$D_{3.1}$	$D_{3.2}$
Corn	+22	+12	+14	+5
Sorghum	+20	+6	+16	+5
Wheat	+11	+2	-9	-23

Water Demand and Supply

As shown in Table 9.4, the imposition of the analog climate on the agriculture of 2030 increases irrigation requirement. Without CO_2 (D_1), wheat requires 11 percent more water and corn and sorghum 22 percent and 20 percent more, respectively. Carbon dioxide enrichment (D_2) reduces that requirement, virtually eliminating the increase in wheat. The increased irrigation requirement is also reduced by adjustments, especially in combination with increased CO_2. Improved irrigation efficiency is partly responsible, although the major effect, especially in wheat whose irrigation requirement is reduced, is attributable to the increased stomatal resistance that reduces ET.

Even if the best of these projections ($D_{3.2}$) were to be realized, water supply problems remain. As noted above, the analog climate implies significantly less water for the MINK states and the river basins of which they are a part. The impacts are particularly great in the Missouri River Basin, where flows at the outflow point are only 69 percent of the long term average. Flows in 2030 would be the same, barring changes in net exports and increased evaporation from manmade reservoirs.

The amount of water stored in the Ogallala aquifer underlying the Kansas and Nebraska High Plains in 2030 is not likely to differ significantly as a result of climate change. Under both the control and analog climate cases, increasing water costs are likely to curb groundwater use to levels approaching sustainable supplies by 2030. Although the quantities of water stored in the aquifer in 2030 will probably not depend on climate, the path taken to reach that level and the sustainable level of pumping will be climate dependent.

The hotter and drier conditions of the analog climate would tend to improve the economics of irrigated, relative to dryland, agriculture. The EPIC simulations of representative farms in Iowa and Missouri suggest that irrigated farming would be more profitable than dryland farming in these states. Net revenues increase by 42 percent to 49 percent in Iowa and by 28 percent to 31 percent in Missouri. These projections consider only baseline (Task A) production costs; allowing for expected increases in energy costs reduces, but does not eliminate, the advantage of irrigation. Pumping depths in Iowa and Missouri are likely to be much less than the 70 m to 90 m typical of the High Plains.

While the hotter/drier analog climate might improve the economics of irrigation agriculture, the increased scarcity of water will constrain its expansion. Under the control climate, water is available for a major expansion of irrigation in the two eastern MINK states without any sizable depletion of groundwater stocks. Under the analog climate however, water use exceeds mean assessed streamflow in all but 2 of the 15 water resource subregions. While groundwater stocks are large in Iowa and Missouri, recharge rates would be lower under the analog climate. Moreover, pumping from alluvial aquifers would tend to draw down surface flows. The net effect of the increased profitability of irrigation and the increased water scarcity on Iowa and Missouri is unknown. But it seems unlikely that irrigation would grow much faster than it did in the period from 1965 to 1985 (3.7 percent per annum).

Economic Impacts of the Analog Climate on Current Baseline Conditions (the B Scenarios)

Effects on Crop Production

The changes in crop yields under the analog climate in the B scenarios would tend to change the costs of producing the crops. This would affect the amount of crop production in two ways: (1) production would be cut back because the previous level would no longer be profitable under the higher costs and (2) some farmers would shift entirely out of a no longer profitable crop and into a profitable one.

In our B_1 and B_2 scenarios only the first factor affects production. We lacked the resources to estimate the cost curves for each crop so we assumed that costs would increase such that the decline in production would be proportional to the decline in yields. Our B_3 scenario reflects both the direct yield effect on production and the conversion of some land in irrigated corn production to dryland production of wheat and sorghum. Consequently, in the B_3 scenario corn production declines more than in proportion to the decline in corn yields, and production of wheat and sorghum declines proportionately less than the decline in their yields.

Table 9.5 shows the changes in the value of production of corn, wheat and sorghum under the B_1 scenario and the B_3 scenario with and without the CO_2 effect. The impact on soybean production is also included. The B_1 scenario represents a "worst case" situation because it excludes the CO_2 effect and adjustments that farmers would make to the analog climate. The B_3 scenario with the CO_2 effect represents a "best case" situation because it includes on-farm adjustments, as well as the CO_2 effect. We include the B_3 scenario without CO_2 because comparison of it with the B_1 scenario indicates the effect of on-farm adjustments in easing the impact of the analog climate.

Table 9.5 indicates that the combination of on-farm adjustments and the CO_2 effect reduces the decline in the value of production of the four crops from $2,662 million to $967 million. The CO_2 effect accounts for 58 percent of the difference and on-farm adjustments for 42 percent.

TABLE 9.5. *Changes in the Value of Crop Production Under the Analog Climate*

Crop	B1 scenario[a] Millions 1982 $[c]	B1 scenario[a] % change cf. 1984/87	No CO_2, $B_{3.1}$ Millions 1982 $	No CO_2, $B_{3.1}$ % change cf. 1984/87	CO_2, $B_{3.2}$ Millions 1982 $	CO_2, $B_{3.2}$ % change cf. 1984/87
Corn	-1,644	-21.3	-1,729	-22.4	-1,236	-16.0
Sorghum	-14	-0.8	+361	+19.7	+139	+7.6
Wheat	-215	-17.1	-35	-2.8	+178	+14.1
Soybeans	-789	-23.0	-542	-15.8	-48	-1.4
Total	-2,662	-18.7	-1,945	-13.7	-967	-6.8

The B1 scenario[a] spans the first two columns; the B3 scenario[b] spans the remaining four columns, subdivided into No CO_2, $B_{3.1}$ and CO_2, $B_{3.2}$.

[a] Assumes no CO_2 effect and no on-farm adjustments—i.e., a "worst case" impact.
[b] Assumes on-farm adjustments—i.e., a "best case" impact.
[c] Calculated from average 1984/87 production valued in 1982 prices.

Effects on the Regional Economy

Because crop production is an integral part of the MINK economy, changes in crop output will affect production in other sectors. These effects can be represented quantitatively by "multipliers," numbers that show the effect on regional production of both the initial change in crop output and the change in the rest of the economy induced by the initial change.

We calculated such multipliers from data provided in IMPLAN, a county-level input-output model of the entire United States developed by the U.S. Forest Service (1989). The IMPLAN model for MINK was built by aggregating the county level data for the region as a whole. We present an account of IMPLAN and the aggregation process in DOE Report (1991) Part VI. Here we summarize the results the model gives in estimating the multiplier effects for the decline in corn and soybean production under the analog climate. We focus on corn and sorghum production because the IMPLAN model indicates a strong multiplier relationship between feed grains and animal production. This in turn has a strong multiplier relationship to the rest of the economy.

Two sets of estimates of the multiplier effects for the decline in corn and sorghum production imposed by the analog climate were used. In one set we assume that the increase in costs of feed grain production caused by the climate-induced losses of yield results in a reduction in export demand for the crops equal to the simulated decline in crop production. Animal producers in MINK are assumed to substitute lower priced grains from outside MINK, so animal production in the region is not affected by the decline in grain output. The other set of multiplier estimates assumes that the full amount of the decline in grain production is borne by animal producers in MINK, exports of the crops being unaffected. Neither of the two underlying assumptions is fully realistic and should be regarded as polar cases setting the limits within which the actual multiplier effects would fall.

Results were obtained for two scenarios: One assumes no CO_2 effect or adjustments by producers (B1 case); the other assumes both the CO_2 effect and adjustments (B3.2 case). Under worst case conditions—no CO_2 effect, no on-farm adjustments, and the total burden of the decline in feed grain production falling on animal producers in MINK—the climate-induced decline in feed grain production would reduce total production of all sorts in the region by $29.9 billion, or 9.7 percent. Under the best case conditions—CO_2 effect with on-farm adaptations and the total burden of the feed grain production decline falling on exports—total regional production would decline by $1.4 billion, or 0.5 percent.

The main reason for the difference

between the two polar cases is that when the feed grain production decline is absorbed internally it reduces animal production, which in turn reduces output of the meatpacking industry. The latter is not only the largest single manufacturing activity in MINK, it also has the largest multiplier effect on the rest of the economy, because it draws heavily on locally produced animal and other inputs.

We believe that the impact of the analog climate on the MINK economy would be closer to the lower than to the higher polar case. There are three reasons. One is that farmers would surely adjust to the changed climate, as in the B_3 scenario. The second reason is that some benefits of the CO_2 concentration increase in the atmosphere, which we have estimated conservatively, would be likely. The third reason is that the assumption that the burden of reduced grain production would fall on exports is more realistic than that it would fall wholly on MINK animal producers. The higher grain production costs under the analog climate would weaken the competitive position of MINK grain outside the region, leading to a decline in its exports. MINK animal producers at the same time could import cheaper feed grain from outside the region. Thus, the impact on animal production in the region, and hence on the meatpacking industry, would be less than in the worst-case outcome. Over the long term, however, some animal production might shift to other grain-producing regions less affected by climate change. This could induce a decline in meatpacking in MINK, because that activity tends to locate in proximity to animal production.

Economic Impacts of the Analog Climate on Future Baseline Conditions (the D Scenarios)

Size of the Regional Economy in 2030

To represent the MINK economy in the year 2030 without climate change we used population and economic projections for the MINK states developed by the Bureau of Economic Analysis (BEA) (U.S. Department of Commerce, 1990). We note here only two aspects of these projections: (1) they show the MINK economy in 2030, measured by real personal income, to be 75 percent larger than it was in the 1984-1987 baseline period and (2) the share of farm income in total regional income, only 3.1 percent in 1984-1987, declines to 2.8 percent in 2030. (Our projections of MINK agriculture, done independently, are consistent with the BEA projections.)

The Agricultural Economy

Wheat, feed grains and soybeans dominate world crop production, and they dominate world agricultural trade in crops even more. MINK agriculture, therefore, is integrally tied to world agriculture, and its future cannot be adequately evaluated except in the context of the future of world agriculture.

We have done this evaluation in a number of steps described in detail in the DOE Report (1991), Part IIA on MINK agriculture. Suffice it to say here, our analysis of prospective global demand and supply conditions for grains and soybeans and of America's and MINK's competitive position in world markets for these crops suggests several plausible scenarios. In the one presented here world trade in grains and soybeans grows in proportion to world demand for them (driven primarily by population and per capita income growth in the developing countries), the U.S. shares of world trade remain at 1980s levels, and MINK's shares of U.S. production remain as they were in the 1984-1987 baseline period.

Under these conditions MINK's production of grains and soybeans would increase 67 percent over the baseline amount, not much less than the BEA projection for regional income (75 percent). In this scenario U.S. demand for grains and soybeans grows with population—about 25 percent from 1984/87 to 2030 (United Nations, 1989). On the plausible assumption that MINK's production is divided between domestic and export demand in the same proportions as in the nation, then most of the increase in the region's production of wheat and soybeans would be for export abroad, well over half the increase in sorghum production would be for that purpose, and a projected 40 percent increase in corn production would be split almost evenly between domestic and export demand.

Impacts on Crop Production

As in the B scenarios, the impacts of the analog climate on 2030 baseline yields depicted in the D scenarios would change the baseline cost of crop production, with consequent effects on the level of production. Also, as in the B scenarios, the changes in production would reflect both the yield effects and the effects of shifts of some land out of irrigated corn production into dryland production of wheat and sorghum.

Table 9.6 shows the changes in the value of crop production from the 2030 baseline. This table corresponds to Table 9.5 in showing both the "worst case" and "best case" situations, as well as a scenario that permits calculation of the difference made by on-farm adjustments. Comparison of the worst case scenarios in Tables 9.5 and 9.6 shows that, without the CO_2 effect or on-farm adjustments (scenarios B_1 and D_1), the analog climate would decrease production of the four crops by 22 percent from the 2030 baseline and by 19 percent from the current baseline (Table 9.5). However, because total crop production would be substantially greater in 2030, the decline in production from that level, $5,126 million in 1982 prices (Table 9.6), would be almost double the decline from the current baseline, $2,662 million (Table 9.5).

Comparison of the "best case" scenarios shows that production actually would increase from the 2030 baseline (Table 9.6), while declining by about 7 percent from the current baseline (Table 9.5). The increase in production from the 2030 baseline results from the favorable effects of the increased

CO_2 concentration and from the assumed success of researchers and farmers in the region in devising technologies to counter the effects of the analog climate. Although we think the emergence and adoption of these technologies is plausible, they would contribute to greater crop production in MINK only if other competing regions lacked access to them. The record shows that new agricultural technologies diffuse rapidly around the United States (and the world). Should the technologies that boost MINK output in our best case D scenario actually become available by 2030, they more likely would result in lower grain and soybean prices than increased production on the scale shown in the scenario.

Effects on the Regional Economy

The production multipliers showing the impact of changes in agricultural production on the rest of the regional economy reflect technical and economic conditions determining MINK's economic structure. To develop projections of the multipliers would require a detailed analysis of these underlying conditions, which is beyond the scope of this chapter. Something useful can nonetheless be said about the regional economic impacts of the D scenarios. The analysis of the B scenarios showed that, if all of the decline in crop production were reflected in reduced exports, the effect would be to reduce total regional production by roughly 0.5 percent to 1.5 percent. Noting this, and allowing for the growth of the MINK economy to 2030 (75 percent according to BEA projections)

TABLE 9.6. *Changes in 2030 Baseline Crop Production Under Analog Climate*

Crop	D_1 scenario No CO_2, no adjustment		D_3 scenario No CO_2, with adjustment		CO_2, with adjustment	
	Millions 1982 $	%	Millions 1982 $	%	Millions 1982 $	%
Corn	-2,462	-23	-1,607	-16	-641	-6
Sorghum	-419	-20	+310	+15	+539	+26
Wheat	-424	-11	+126	+3	+639	+17
Soybeans	-1,821	-26	-754	-11	+149	+2
Total	-5,126	-22	-2,029	-9	+686	+3

suggests that, even if the crop production multipliers were to double from the current to the 2030 baseline, the decline in crop production in the worst-case D scenario would have a small percentage impact on the regional economy, if all the decline occurred in crop exports.

However, if the decline in crop production resulted in an equal reduction of feed grain supplies to animal producers in MINK, and if the crop production multipliers were to increase, then analysis of the B scenarios suggests that the worst case D scenario would imply a decline in total regional production of more than 10 percent—no longer a trivial amount. This worst case result for the economy as a whole probably is less likely than the best case result, for the same reason as in the B scenario: The decline in crop production is more likely to show up as reduced exports than as reduced feed grain supplies to local animal producers.

The worst-case result in the D scenarios also assumes that animal production in MINK remains closely related to feed grain production, and that the animal-meatpacking relation remains important. The high multiplier effect of the decline in crop production in the worst case result for the regional economy assumes continuation of the strong locational relationship in MINK among feed grain production, animals and meatpacking.

Of course if crop production in MINK should increase, as in the D_3 scenario with the CO_2 effect and on-farm adjustments, then the impact of the analog climate on the regional economy would be positive, at least as far as agriculture is concerned.

Summary and Conclusions

A methodology has been developed to allow assessment of the economic impacts of climate change on the regional scale and to assess the efficacy of responses to climate change. In the study reported above we have imposed an analog of the world climate change anticipated to follow greenhouse warming onto the four-state MINK region as it is today and as we project it may be 40 years from now when the climate change may actually be upon us. Under both present and future circumstances the impacts of the

climate change analog on crop production, water demand and supply are calculated with and without farm level adjustments and with and without an increase in atmospheric CO_2 concentration of 100 ppm (above the current concentrations of about 350 ppm).

The response of crop yield and evapotranspiration to the climate change-CO_2-adjustment cases was simulated with the EPIC model, which is especially well adapted to deal with increased CO_2 and increased aridity. Changes in surface water supplies were calculated by relating the flow in the 1930s at a number of gauging stations unaffected by human activities to the flow during the control period (1951-1980) at the same stations.

Were the climate change to occur now, yields of grain crops, except for wheat, would be lowered significantly. The reduction would be tempered by CO_2, and simple adjustments would lessen the yield reduction further still. Under the best of these scenarios, however, the productivity of the region's agriculture would be significantly diminished.

The number of days of water stress for crops would increase under the analog climate, indicating an increased requirement for irrigation to avoid losses of yield. However, surface water flows would be considerably reduced, especially in the Missouri River Basin, so that the increasing irrigation demand would probably not be met with greater withdrawals of surface water for which there are already many competing demands. In much of the western MINK states irrigation is based on depletable groundwater sources. A change in climate would probably result in a faster rate of depletion of these sources.

Crop productivity and irrigation requirements in the absence of climate change were projected to the year 2030 by means of EPIC simulations. This was done by adjusting EPIC to consider a set of foreseeable technologies. Yields of the grain crops considered in the paper will be 70 percent 90 percent greater in 40 years according to EPIC. Irrigation requirements in the absence of climate change will not be much different in 2030 from the current situation, but by that time additional constraints on water supply are likely.

When the analog climate is imposed upon the MINK region of 2030 with no farm-level adjustments and no CO_2 effect, yields are reduced in percentages not very different from those under the same conditions in the present situation. However, total production remains higher in 2030 than at present. The CO_2 effect and simple agronomic adjustments moderate the analog climate effects as before, but two additional adjustments that might be driven by an awareness of oncoming climate change—e.g., drought resistance and irrigation efficiency improvements—tend to counteract the climate induced losses more effectively. Indeed, after allowance for all of these improvements and the CO_2 effect, crop production in MINK might even be a bit higher with the analog climate than without it.

The increased requirement for irrigation water under climate change would be still more difficult to meet in 2030, because surface flows will be lower and the aquifers underlying much of the region will by then be depleted or regulated for use at no more than sustainable rates.

The climate-induced declines in crop production would reduce total MINK production because of the linkages between crop production and the rest of the economy. If the decline in crop production resulted mostly in a reduction of crop exports, the decline in total regional production would be on the order of 1 percent to 2 percent. However, if the crop production decline resulted mostly in reduced feed grain supplies to local animal producers and hence of supplies of animals to the meatpacking industry, total regional production could fall 10 percent or somewhat more. The actual decline in total production likely would be closer to 1 percent to 2 percent than to 10 percent. Should crop production rise, as it does in one scenario, total regional production would be higher under the analog climate than without it.

We caution the reader not to think of the foregoing as predictions of what will be happening in the MINK region over the next 40 years, but rather as illustrations of what might happen given the region's agricultural and water resource vulnerabilities to climate change and its economic structure.

Our methodology provides a framework for analysis that overcomes important limitations of prior studies. We recognize that our data sources, models, and analytical tools are imperfect, and that better ones need to be developed. There is reason for confidence, however, that such improvements are possible and that their development will keep pace with or lead advances in climate modelling of the kinds that are needed to provide reliable predictions of the regional distributions of climate change. As confidence grows in GCM predictions, their results can readily be employed in the MINK study framework.

References

Bolin, B., B.R. Doos, J. Jager, *et al.* Warrick (Eds.), 1986. *SCOPE 29: The Greenhouse Effect: Climatic Change and Ecosystems.* John Wiley, New York.

Campbell, D.C., 1984. The Pick-Sloan program: A case of bureaucratic economic power. *Journal of Economic Issues.* **18:**449-56.

Cure, F.D. and B. Acock, 1986. Crop responses to carbon dioxide doubling; A literature survey. *Agriculture and Forest Meteorology*, **38:**127-45.

Harbeck, G.E. Jr., 1962. *A Practical Field Technique for Measuring Reservoir Evaporation Utilizing Mass-Transfer Theory.* Geological Survey Professional Paper 272-E. U.S. Government Printing Office, Washington D.C.

Intergovernmental Panel on Climate Change (IPCC), 1990. *Working Group I: Scientific Assessment of Climate Change.* WMO/UNEP, Geneva, Switzerland.

Kimball, B.A., 1983. Carbon dioxide and agricultural yield: An assemblage and analysis of 430 prior observations. *Agronomy Journal* **75:**779-88.

———, 1986. Influence of elevated CO_2 on crop yield. In: *Carbon Dioxide Enrichment of Greenhouse Crops.* H.Z. Enoch and B.A. Kimball (Eds.). CRC Press, Boca Raton, Fla., pp. 105-15.

Lamb, P.J., 1987. On the development of regional climatic scenarios for policy-oriented climatic-impact assessments. *Bulletin of the American Meteorological Society*, **68:**1116-23.

Manabe, S., and R.T. Wetherald, 1986.

Reduction in summer soil wetness induced by an increase in atmospheric carbon dioxide. *Science*, **232**:626-28.

Mooney, H.A., J. Ehleringer, and J.A. Berry, 1976. High photosynthetic capacity of a winter annual in Death Valley. *Science*, **194**:322-24.

Palmer, W.L., 1965. *Meteorological Drought*. Research Paper No. 45. U.S. Department of Commerce, Weather Bureau. U.S. Government Printing Office, Washington D.C.

Penman, H.L., 1948. Natural evaporation from open water, bare soil and grass. *Proceedings of the Royal Society London Series A:* **193**:120-45.

Rosenberg, N.J., 1982. The increasing CO_2 concentration in the atmosphere and its implication on agricultural productivity. II. Effects through CO_2-induced climate change. *Climatic Change*, **4**:239-54.

———, B.A. Kimball, P. Martin, *et al.*, 1990. From climate and CO_2 enrichment to evapotranspiration. In: *Climate Change and U.S. Water Resources*. P.E. Waggoner (Ed.), John Wiley, New York, pp. 151-75.

Ruttan, V.W. 1989. Biological and technical constraints on crop and animal productivity: Report on a dialogue July 10-11, 1989. Staff Paper P89-45, Department of Agricultural and Applied Economics, University of Minnesota, St. Paul (95 pp, mimeo).

Schlesinger, M.E., and J.F.B. Mitchell, 1985. Model projections of the equilibrium climatic response to increased carbon dioxide. In: *Projecting the Climatic Effects of Increasing Carbon Dioxide*. M.C. MacCracken and F.M. Luther (Eds.). DOE/ER-0237, U.S. Department of Energy, Carbon Dioxide Research Division, Washington D.C., chapter 4.

Schneider, S.J., and N.J. Rosenberg, 1989. The greenhouse effect: Its causes, possible impacts and associated uncertainties. In: *Greenhouse Warming: Abatement and Adaptation*. N.J. Rosenberg, W.E. Easterling III, P.R. Crosson, and J. Darmstadter (Eds.),

Resources for the Future, Washington D.C., chapter 2.

Singer, M., 1987. *Passage to a Human World*. Hudson Institute, Indianapolis, Indiana.

Stockle, C.O., J.R. Williams, N.J. Rosenberg, *et al.*, 1992. Estimating the effect of carbon dioxide-induced climate change on growth and yield of crops. I. Modification to the EPIC model for climate change analysis (in press).

Thornthwaite, C.W., 1948. An approach toward a rational classification of climate. *Geographical Review*, **38**:55-94.

United Nations, 1989. *World Population Prospects 1988*. Department of International Economic and Social Affairs, New York.

U.S. Congress, Office of Technology Assessment, 1986. *Technology, Public Policy and the Changing Structure of American Agriculture*. OTA-F-285. U.S. Government Printing Office, Washington D.C.

U.S. Department of Agriculture, 1981. *Land Resource Regions and Major Land Resource Areas of the United States*. U.S. Government Printing Office, Washington D.C.

———, 1987/1988. *Agricultural Statistics 1987 and 1988*. U.S. Government Printing Office, Washington D.C.

U.S. Department of Commerce, 1987. *Statistical Abstract of the United States*. Bureau of the Census, Washington D.C.

———, 1990. *Regional Projections to 2040. Vol 1: States*. Bureau of Economic Analysis. U.S. Government Printing Office, Washington D.C.

U.S. Forest Service, 1989. *Micro IMPLAN Release 89-03 Help File*. U.S. Department of Agriculture, Forest Service, Land Management Planning. Fort Collins, Colo. (March).

Williams, J.R., C.A. Jones and P.T. Dyke, 1984. A modeling approach to determining the relationship between erosion and soil productivity. *Transactions of American Society of Agricultural Engineers*, **27**:129-44.

10. GLOBAL WARMING AND FORESTS OF THE GREAT LAKES STATES: AN EXAMPLE OF THE USE OF QUANTITATIVE PROJECTIONS IN POLICY ANALYSIS

Daniel B. Botkin

Part of the reason that we seem to be so unsuccessful in dealing with large scale environmental issues is that we operate under the wrong worldview, the wrong beliefs about how natural ecological systems function. Over the years, our management policies developed as if nature undisturbed achieved a single state of equilibrium that would continue indefinitely, and which was both "good" for all life and "desirable" to people. Our policies developed as if biological resources could be harvested at a constant rate indefinitely, without variation, as if nature ran like a diesel engine (Botkin, 1990). We believed that natural ecological systems were linear when they are nonlinear; that they were equilibrium systems when they are nonequilibrium.

From the perspective of scientific, quantitative projections, this set of beliefs has led us to apply highly unrealistic and overly simplistic models as a basis for policy development. Thus, the correct approach to management of our natural resources at a regional level requires two fundamental changes: a change in our deepest beliefs about nature and, as a consequence, a development of realistic, quantitative methods to allow us to make projections of likely effects of our activities. In spite of all of the talk about advances in the understanding of environmental issues, we still tend to think of the environment at global and regional levels in outdated, equilibrium terms, and at the same time we lack realistic and accurate models of ecological phenomena.

It has become fashionable to speak of "sustainable use" of resources and "sustainable development" of the land. Such idealized sustainability has never been achieved. Sustainability is a simple matter for an equilibrium system, but may not be achievable in the sense of a continual output of products, with no variation, for complex, nonlinear, nonequilibrium ecological systems. Thus, as we move away from the old ideas, we will need to recast the concept of sustainability.

No issue illustrates our dilemma—the lack of ideas and techniques—more than global warming, and no region better represents the dilemma we face in management of natural biological resources during global warming than the Great Lakes region of the United States. This region has been subjected to intense land-conversion during the past two centuries. Because middle latitudes are projected to be altered most by global warming, this region may suffer some of the greatest impacts of future environmental change. The possibility of global warming remains controversial. The question is: How should we deal with this contingency in our planning? My perspective is that the consequences for natural resources, some of which are explained in this chapter, are so severe that it is only prudent that we plan for global warming now; that we project the impacts of global warming on natural resources as accurately as possible and develop responses while there is still time to plan.

Environmental Changes Since European Settlement

Ever since European settlement, the Great Lakes region of the United States has undergone rapid, dramatic environmental changes. Prior to European settlement, forests covered most of Ohio, Indiana, Michigan, and Wisconsin, the eastern half of Minnesota, and about half of Illinois. At first, soon after the founding of the United States, these forests were considered to be encumbrances in the way of progress. The preface to the U. S. Census of 1810 referred to "bothersome trees" that "encumber a rich soil. . . and prevent its

154

cultivation" (Perlin, 1989). By 1880, much of Ohio and Indiana was cleared, and the undesirable effects of this clearing were becoming evident. The famous American forester Charles Sargent wrote that deforestation of these states might lead to "widespread calamity which no precautions taken after the mischief has been done can avert or future expenditure prevent" (Sargent 1884). By 1920, little of the original forest remained uncut (Paulik and Wright, 1932).

The environmental impact of this logging was worsened by poor subsequent land-use practices. Only the most valuable part of a tree, the central trunk or "bole," was transported to the mills. "Slash"—the smaller limbs, branches, twigs, and leaves—was left on the ground to decay or burn. Fires that were set carelessly were allowed to burn. Slash provided abundant fuel and promoted intense fires that cleared large areas. Sometimes these fires were hot enough to burn away the organic matter from the soil, ruining the land's fertility. Against this rapid Paul Bunyanesque clearing of the magnificent forests, nature seemed constant. Human beings appeared to be the only disturbing factor among the millions of trees.

Natural and Human-Induced Changes

Natural changes, if they occurred, seemed invisible against the widespread clearing of forests and prairies, and destruction of wildlife. In the nineteenth and early twentieth century, those who studied natural history concluded that without human influence nature was constant unless disturbed externally, and capable of simple recovery from disturbance. This belief was perhaps best expressed by George Perkins Marsh, the intellectual father of conservation in the United States, who wrote in 1864 in his landmark book *Man and Nature*, that "Nature, left undisturbed, so fashions her territory as to give it almost unchanging permanence of form, outline, and proportion, except when shattered by geologic convulsions." He wrote further that "in countries untrodden by man," all factors balance one another, so that "the geographical conditions may be regarded as constant and immutable" (Marsh, 1864, pp. 29, 35).

North of the massive logging in Michigan lay Isle Royale, a large, remote island in Lake Superior, too inaccessible and too deficient in economic resources to be victimized by large-scale land clearing. Prior to European settlement, American Indians had visited the island only occasionally to mine native copper. This remote island, a remnant of undisturbed nature, became a favorite place for ecological and natural history study. Here, left to her own devices, nature seemed to achieve that permanence of form and substance imagined by George Perkins Marsh.

In 1909, near the end of the great clearcutting of the white pine forests of Michigan, William S. Cooper, a biologist from the University of Chicago, traveled to Isle Royale to study its forests and their development following natural disturbances from windstorms and fires. The result was one of the classic studies of forest ecology of the early twentieth century (Cooper, 1913). Viewing the forests of balsam fir, white spruce, and white birch, interspersed with bogs and stands of sugar maple, Cooper concluded that he had found on this wilderness island "the final and permanent vegetational stage, toward the establishment of which all the other plant societies are successive steps," and "both observational and experimental studies have shown that the balsam-birch-white spruce forest, in spite of appearances to the contrary is, taken as a whole, in equilibrium; that no changes of a successional nature are taking place within it," even though "superficial observation would be likely to lead to exactly the opposite conclusion."

Cooper's idea of forest succession—a fixed pattern of development that led to an equilibrium state, in spite of, as he put it, "appearances to the contrary"—was the widespread explanation of forest development throughout most of the twentieth century, and it remains a dominant idea in the management of forest resources. This classic forest succession to a single climax leads to a simple solution to management: To obtain and sustain a natural forest—a wilderness area—simply remove a disturbing factor and a forest will recover to its original abundance and diversity. With global warming, one would need simply to wait for the warming episode to pass, and the forests and their wildlife would return to their former abundances. The idea of classic forest succession also leads to a simple solution to the management of economically important timber resources. Succession to a fixed equilibrium implies that forests can be harvested at a constant rate to provide a continual, maximum yield. Each acre can be cut at a fixed interval; all acres are logged using the same harvest interval, but some acres are logged one year, others the next, and so forth, to provide a constant sale of timber products. Variations in growth as climate changes, or in response to insect outbreaks or disease epidemics, have been simply ignored.

These ideas about natural forests lead to an apparent, but false, conflict between conservation and utilization of natural resources. On one side are those who have viewed the destruction of forests and seek preservation of the remaining stands. From their perspective, if nature undisturbed is constant and this constancy is good, then there is only one nature and that is the nature in its mature, fixed, permanent condition. By definition, any harvesting of timber creates an unnatural and undesirable forest stand.

On the other side are those who see this same equilibrium forest as a well running machine that can and ought to be maintained at its constant power output. Leaving nature alone, allowing death and decay of trees, is simply an inexcusable waste of an important resource.

As long as all of nature is seen as one thing and in one condition, these two uses of forests appear in opposition. But this perception is false. The true qualities of nature allow for variation, variety, change, and complexity, and offer opportunities for both conservation and wise utilization of natural resources, if we only take the effort to measure what we must know about the woodlands.

The Real Dynamics of Nature

In the last two decades, rapid advances have been made in our understanding of natural ecological systems, and these paint a very different picture from the constant and simple stability of nature as perceived by George Perkins Marsh and William S. Cooper. We know now that forest succession takes place—that a repeatable pattern of development of vegetation follows clearing of the land—but it does not lead to the static equilibrium condition imagined in earlier times.

About 10,000 years ago, the Great Lakes region was covered by the last continental ice sheet. Following the retreat of the glaciers, tundra plants occupied the area, as revealed in studies of the land to the west of Isle Royale, in what is now Superior National Forest of Minnesota and its million acres of designated wilderness called Boundary Waters Canoe Area (Heinselman, 1973). The first trees to arrive were spruce. About 9,200 years ago the spruce forest was replaced by jack pine and red pine, trees characteristic of warmer and drier conditions, suggesting a warming and drying of the climate. Paper birch and alder immigrated into this forest about 8,300 years ago; white pine arrived about 7,000 years ago, and then there was a return to spruce, jack pine, and white pine, suggesting a cooling of the climate. About every thousand years a substantial change occurred in the vegetation of the forest, reflecting in part changes in the climate and in part the arrival of species that had been driven south during the ice age and were returning slowly.

The eighteenth-century voyageurs saw a forest that may have appeared permanent to them but, in terms of the lifetimes of trees, these forests were a relatively recent phenomenon, probably too recent for the forests to have settled into a state of equilibrium, if indeed such a condition could be attained even under a constant climate. If climatic change caused variation in the forests over a thousand years, other factors have altered the forests at shorter time scales. In the Boundary Waters Canoe Area, fire has been a common and persistent feature, as revealed by written accounts of fires and fire scars in trees during the past 300 years. In jack pine forests that grow on sandy upland plains, fire occurs about four times a century; on wetter sites where black spruce is common, fire occurs much less often, once in 150 to 200 years (Heinselman, 1981). The frequency of fire has varied over the past four centuries, as reconstructed through written histories, fire scars imbedded in tree rings, and carbon in lake deposits (Heinselman, 1973). As in many parts of North America, the frequency of fires in the Great Lakes region probably was increased by the American Indians, while fire suppression was the rule in the first half of the twentieth century.

New space-age technologies provide observations that reinforce the new dynamic idea about forests of the Great Lakes region. Recently, we used Landsat imagery taken 10 years apart, in 1973 and 1983, to measure changes in successional status of vegetation in Superior National Forest of northern Minnesota (Hall *et al.*, 1991). In these images we could distinguish five stages in forest succession: clearings; very young, regenerating stands; somewhat older aspen stands; even older mixed aspen and conifers; and oldest of all, pure conifers. Using computer analysis and overlaying 1973 and 1983 images, we could determine what changes occurred for each of 250,000 Landsat units, each covering an area 80 m square. About half of the oldest stands changed from one type to another within the 10 years, while younger stands changed more rapidly. The forest was revealed as highly dynamic over a period of 10 years at this small spatial scale.

Isle Royale, that epitome of undisturbed, supposedly unchanging nature, has also changed over time. Its ice age history would have paralleled that of the nearby Boundary Waters Canoe Area. In the twentieth century, wildlife populations on the island have undergone great booms and busts. At the time William S. Cooper began to study the island, moose had arrived only recently. Cooper noted the great impacts this large herbivore was having on the vegetation, especially on water lilies and yew, two of the moose's favorite foods. The moose population increased rapidly into the 1930s, declined, and increased again. The population reached one of the highest densities of moose anywhere in the world, leading to great changes in the vegetation. The land became open, vegetated by many small stems showing signs of repeated browsing. The condition of the vegetation thereafter varied over time with changes in the moose population.

Wolves arrived at the island soon after World War II. Instead of growing to a constant number and stabilizing the moose population, as it had been believed predators would do, the wolf population fluctuated greatly, approximately doubling and halving its size over the years (Botkin *et al.*, 1973; Jordan *et al.*, 1971).

All of these factors contributed to changes in the vegetation on the island: actions of animals, variations in the external environment, and human activities. But even without these changes, it is likely that forested ecosystems would undergo changes. Studies in other parts of the world suggest that over the long term, chemical elements required for vegetation growth are leached by water from the upper soil to lower layers where these elements cannot be reached by tree roots. In comparatively stable environments with high rainfall, vegetation appears depauperate: The stage dominated by huge trees with maximum abundance and diversity is eventually replaced by grasses and shrubs. As an example, a study of a series of sand dunes in Australia representing a sequence of 100,000 years showed that in the oldest dunes mineral nutrients leached far down into the sands, too far for tree roots to reach, and the dominant vegetation was a few species of grasses and shrubs (Walker *et al.*, 1981). The flux of chemical elements, a part of the internal dynamics of the ecosystem, forced the vegetation away from its classic ecological climax. Thus, there are three mechanisms that tend to force natural ecological systems away from constancy: long-term climatic and other environmental change; human actions; and internal dynamics of ecosystems.

These new observations, making use of historic records, fire scars, pollen deposits, remote sensing, and simple, direct observations, reveal a very different forest from that of nineteenth-century writings: a forest that is continually changing at many temporal scales. To be successful, any new approach to management of forests, even under normal climate variations, must take these new insights into account.

Here is the essence of a dilemma that still confronts and confuses us in our attempts to manage natural resources in the Great Lakes region and elsewhere in North America and throughout the world: Against a foreground of rapid and large human-induced changes, slower natural variations seem to vanish from view. Our management of these resources has been predicated on a constancy and simple kind of stability of nature. Today, as we realize that we may have set in motion, through global warming, environmental changes that will dwarf those of nineteenth century human activities, we still pretend that the environment, unless directly altered by our hands, will maintain a constant condition, or if disturbed from that constant condition, will return to its former state of equilibrium. A new approach is needed; an example of such an approach is given in the next section.

A Small Bird Illustrates A Large Change in Approaches

One of the clearest examples of key aspects of the new approach to management of natural resources is represented by the history of attempts to save the Kirtland's warbler, an endangered species that nests only in jack pine woodlands where these grow in central Michigan on a coarse sandy soil, called Grayling sand. The warbler has long been of interest and concern to naturalists and conservationists. It was once proposed as the state bird of Michigan, and was the first songbird subject to a complete census in the late 1950s when somewhat more than 400 singing males were counted. A decade later the population had declined abruptly, and by 1970 only 200 males were counted. Ornithologists soon realized that the cause of the warbler's rapid decline was the suppression of fire that had become standard policy for forests during the mid-twentieth century.

The problem for the warbler is that jack pine needs fire. It has evolved in fire-prone forests and is adapted to fire. Its cones have a glue that prevents the scales from opening and releasing their seeds. The glue is loosened only by fire. The scales warp allowing their seeds to fall after they are wetted by the first rain that follows the fire—an optimum time for seeds to germinate.

Jack pine is a shade-intolerant species, one that grows well only in bright light of forest openings. Even planted jack pine trees cannot grow in the shade of other trees. If a mature forest develops, the pine cannot grow and the trees die for lack of light. This is what was happening in the 1950s and 60s throughout the warbler's small nesting region, because forest managers operated under the belief that all fires were bad and should be suppressed.

In the 1960s, the Michigan Department of Fish and Game, the U.S. Fish and Wildlife Service, and The Audubon Society began a cooperative program to save the warbler. An area of approximately 38,000 acres was set aside near Roscommon, Michigan, where fires are set periodically on small patches to promote reestablishment of jack pine, and to

maintain a patchwork landscape with jack pine woodlands in various stages of growth. This has been a good plan based on the real needs of the warbler, not on a hypothetical idea of the constancy of nature, and under the present climate appears it effective in maintaining the warbler.

Forests of the Great Lakes States and Global Warming

What Might Happen Under a Business-as-Usual Approach

If global warming occurs, our natural resources will face a crisis that cannot be solved under the old paradigm so completely divorced from both human practice and the real characteristics of the nature it purports to describe and explain. Under human impacts of the past century, the hypothetical sustainability of natural resources predicted by the myth of natural constancy and stability was never realized. To save the living resources of the Great Lakes states, we need a new paradigm for nature, and a new basis for theory from which we can project the effects of our actions on the abundances of our living resources.

How then are we to proceed? We need a way to project future conditions wedded to the observed, real, dynamic characteristics of populations and ecosystems. An example of such an approach is provided by the JABOWA computer model of forest growth, which we have used to analyze possible effects of global warming on forests of the Great Lakes States. In 1970, the author developed this model along with two colleagues, James F. Janak and James R. Wallis. Since that time, JABOWA has become widely used for the study of forests. There are about 17 versions applying the model to forests from New Zealand to Norway and from the tropics to the arctic timberline. We have shown that this model is realistic and accurate in its projections. It does not assume or require that nature achieve a constancy or stability believed in by George Perkins Marsh or W. S. Cooper. Instead, it is based on a long history of study of the growth of trees in laboratories and of the natural history of forests. The model projects the growth of individual trees on small plots, taking into account mean monthly temperature and rainfall, soil depth, water holding capacity, depth to the water table, and soil fertility. Trees compete for light and taller trees shade smaller ones. Species differences in seed production and seedling establishment, in longevity, in tolerance to low light and low fertility, and to drought and flooding, are taken into account.

The projected effects of global warming on forests of the Great Lakes states are surprisingly severe and rapid. The most alarming projection that we have made is for the habitat of the Kirtland's warbler. If global warming occurs at the rate projected by the global climate model developed by James Hansen of Goddard Space Flight Center, the jack pine forests in the warbler's preserve should now be growing slower, on average, than they have in that region in the past. Stands burned around the turn of the next century, or by the year 2010, would not grow back to jack pine (Botkin et al., 1991).

At first, other species—white pine and red pine—might come in to these areas, but by the latter part of the twenty-first century the climate would become too warm and dry for all native species. What are now woodlands on sandy soils in Michigan might look like the worst of the areas cut over during the great white pine logging of the nineteenth and early twentieth century—"stump barrens" as they are known—areas where intense fires burned away organic material in the soil. For this and other reasons not yet entirely clear, but perhaps influenced by long- term climate change, these stump barrens never reforested. Instead, among the grey stumps and rotting remnant logs grow reindeer moss, blueberry and huckleberry shrubs, bracken fern, and grasses. If global warming occurs as forecast, and if the best existing management practices simply continue as they have been applied in the past 15 years, Kirtland's warbler is likely to become extinct and the jack pine woodlands in its habitat replaced by barrens.

Global Warming and Use of Forests

The demise of a single endangered species may seem a small incident in the larger scheme of things, and many in our society might dismiss a concern with the impact of global warming on the Kirtland's warbler as unimportant against the massive changes that might occur in the environment. But the decline of Kirtland's warbler's could be an early symptom—an early warning, like the canary in the coal mine—of more general effects of global warming on the forest resources of the Great Lakes region.

To explore the effects of global warming on commercial forests of the Great Lakes states, we have made projections of forest conditions for the southern part of Superior National Forest of northern Minnesota (Botkin et al., 1989; Botkin and Nisbet, 1990). Here, commercial forestry is practiced in the transition between boreal and northern hardwoods. We examined forest growth both from clearings (initial conditions of no trees) and of old age (400 year) forest stands for four soil types: deep relatively dry soils; deep relatively wet soils; shallow wetland soils; and shallow dry soils. These soil types provide a broad range of forest conditions, from old age cedar bogs and old age balsam fir stands, to regrowth of aspen on thin sandy soils.

As in the case of the habitat of the Kirtland's warbler, the projected climatic changes lead to surprisingly large alterations in the vegetation of Superior National Forest. Biomass of present dominant species could decline to one-fifth of the current amount sometime between 2010 and 2040, with the character, quantity, and timing of the changes dependent on soil moisture conditions. Although slower than the projections for jack pine, these changes are rapid compared to the natural rates of change in these forests. In a true bog on a shallow wet soil, a forest undisturbed for 400 years and dominated by white cedar would decline in the next century to an almost treeless, shrub-dominated bog. On a deep, drier, but fertile sandy upland soil, the abundance of white birch in a forest undisturbed for 400 years would decline to about 10 percent of its 1980 abundance by about the year 2020, to be replaced by a sugar maple forest. Total biomass would decline about one-half by about the year 2070. On shallow dry upland soil cleared in 1980, trembling aspen would dominate the forest if current climatic conditions continued, but this species would be replaced by a sugar maple forest under the global warming. A 400-year-old stand dominated in 1980 by balsam fir on a deep, fertile, moist soil would lose two-thirds of the balsam fir abundance by the year 2010. Sugar maple would replace fir as the dominant species. The soil, previously too wet for optimal tree growth, would be warmed and dried, and the site would become well-watered but well-drained by 2010. In contrast to the other sites that suffered a decline in total biomass, on this site the total biomass would nearly triple, because global warming would improve the soil conditions for total tree growth. These impacts are projected by the model, even if increases in carbon dioxide in the atmosphere stimulated tree growth; this fertilization effect would be small in comparison to the effect of warming of the air and drying of soils.

The aesthetic quality of the forests including those of the Boundary Waters Canoe Area and Isle Royale National Park would change dramatically. As the transition from boreal to northern hardwoods occured, massive die-offs of mature trees would occur. The understory would become a thicket of maples and other deciduous species. The pleasant forests of today, open enough for hikers to pass through easily, would become unpleasant thickets of shrubs, seedlings, and saplings, exposed above and hot under the sun. The canopy trees, damaged by too much heat and too little water, would become subject to insect attack and disease outbreaks. Complaints might be raised about mismanagement of the wilderness, and speculations might attribute the forest decline to local pollutants and to direct human actions that somehow increased the abundances of insect pests and tree disease organisms, while in reality these could be associated with the lower vigor of the trees, and could be merely secondary responses to global warming. Outbreaks of fire would increase as the dying trees would provide abundant fuel, producing intense and destructive fires rather than light and beneficial ones.

The effects on commercial forestry could be severe. The change from conifer forests, useful for construction timber, pulp, and paper, to hardwood forests, valuable for furniture, might require major retooling of the forest industries. Presently the hardwoods cannot be used to make pulp and paper. New technologies permit that production, but would require major capital outlay.

Global warming might lead to a number of decades during which the boreal forest species were declining and northern hardwood species had not yet grown to maturity. From our viewpoint, the forests would be neither aesthetically pleasing nor commercially useful. Even when fully developed, the forest induced by global warming might be unlike any forest of today. Sugar maple would be much more abundant than in present northern hardwood forests, and as a result the biological diversity of the transformed forest would be less than in present northern hardwoods.

The model suggests another problem that might arise would be continued climate change after the year 2070. If CO_2 concentrations continued to increase, the climate would continue to warm. Conceivably, even large contiguous regions similar to the Boundary Waters Canoe Area might continue to undergo such rapid climatic change that seedlings that did become established could not reach maturity before the climate changed to conditions unfavorable for their growth. As a result, large areas might remain treeless, dominated by shorter-lived annual and perennial herbs and shrubs, as was projected for the jack pine stands in Michigan. We have found that the computer model predicts similar rapid and strong changes in forests at other locations in the Great Lakes states.

Some Possible Solutions

Maximizing Timber Yield Under Global Warming: An Undesirable Solution Carries A Warning

With all the attention focused on global warming by conservationists and environmentalists, surprisingly little consideration has been focused on how this large-scale environmental change might alter the harvesting policies of forest industries, and of government agencies that determine policy on federal and state lands.

To consider how the forest industry might react to global warming in the Great Lakes states, consider first what a "naive" forester might be motivated to do—perhaps someone we could refer to as an economist's idea of a forester—one with a single motivation, which is to maximize the yield of harvestable timber from the forester's land. Using the JABOWA forest model, we have projected how total timber yield would change under different harvesting policies: those that determine when the initial logging would be done, and those that determine the period between cuts, known as the harvest rotation. We considered the same global warming scenario described earlier, the one projected by the NASA general circulation model (Botkin and Nisbet, 1990). The results are simple but surprising: The greatest harvest to be obtained between today and year 2070 is the harvest one can obtain now. This statement remains true throughout the period. Forests after today will only decline in total organic matter. Some may recover in the future, as species adapted to warmer and drier habitats migrate onto the land or are planted, but regrowth will be slow compared to human economic interests.

This result is ironic. As soon as environmentalists accomplish their goal—persuade society that global warming is a serious issue that must be treated as though it were quite likely—a single-purpose forester would do exactly what the environmentalists would least like him to do. He would go out and cut all of his existing timber stands and try to sell the forest products before his competitors realized the same thing and flooded the market with timber.

It is important that we integrate such knowledge into our analysis of how to respond to global warming. Given these simplistic motivations, we have to ask: What policies would best achieve a reasonable use of forests, that is, maintaining forests for future production while providing wood products needed by our nation and for international trade? Clearly, a business-as-usual approach—assuming that forests left alone without direct human intervention will grow back to their former abundance—can no longer be treated seriously under global warming, if it ever could have been treated seriously. We need to develop policy mechanisms that reward commercial foresters who delay logging so as not to flood the market with timber. Perhaps greater emphasis could be placed on selective logging, in which the trees most sensitive to climatic change were harvested first. Trees might be removed selectively to open up the canopy and provide more light to the remaining vegetation, which might allow the remaining trees to persist longer. This might delay the overall die-back, and reduce the undesirable secondary

effects, such as increases in surface erosion. The logging of dying trees as the forests decline might be done in conjunction with a program to plant species that are better adapted to future conditions. Exactly how these inducements can best be accomplished is a task for economists and others, but in these activities the type of computer simulation I have described can be a useful tool.

A program of tree planting could be started now to introduce species that are better adapted to the projected warmer and drier conditions. However, the selection of species must be done carefully. We have begun to apply the forest model to help in this selection. We have added to the model three species of pine found in the southeastern part of the United States: loblolly, short-leaf, and Virginia pine, with the idea that one of these species might be able to replace jack pine on the sand plains of Michigan. Surprisingly, we found that, although the climate becomes too warm and dry by year 2070 for any of the native pines of Michigan, it does not warm up enough for the southern pines. Through this application of the model we have avoided a potentially costly and wasteful tree planting program. The model, along with other models and with expert advice, should be applied to the question of which trees should be planted.

Biological Conservation in a Global Warming Environment

The prospect of future global climate change challenges us to reconsider fundamental aspects of our conservation ethic. If the natural environment for most areas will change drastically during the next 50 years, we will be forced to reconsider the meaning of "wilderness" and to decide what we mean to conserve when we designate areas as legally protected wildernesses. The failure of a program that only lets nature take its own course will become obvious. As the author of *Discordant Harmonies: A New Ecology for the 21st Century* argues in his book, in the future we will need three kinds of natural areas: areas set aside for recreation, areas for biological conservation, and areas for baseline measurements and scientific research (Botkin, 1990). In the past we imagined that these three kinds of natural areas were one, but in fact each is different from the others and requires different policies.

Consider an attempt to maintain a wilderness recreation area, such as the Boundary Waters Canoe Area or Isle Royale National Park, when the climate rapidly changes the composition of forests of these areas. Left alone, these will become as no person has seen them before. Suppose we agree to develop an active policy to maintain the look of a wilderness for these two areas. As mentioned, these areas were "naturally" covered by ice, tundra, and a variety of forests unlike the present ones until only a few centuries ago. Which of these past conditions would we choose as our management goal? Laying out the possibilities in this way, the answer becomes almost obvious. It is my belief that the popular American idea of wilderness extends past the usual definition of a place "untrammeled by human action" to include the look of North America as seen by its first explorers. In the Great Lakes region, this wilderness would be represented by the forests as they appeared to the voyageurs in the eighteenth century. Thus, one type of natural area needed for future recreation would be a landscape with the appearance of eighteenth century forests. Under global warming, a landscape with this look would require intense behind-the-scenes management, assuming that would be possible.

The second kind of natural area would be land set aside to conserve specific endangered species or rare ecological communities. As shown by the example of the Kirtland's warbler, this kind of natural area may require intensive management, sometimes, as with the people-shy warbler, requiring the exclusion of humans during the bird's breeding season.

Only the third type of natural area--one set aside for baseline measurement and monitoring—would include land completely undisturbed by direct human action. And, ironically, this type of area would probably change into one unknown previously. It would be quite unnatural in the sense that it would not look much like the landscape visited by the voyageurs, or be familiar to the native American Indians before them, or to any human inhabitants through the end of the twentieth century. But such areas will be important to serve as bench marks against which we can observe the effects of our other direct management policies.

To save biological resources and use them wisely, we must change our concepts and our quantitative methods. Instead of believing that the natural condition is one of uniformity and constancy, we can make use of the real heterogeneity of the landscape, seeking a variegated landscape with conservation areas and intensive-use areas mixed together.

Some observations about land use in the Great Lakes region will clarify what needs to be done. These only become important once we accept the two ideas discussed earlier: that the "desired" condition of nature is up to us, and that natural ecological systems are not in equilibrium and the landscape is therefore highly variable. Thus, it should not be surprising that such simple facts have been ignored in many past discussions of approaches to conservation and the use of biological resources. As long as nature is assumed to remain in a single equilibrium, there is no management need to know what that equilibrium is.

Many nations that are rich in biological resources, as is the United States in general, and the Great Lakes states in particular, allocate 10 percent or more of their land to parks and biological preserves. Kenya, for example, has about 10 percent of its land in parks and natural areas; Costa Rica about 20 percent. In contrast, the six Great Lakes states—Michigan, Minnesota, Illinois, Indiana, Ohio, and Wisconsin, a total area of more than 800,000 km^2 (which approaches the area of France and Germany combined)—allocate less than one-half of 1 percent to parks, and less than 1 percent to designated wilderness (Table 10.1). In a program of wise management of our biological resources, it seems only prudent to allocate a reasonable portion of the land to each of the necessary functions, and it is suggested here that a program of environmental management for the Great Lakes states should allocate approximately 10 percent of the land area to biological conservation. Less than 8 percent of the area of the six states is in state and national forests, and slightly less than 9 percent is in lands that could be managed by the federal government or state agencies for conservation, recreation, water supply, and timber resources. Much of the 10 percent could be obtained from these lands alone, but some of the biological conservation lands could also

TABLE 10.1. *Land Use Area of the Great Lakes States*

(A) Areas in square kilometers

State	Total Area	State Forests	National Forests	State Parks	National Parks	Wilder-ness	Protected Area
Michigan	147,511	15,000	10,927	1,011	2,314	383	29,635
Minnesota	206,030	16,000	10,927	818	882	4,388	33,015
Illinois	144,120	1,113	1,033	–	–	–	2,146
Indiana	93,064	–	760	219	–	86	1,065
Ohio	106,201	699	769	801	--	–	2,269
Wisconsin	140,964	1,900	6,070	268	–	183	8,421
Totals	837,890	34,712	30,486	3,117	3,196	5,040	76,551

(B) Percentages of total

State	Total Area	State Forests	National Forests	State Parks	National Parks	Wilder-ness	Protected Area
Michigan	147,511	10.17	7.41	0.69	1.57	0.26	20
Minnesota	206,030	7.77	5.30	0.40	0.43	2.13	16
Illinois	144,120	0.77	0.72	0.00	0.00	0.00	1
Indiana	93,064	0.00	0.82	0.24	0.00	0.06	1
Ohio	106,201	0.66	0.72	0.75	0.00	0.00	2
Wisconsin	140,964	1.35	4.31	0.19	0.00	0.13	6
Totals	837,890	4.14	3.64	0.37	0.38	0.60	9

be composed of private lands for which certain easements or tax incentives could be provided, or for which land trades could be made.

Since, from the new perspective, such conservation areas need not be uniform nor uniformly pristine, nor allocated to a single use, such a program might be accomplished with surprisingly little undesirable effects on agriculture, forestry, and human settlements. Where possible, the 10 percent of the land allocated for biological conservation could also provide urban water supply and recreation and, especially where early successional conditions were important, might be combined with certain logging practices.

Having provided a general outline for proportionment of the land for each use, the next question arises: How can we establish priorities for land use and allocate various parcels to specific uses? I propose a new scheme that makes use of what I will call relative land-use elasticity, by which I mean how easy it is to move one use of the land from one location to another. The scheme I suggest assumes that recreation and commer-

cial forestry are the most elastic activities, in the sense that operations of a timber company can be readily moved around, and the destination of a vacationer can be readily changed. Biological conservation and urban water supply are the least elastic uses, because climate, local habitat conditions, and intrinsic characteristics of plant species operate together to dictate where a species can persist, and urban water supply can be obtained from new locations only at great expense. We should determine the needs for the least elastic activities first, so that we should allocate land use by geographic areas in the following order of priority: biological conservation; urban water-supply; commercial timber harvest; recreation. Of course, any area can have multiple uses and uses should be combined where feasible, with each use added according to the priority given here.

The implication is clear: We must reverse the past priority for land-use allocation. In the past, first the land was stripped of its existing commercially useful biological resources; then the land was settled, water obtained where convenient to the settlements, and eventually recreation areas established, usually on land considered useless for other purposes. Last of all, areas were allocated to biological conservation. The scheme I have described could be applied to government-controlled land only, or it could be applied independent of land ownership, as long as it is understood that a system of fair compensation would be developed for any private lands for which new uses were considered important. Determining priorities of land use for government-controlled areas in the Great Lakes states would, by itself, be an important step.

The Role of Science in Management of Natural Resources

The new perspective on nature leads to a new understanding of the role of scientific research in land and biological resource management. When I began research at Isle Royale National Park about 1970, I found that the National Park Service staff considered scientific studies as neutral and amusing additions at best and, at worst, as an interference with the real purpose of the park, which was seen as recreation. The use of the park for biological conservation, and the need of that conservation for research, was little recognized. The utility of research to sustain forests and lakes and wildlife of the island was little appreciated. This attitude has been widespread in the approach to land management in the United States. It derives in part, I believe, from the old faith in the balance of nature. As long as one believes that undisturbed nature achieves a single constant condition, and that this condition is the most desirable one, then scientific study is not necessary. Nature knows best and may maintain whatever conditions are best for her. We need only to feel good about nature, want her to succeed, and leave her alone.

But as emphasized, natural ecological systems change continually and they will change dramatically under global warming; hence knowledge becomes essential to successful management. Like it or not, we find ourselves in the pilot seat of spaceship Earth, and we must discover the dials that we must read if we are going to know how to turn the controls to guide our planetary life support system into the future. We must know what kinds of changes to expect, their natural rates

of change, and their causes. We must be able to project what could happen to our forests, and project how various management options will alter the future of these forests. We must monitor our forests and compare projections with real changes. Science must be an intrinsic part of management.

Nature As Our Guide to Change

At a more general level, our new perspective allows us to return to nature for guidelines. Now that we know that nature changes at essentially every scale of time and space, we can use nature as our guide. In the past, prior to the rise of technological civilization, forests changed at certain rates and in certain ways, under certain kinds of environmental variations. These changes and variations were repeated over very long periods, so that species were able to evolve and adapt to them. Thus, an environment that changes at these "natural" rates and in these natural ways is likely to be benign for forests and for all living things. In the forests of the Great Lakes states, migration of tree species has occurred slowly, at a rate of about 20 km/century, while disturbance from fire, as mentioned earlier, occurred approximately four times a century on dry sites and less than once a century on moist sites. These rates of disturbance and rates of migration can be our guidelines. Anything that we can do to slow the rate of global warming, so that climate change better matches the natural migration capacities of tree seeds, will ameliorate the situation.

Conclusions

Past management policies developed as though undisturbed forests achieved a single state of equilibrium that would continue indefinitely, that was both "good" for all life and "desirable" to people, and as though biological resources could be harvested at a constant rate indefinitely, without variation, as if nature ran like a diesel engine. We believed that natural ecological systems were linear when they are nonlinear; that they were equilibrium systems when they are nonequilibrium. In this chapter we described how we might take a new approach to management of forest resources in the Great Lakes states, accepting the dynamic, variable condition of these ecological systems whether disturbed or undisturbed by human actions. The application of perhaps the most successful ecological computer simulation developed in this century was described. It addresses the problems associated with the management of these forests in the future and the development of land use policies under a global warming climate. The new perspective and new tools provide positive approaches to complex problems and lead us to take a fresh look at some simple facts about the region.

The combination of a new paradigm and new methodologies suggest a strategy for action:

- Allocate land according to land-use elasticity; allocate to biological conservation a reasonable fraction of the land, on the order of 10 percent of the land in the Great Lakes states.
- Develop three kinds of natural areas, one for recreation, one for biological conservation, and one for baseline measurement and monitoring.
- Use the new kind of computer simulation as part of a management program for timber production so that appropriate responses under a global warming climate can be made and options that will fail because of global warming may be avoided.
- Seek to reduce change in forests of the Great Lakes states to natural rates.
- Use logging techniques that impose natural kinds of alterations.

This chapter has suggested the beginning of a new approach. Much more has to be done; we have only touched the surface of the potential for new ideas and new methods. With this new approach we can be more optimistic about our capacities to achieve economic utility and biological conservation on the same landscape. With the Great Lakes states as a prototype, the essence of the approach suggested here could be applied widely.

References

Botkin, D.B., 1990. *Discordant Harmonies: A New Ecology for the 21st Century.* Oxford University Press, New York.

————, P.A. Jordan, A.S. Dominski *et al.,* 1973. Sodium dynamics in a northern terrestrial ecosystem. *Proceedings of the National Academy of Science* **70**: 2745-48.

————, R.A. Nisbet, and T.E. Reynales, 1989. Effects of climate change on forests of the Great Lakes states. In: *The Potential Effects of Global Climate Change on the United States.* J.B. Smith and D.A. Tirpak (Eds.). USEPA, Washington D.C., pp. 2-1 - 2-31.

————, and R.A. Nisbet, 1990. Response of forests to global warming and CO_2 fertilization. *Report to EPA* (unpublished).

————, D.A. Woodby, and R.A. Nisbet, 1991. Kirtland's warbler habitats: A possible early indicator of climatic warming. *Biological Conservation,* **56**:63-78.

Cooper, W.S., 1913. The climax forest of Isle Royale, Lake Superior, and its development. *Botanical Gazette,* **55**:1-44, 115-40, 189-234.

Hall, F.G., D.B. Botkin, D.E. Strebel *et al.,* 1991. Large scale patterns in forest succession as determined by remote sensing. *Ecology,* **72**:628-40.

Heinselman, M.L., 1973. Fire in the virgin forests of the Boundary Waters Canoe Area, Minnesota. *Journal of Quaternary Research,* 3:329-82.

————, M.L., 1981. Fire and succession in the conifer forests of northern North America. In: *Forest Succession: Concepts and Applications.* D.C. West, H.H. Shugart and D.B. Botkin (Eds.). Springer-Verlag, New York, pp. 374-405

Jordan, P.A., D.B. Botkin, and M.I. Wolfe, 1971. Biomass dynamics in a moose population. *Ecology,* 52:147-52.

Marsh, G.P., 1864. *Man and Nature* (reprinted and edited by D.Lowenthal, 1967). Belknap Press, Cambridge, Mass. pp. 29, 35.

Paulik, C.O., and J.K. Wright, 1932. *Atlas of the Historical Geographic of the U. S.* Carnegie Institution, Washington, D. C. (reprinted 1975, Greenwood Press, Westport, Conn.).

Perlin, J., 1989. *A Forest Journey: The Role of Wood in the Development of*

Civilization, W. W. Norton, New York.

Rosenfeld, A.H., and D.B. Botkin, 1990. Trees can sequester carbon, or die, decay, and amplify global warming: Possible positive feedback between rising temperature, stressed forests, and CO_2. *Physics and Society*, **19**:4.

Sargent, C., 1884. Report on the forests of North America. *Tenth Census*, **9**:509.

Sandoz, M., 1978. *The Buffalo Hunters: The Story of the Hide Men*, University of Nebraska Press, Lincoln.

Walker, J., C.H. Thompson, I. F. Fergus *et al.*, 1981. Plant succession and soil development in coastal sand dunes of subtropical eastern Australia. In: *Forest Succession: Concepts and Applications.* D.C. West H.H. Shugart and D.B. Botkin (Eds.). Springer-Verlag, New York, pp. 107-31.

11. THE IMPACT OF CLIMATE CHANGE ON RICE VARIETY SELECTION IN THAILAND

Suwanna Panturat and Amos Eddy

This chapter is concerned with the impact of global warming on rice production in Thailand. Rice is Thailand's principal staple and the major agricultural export crop, accounting for approximately 18 percent of the agricultural export income for the year 1988 (Chuprakob, 1989). Approximately 70 percent of the population depends on agriculture. Forty percent of Thailand's total land area is devoted to growing crops, of which 62 percent is rice. About 70 percent of the planted area is rainfed and single cropped during the wet season. Yields of the upland rice crop are low. In the lowlands, yields are usually much better, and often double that of the upland rice. The export of rice, maize, cassava and sugar cane is a significant source of foreign exchange (Isaranurak and Rata, 1988).

Although rice is grown in virtually every one of the 73 provinces in Thailand, the bulk of the productivity is in the Chao Phraya River Basin surrounding and north of Bangkok. Some half dozen major varieties of rice have been identified by Huke (1982), ranging from dryland in the Northeast region to deepwater in the southern portion of the Central region, the former subject to severe droughts, and the latter to extensive flood damage. Comprehensive studies, such as that reported by Janatwat and others (1985) for northeastern Thailand, have related rice production problems to interactions between the variability of the atmospheric water delivery system and the ability of the soil to retain this water supply for later use by the crops.

The assessment of winners and losers in the context of global warming is still a controversial issue among scientists and policy-makers. No one can be certain of the answer. The amount of precipitation delivered on a given day to a given location could remain the same whether the precipitation itself were to be delivered by a rain-producing severe thunderstorm, or by a steady warm frontal rain falling over many hours. Precipitation in Thailand comes mainly from monsoonal troughs and tropical cyclones. There has been a suggestion that the monsoon season has changed during the past decade, coming later to Thailand and producing less rainfall. The average rainfall of the past six years, for example, is 5 percent to 18 percent lower than the long-term average (with the greatest decrease in the northeastern region) (Isaranurak and Rata, 1988).

Rainfed farming is a high risk enterprise, particularly in the monsoon type of climate, where usually heavy rains are received for periods varying from three to eight months, followed by extended dry periods of four to nine months. In many areas there is a high degree of interannual variability, not only of the amount of rainfall, but of the timing of the rainy season. Frequently, there are long spells of dry periods within the rainy season, which, if they occur at the critical stages of growth of a crop, may result in reduced yields or even in a complete loss of the crop. In 1987 Thailand experienced one of the worst droughts in many years, with about 960,000 ha of cropland adversely affected. Rice production decreased by about 8.2 percent from the previous season (Isaranurak and Rata, 1988).

For a given soil, cultural practice, and varietal genetics, the yield (both grain and straw) of the rice also depends on temperature and solar radiation. Plants require solar radiation for their photosynthesis process and are sensitive to relationships between radiation intensities at specific wavelengths. Their phenology also reacts to the length of the day.

In Thailand the southwest monsoon, with its extensive cloudy periods, decreases available solar radiation during the major rice season, thereby decreasing yields. In contrast, the northeast monsoon provides cloud-free sunny weather during the second rice season, thereby making possible higher rice yields (given irrigation). In the lowlands, flooding frequently reduces yields and acts as a disincentive for the use of fertilizer and other cash inputs. On the flood-free terrace lands the risk of drought and an early end to the rainy season are the main disincentives. Fortunately, when floods reduce production in the lowlands, the better rains usually mean a higher yield on the terraces.

The principal question for the present study is: With respect to food production, what would be the response of policy planners to changes in the climate, in order to maintain adequate levels of food production for both local consumption and world export? The answer could be made up of some or all of the following elements:

- alternate crops and alternate cropping practices;
- invest in new irrigation canal systems;
- invest in new reservoirs;
- encourage farmers to modify their tillage practice, fertilizer use, irrigation methods and other farming techniques; and
- recommend to the rice breeder to devise new varieties of rice that will improve the yields for the climate change scenarios.

This chapter focuses on the last element as input to policy exercises conducted at national or regional levels. To accomplish this objective, all information needed for model input was collected from seven sites selected from all four regions of Thailand: Northern, Northeastern, Central and Southern.

Both the BASE climatological data (raw data from the Thailand Meteorology Department) and the GISS (Goddard Institute for Space Studies) climate change data have been used as inputs to the state-of-the-art plant process, CERES-RICE (Version 2). CERES-RICE was developed by Godwin and Singh (1989) of the IFDC (International Fertilizer Development Center) for the simulation of both lowland and upland rice growth and development. This model makes explicit use of precipitation, temperature and solar radiation on a daily basis in estimating changes in biomass and phenological development of the crop.

The genetics are represented by an 8-vector of coefficients that control the length of the phenological stages, photosensitivity and growth properties.

The plant process computer simulation model is used to examine potential impacts of changes in ambient weather conditions, and of changes in varietal characteristics of the plant, on rice yield under specific soil and cultural practice factors. Different types of climate, soil and the cultural practice of the farmer require different rice varieties to obtain optimal yield. Consequently, to optimize the yield, the farmer should use the variety with genetic characteristics that can make the best use of the other three factors mentioned above.

Rice Climate

GCMs (Global Climate Models) have found an increasing use in climate impact analyses (Panturat, 1987, Eddy and Sladewski, 1988; Chen and Parry, 1987). Outputs were obtained from NCAR (National Center for Atmospheric Research), GISS (Goddard Institute for Space Studies), GFDL (Geophysical Fluid Dynamics Laboratory), and the OSU (Oregon State University). However, data from the GISS model were used in this study,[1] because GISS was the one model that incorporated a diurnal cycle. Using data obtained from Dave Rind, GISS, the following differences were evident for climate change scenarios with and without a diurnal cycle:

- the rain over the warmer land increased by up to a few mm per day without a diurnal cycle;
- low-latitude temperature in winter was about 10 °C warmer over land without a diurnal cycle, while high latitude temperatures were a few degrees warmer;
- near the equator, there was a 50 percent increase in low clouds over land for no diurnal cycle;
- more monsoon activity and greater rainfall occurred with no diurnal cycle.

---

[1] Long-term monthly mean values of temperature and precipitation for $1xCO_2$ and $2xCO_2$ GISS model runs for Southeast Asia were supplied by Roy Jenne of NCAR for the UNEP project on Socio-Economic Impacts and Policy Responses Resulting from Climate Change in Southeast Asia.

The climate change scenarios are developed by modifying the daily rainfall amounts using the basic observed daily data for the period (1955-1976). The rainfall days of occurrence are not changed. This is a very important constraint because, if the increase in rain were to occur by increasing the number of days of rain instead of by increasing the intensity, then the associated changes in solar radiation reaching the plant canopy could have as much impact on yield as would the changes in water availability. It is the ratios of $2xCO_2/1xCO_2$ for monthly precipitation that were used to produce the GISS precipitation scenario (shown in Figure 11.1). Figures 11.2 and 11.3 show the (average) consequence of applying similar ratios to the daily BASE temperatures and radiation. Since the GISS precipitation values are obtained by multiplying the BASE daily

FIGURE 11.1. Monthly mean precipitation (mm). Solid bar = Base case; open bar = GISS.

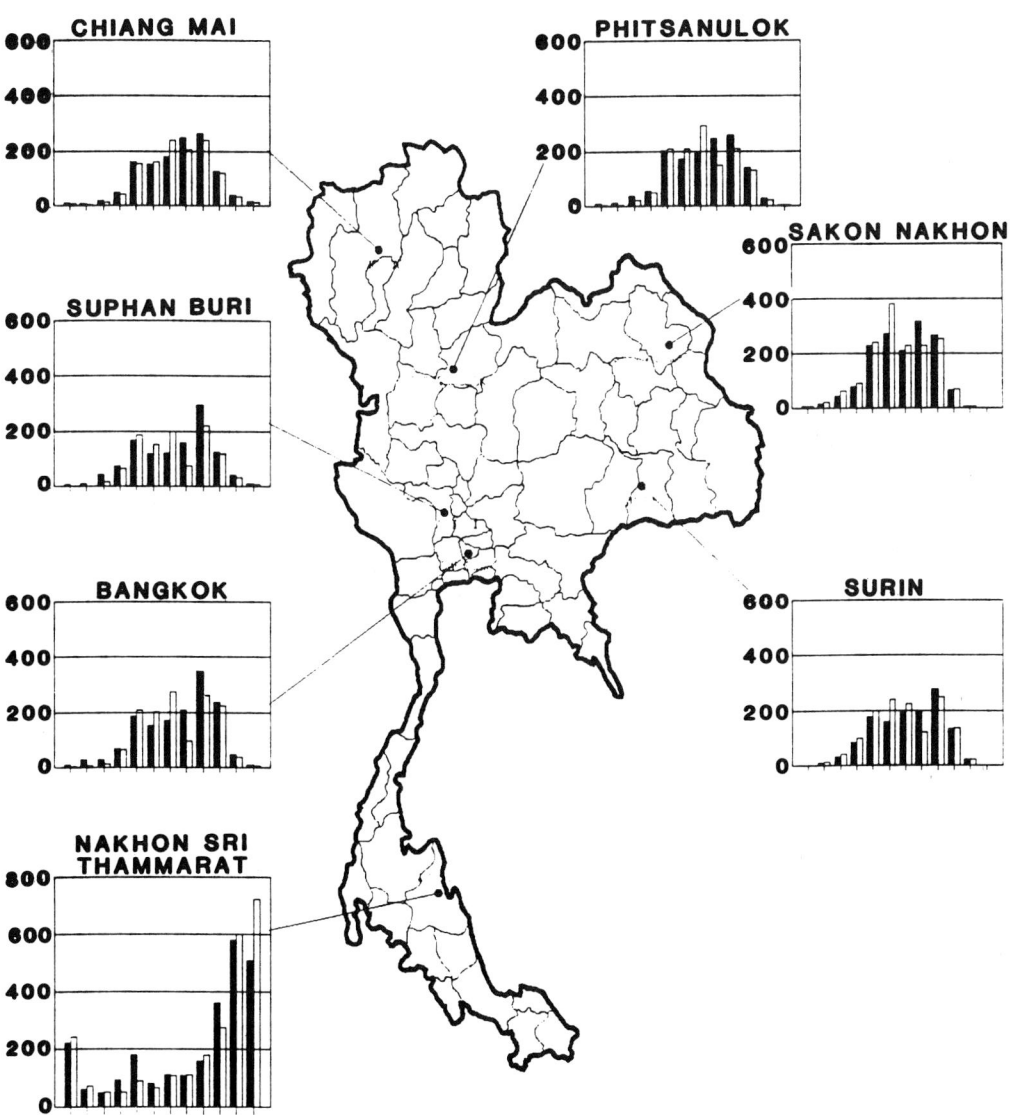

precipitation by a ratio, the variance as well as the mean values will be changed. Since these ratios are a function of latitude and longitude, this change in variance will vary from station to station.

Figures 11.1 to 11.3 represent monthly mean values of the minimum climatological data set required for an understanding of some of the results to be shown in the following climate change impact estimates. All seven selected sites are represented with monthly mean precipitation as well as monthly mean temperature for both the BASE and the GISS scenarios; whereas monthly mean radiation values for both the BASE and the GISS scenarios are represented by seven sites from all four regions close by the selected sites.

The precipitation pattern in Figure 11.1 shows an annual cycle for all seven sites. The precipitation amount obtained from the

FIGURE 11.2. Monthly mean temperature (°C). Solid bar = Base case; open bar = GISS.

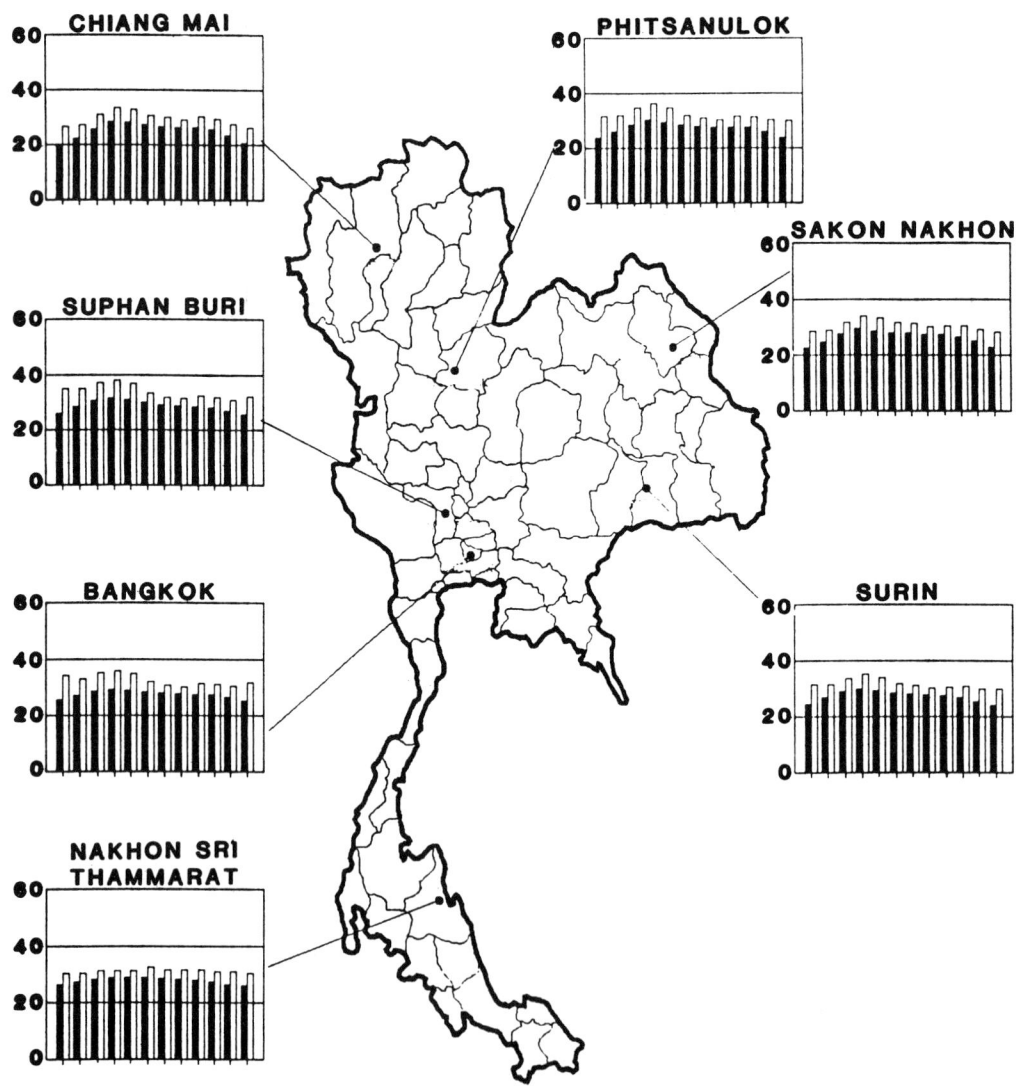

GISS scenario is mostly equal to or less than that obtained from BASE, except for the month of July from those sites in the Northern and Central regions, June from the Northeast region and December from the Southern region. Mean temperature obtained from GISS in all seven sites is significantly greater than that of BASE. All sites except Nakhon Sri Thammarat show the temperature to increase greatly during the dry season (January to April), while at Nakhon Sri Thammarat the temperature increase is nearly the same for every month. Mean radiation values obtained from GISS are significantly greater for the months of April and May for all sites, except Chantaburi, than that obtained from BASE, while at Chantaburi the mean radiation values obtained from GISS are greater than those obtained from BASE during the dry season (January to April).

FIGURE 11.3. Monthly mean radiation (LY/day). Solid bar = Base case; open bar = GISS.

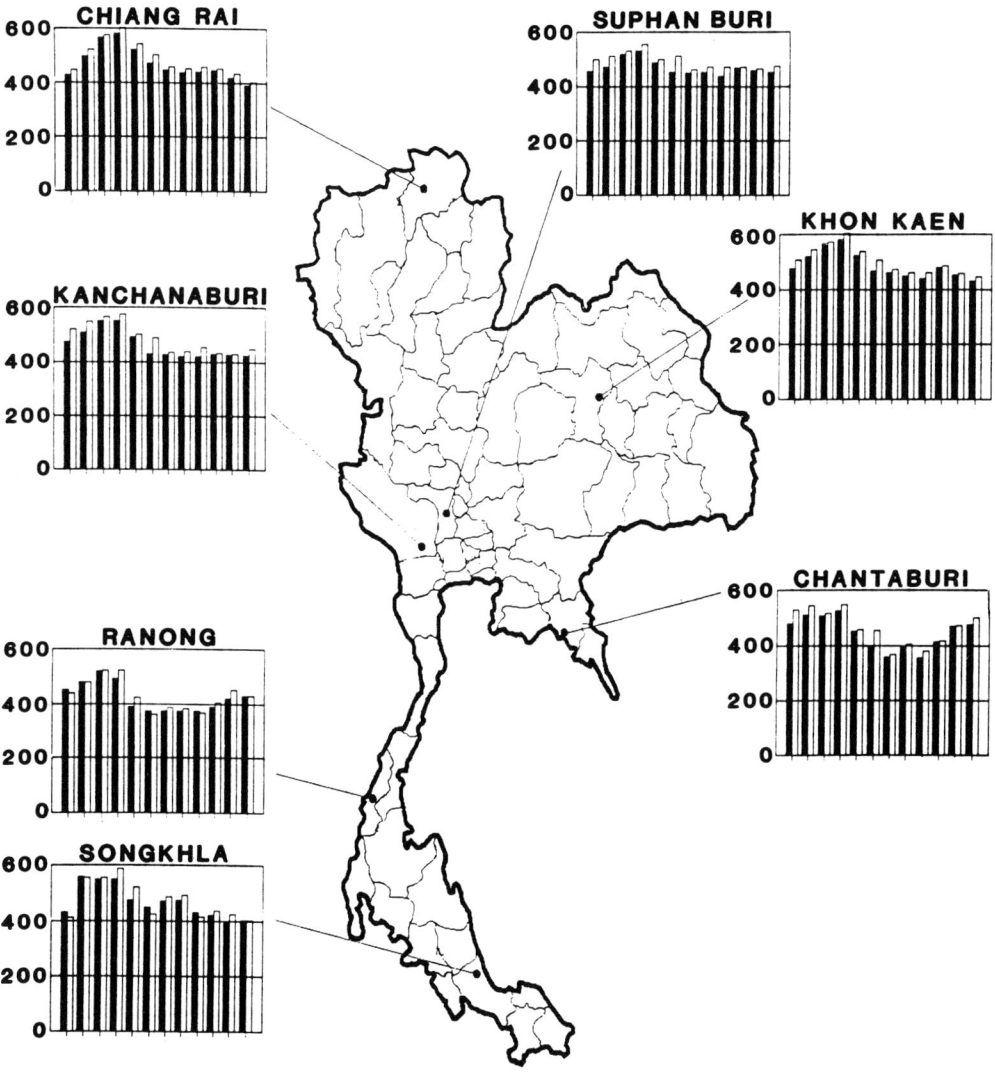

The Rice Model

The CERES-RICE model has been tested on minimum data sets gathered from experiments in various locations of the world. Such experiments were conducted on upland direct-seeded rice and flooded transplanted rice (IBSNAT 1990). The CERES-RICE V2.00 simulation model developed by Godwin and Singh (1989) can be used to represent transplanted or direct seeded, rainfed or irrigated, and upland or lowland cultural practices. Bunding can be designed for the paddies; fertilizer application can be imposed with respect to frequency, timing, type, depths and amounts; planting dates, sowing depth, and plant population densities can be specified; and, the straw and root biomass associated with antecedent harvesting practice can be incorporated into the soil nutrient profile. Many other parameters must be set in the model when it is being tuned to a particular site with its specific history of cultural practice and productivity.

Also, of particular relevance to the present study, the genetic characteristics of the crop that is to be simulated must be specified in terms of an 8-vector of coefficients to be discussed in detail below.

One of the strong points of this model is its simulation of the nitrogen cycle in a manner appropriate to anaerobic conditions associated with lowland (flooded) rice, but to aerobic conditions associated with upland rice. Weak points of the model include its inability to account directly for the negative impacts of pests and pathogens, lodging associated with strong winds, and losses associated with harvest practices.

In summary, the versatility of the model with respect to cultural practice controls, the straightforward procedure associated with data input, and the complete, organized and extensive output data sets all combine to make this algorithm exceptionally valuable for applications such as the one presented here.

The Genetic Coefficients Related to the Phenology

There are eight genetic coefficients used by the model to define the variety chosen for simulation. This 8-vector (P1, P2R, P5, P20, G1, G2, G3, G4) is related to the growth and development of the rice plant, as illustrated in Figure 11.4 (adapted from De Datta, 1981). In the present study we have used a base temperature of 8°C, together with the daily mean temperature, to calculate the daily heat supply provided by the atmosphere to the rice plant. The coefficient P1 represents the cumulative heat (degree days above 8°C) required by the plant to progress through the vegetative phenological stages and to arrive at the reproductive stage. Thus, the calendar time required to pass through this vegetative period is a function of ambient temperature. Similarly, the coefficient P5 represents the number of degree days required for the variety selected to pass through the grain filling and maturation stages. It should be noted that the carbohydrate increase produced by the model during the grain filling period is a function of radiation. Consequently, if the temperature during this stage is increased (thereby decreasing the number of calendar days required) while the radiation is not, then the result will be a decrease in model-estimated yield.

The two coefficients P20 and P2R are used together to represent the extent to which the variety is photosensitive. P20, which characterizes the variety selected, is defined as the number of hours of civil daylight per day (DL) at panicle initiation, and is used together with P2R to define the length of the flowering period. There are three ways in which the model can be used together with a nonlinear programming (NLP) technique to select an "optimal" 8-vector of genetic coefficients.

- If a crop with unknown varietal coefficients has been grown for several years in an area where one has a record of (straw and grain) yields, cultural practice, weather and soil profile characteristics, then one can use the NLP algorithm to select that set of genetic coefficients which will minimize, for example, the variance of model-estimated straw and grain yields about their observed values. This optimization would be subject to constraints imposed by plant physiologists familiar with the area.

- If one wishes to select from a menu of available varieties the variety that would optimize the straw and/or grain yield for a given soil, climate and cultural practice, then one need only use the model to enumerate the model estimates one variety at a time for the period of climatic record. You would then use the mean yield and its variance estimated by the model for each variety to study the relative risks before choosing the "best" variety.

Suppose one wishes to develop a variety-defining set of genetic coefficients that would optimize yield for a given (possibly new) climate scenario under realistic constraints imposed by a plant breeder. In this case, the desired (realistic) cultural practice could be specified for the chosen site (and soil), and the

FIGURE 11.4. The genetic coefficients related to phenology: P1 = degree days above 8°C required to progress through the vegetative phenological stages; P5 = degree days above 8°C required to progress through the grain filling and maturation stages; P20 = optimum photoperiod in hours; DL = civil daylight per day at panicle initiation; P2R = photosensitivity in degree days per hour; REP = not photosensitive degree days = P2R*(DL-P20); G1 = potential grain number; G2 = kernel weight (grams/kernel); G3 = tillering factor; G4 = 1 non-japonica for high temperatures, <1 japonica for low temperatures (adapted from De Datta 1981).

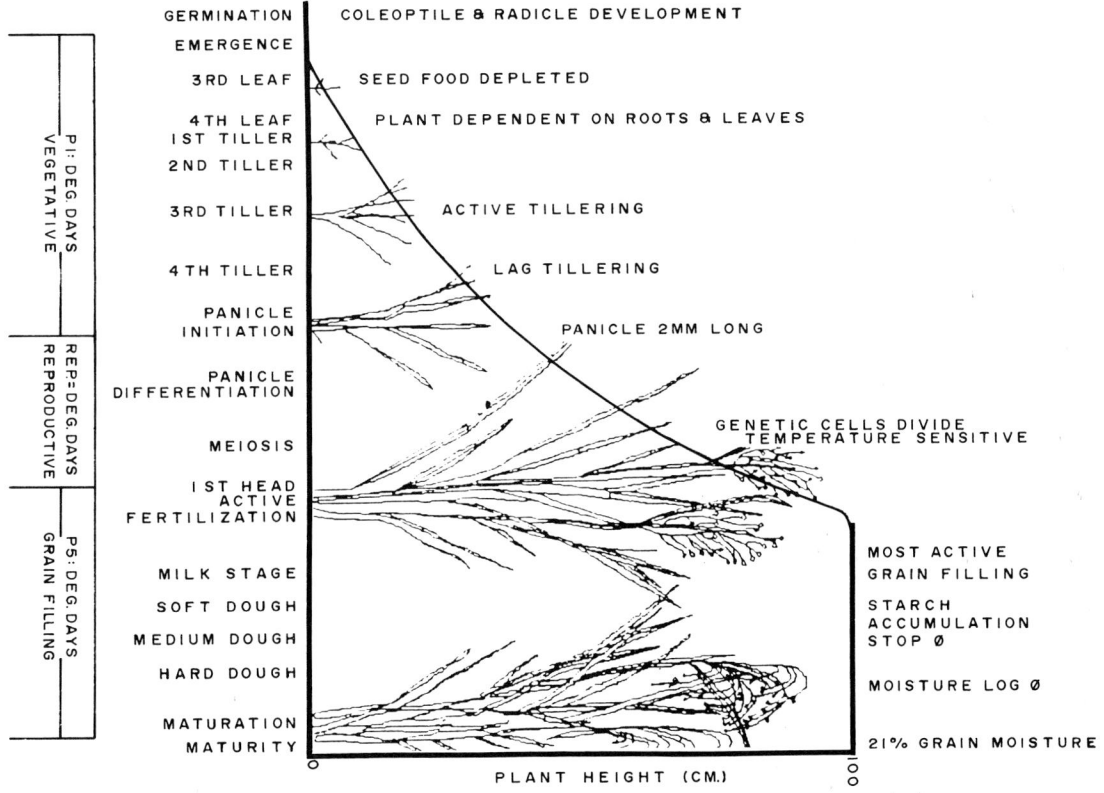

NLP algorithm could be run using the given climate scenario, subject to the imposed genetic coefficient constraints. Optimization criteria could be profit maximization or risk minimization depending, for example, on factors such as whether the rice is to be grown as a cash crop or a food crop.

It is this third use of the rice model yield optimization technique that we will now illustrate.

Optimizing the Genetic Coefficients

Pitsanulok, in northern Thailand, has been chosen as the site to illustrate the methodology used to test the value of changing a subset of the genetic coefficients in order to improve the yield characteristics of the rice grown in that area. We would like to select a variety of rice (or to breed a new one) that would increase the long term (say, 20-year) mean rice yield while at the same time decrease its variance. In this sense, we would like to discover what set of genetic coefficients would be best suited to the current climate, and also what set would be best suited to a future climate (in our case, the GISS $2xCO_2$).

Figure 11.5 will be used to illustrate some of the important features of our approach to achieving such "varietal characteristic optimization." Each point on the 5x5x5 three-dimensional array represents a mean value of rice yield associated with a 20-year run of the rice model. Thus, by comparing the output for the BASE case with that for the GISS case, the yields obtained for 2,500 years of rice "grown" under the BASE climate scenario can be compared with 2,500 years of rice "grown" under the GISS climate scenario.

The three axes represent deviations of the three genetic coefficients (P20, P1, P5) from a set of base coefficients shown in the box on the left side of Figure 11.5. These coefficients are used to describe the RD6 variety that is grown in the area. Entries shown in the arrays have been normalized as follows: at the center point (0,0,0), P20 = 12.0 hours, P1 = 1,106 deg. days, P5 = 382 deg. days. Each of the 125 20-year mean yields has been divided by the mean yield at the center point

(0,0,0) and then multiplied by 100.

Although the model input parameters are given in the figure, there are two additional farm-practice-related model constraints that vary with P20, P2, and P5. Since P20 governs the flowering date in this photosensitive variety (RD6), and since P1 specifies the length of the vegetative period (see Figure 11.4), the sowing date has been made compatible with these two variables at the beginning of each 20-year run. It is the 20-year mean value of P1 for the climate scenario input that is used together with the date implied by P20 to calculate the average sowing date. The fertilizer application schedule is also controlled by the sowing dates and the date of the end of the vegetative period. Consequently, this set of input parameters is also recalculated at the beginning of each 20-year run segment of the program. Each run of 2500 crop years, such as is illustrated in Figure 11.5, required about 15 hours of running time on a 10 MHz 80286/287 PC.

The following information can be drawn from the presentations of rice model output such as that shown in Figure 11.5, given the soil, cultural practice and climate assigned to this site in northern Thailand.

- Greater mean yields under both the GISS and the BASE climate scenarios could be expected from a variety that flowered earlier and had a longer grain filling period than that of RD6 (as represented by our basic set of P1, P20, P5 coefficients). This is because a larger value of P20 leads to an earlier sowing (and hence, transplanting) date, which then makes better use of the rainy season (see Figure 11.1).

- Changing the length of the vegetative period seems not to have had much impact on the 20-year mean yields; however, the yield variability does seem to increase as the time between transplanting and panicle initiation increases.

- The rate of increase in mean yield, as one increases P20 and P5, is greater in the case of the GISS climate scenario than it is in the case of the BASE climate scenario. One reason for this is the availability of enhanced radiation earlier in the growing season during grain filling (see Figure 11.3).

Clearly, Figure 11.5 represents simply

some of the attributes of a 3-space starting vector for an NLP (nonlinear programming) solution to an optimization algorithm. Such an algorithm also requires an objective function to evaluate convergence toward a solution. To provide for an incorporation into an objective function of the relative merits of mean yield increase and yield variability decrease, we have used (at present in a subjective manner) the coefficient of variation of model output yield estimates as a criterion for selecting the coefficients that would define the improved varieties to be examined in the following section. The more objective NLP optimization will comprise the next and much more extensive step in our research.

FIGURE 11.5. Model estimates of the impacts on rice production given specific changes in the genetic coefficients, as indicated. Sowing date and fertilizer application dates were determined from modified P1 and P20.

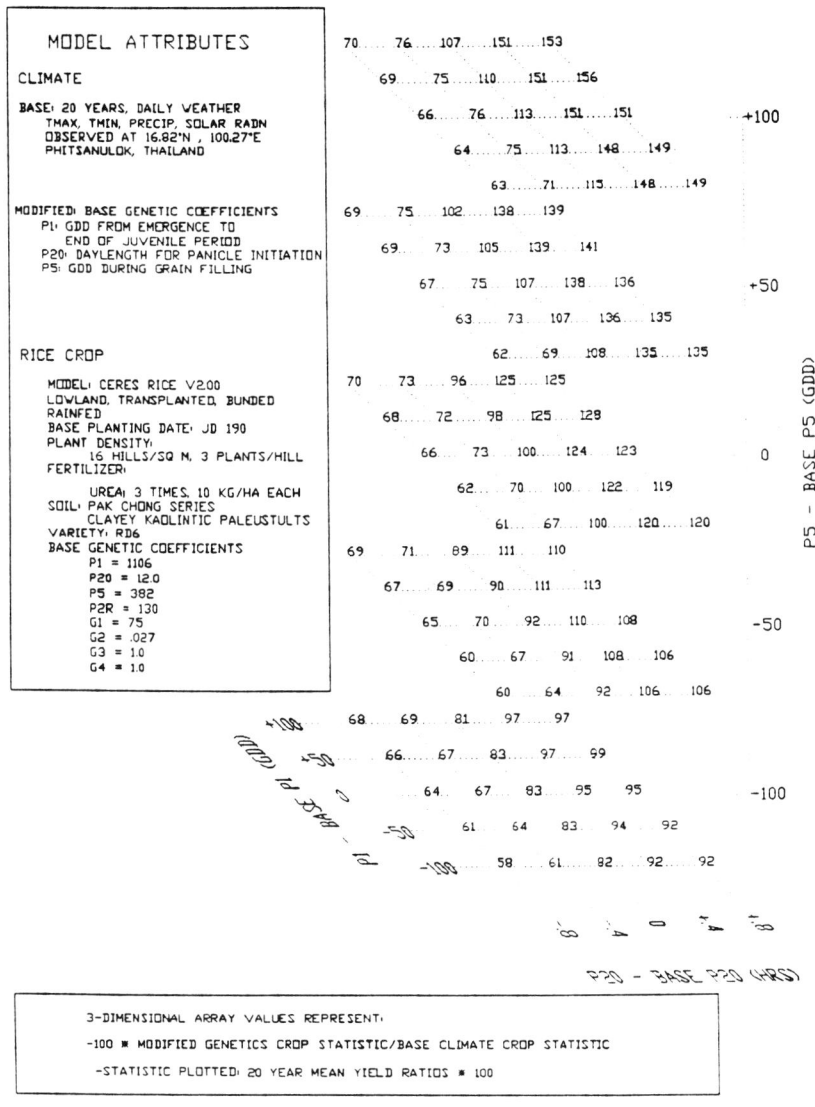

Illustrative Results

The methodology described in the previous section was used to obtain the model output yield data that went into the production of Figures 11.6 and 11.7.

In order to constrain the variability within and among stations (insofar as this was possible) to the consequences of climate change, we have kept the following features common to all runs.

- the soil profile,
- the transplanting option,
- the bunded paddy option, and
- most of the cultural practice options (the exceptions being sowing, and so, trans-

planting, and fertilization dates).

The model output yields and their standard deviations were calibrated using the 16-year time series (1973-1988) of province-level reported yields (production/harvested area) for the province in which each model site was located. Thus, in the case of the BASE climate scenario runs for Chiang Mai, the 20-year model run mean for the basic genetics (point 0,0,0) was scaled to be 3,301 kg/ha and its standard deviation to be 290 kg/ha (see Figure 11.6). The improved variety for this same site, again under the present (BASE) climate, was found to have a 20-year (scaled) mean yield of 5,182 kg/ha and a standard deviation of 267 kg/ha.

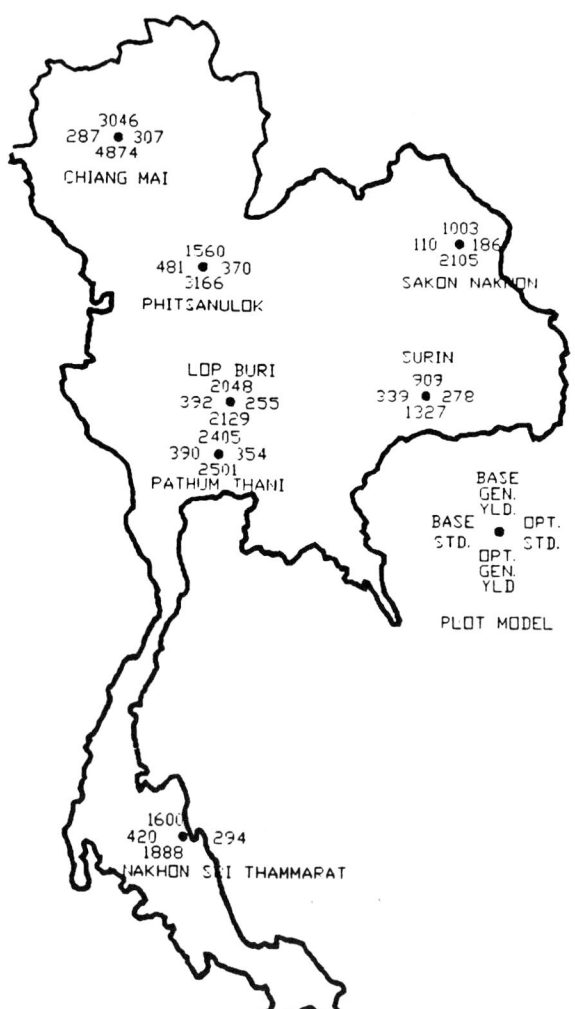

FIGURE 11.6. Base case climate data yields (kg/ha). Model output scaled via observed mean yields (1973-1988).

Figure 11.7 shows the same type of results for the GISS scenario runs.

Further, following the Chiang Mai results, if the same variety and cultural practice "currently" applied were to find continued application into a GISS climate regime, then the mean yield could be expected to drop to about 92.3 percent of its present value. However, if the "improved" variety of rice that we designed for the GISS climate (with its associated changes in sowing date and fertilizer application dates) were to be planted when a GISS scenario prevailed, then the mean yield would increase to about 148 percent of its present value.

Figures 11.6 and 11.7 show some interesting variations in the impact of our climate change scenario across Thailand. Continued use of the currently planted varieties and cultural practice into a GISS climate would find little change in the mean yields in the Central and Southern regions, whereas such continued use in the Northern and Northeastern regions would result in significant decreases in yields. Although it might be tempting to draw conclusions concerning the geographical variations in climate change impacts based on the above results, considerable caution must be exercised. The impacts shown can be as readily produced by differences in sowing date and in the genetic coefficients as they can in the changes between climate scenarios. This means that the actual climate change impacts must comprise a careful composite of all va-

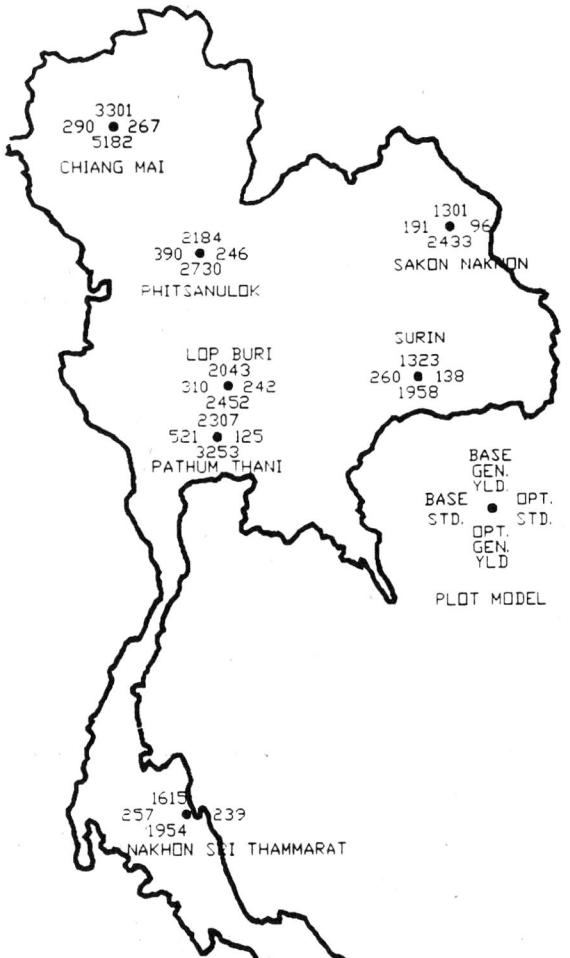

FIGURE 11.7. GISS climate data yield (kg/ha). Model output scaled via observed mean yields (1973-1988).

rieties of rice grown in an area and under the actual cultural practices employed, in order to approach an expression of potential reality. Our present study is a first step in this direction; but it does show that the job can be done.

Table 11.1 shows the varieties of rice currently used at the sites selected, as well as the ones chosen for our study.

Aggregating to the National Level

The sites discussed above will now be used to represent over 90 percent of the rice productivity in Thailand. This statement represents the result obtained from the following rather simplistic procedure.

- The sites selected were constrained to:
    - be representative of the cultural practice, soils, and varietal characteristics of the surrounding area,
    - be representative, from a climatological point of view, of daily weather data from a nearby station of the Thailand Meteorological Department and covering at least a 20-year time period, and,
    - be numerous enough to represent adequately all of the four regional subdivisions of the nation.
- Complete time series (1973-1988) for major rice yield (calculated from production and harvested area) were obtained (e.g., Chuprakob, 1989) for all 73 provinces and were grouped into the following four commonly designated regions: Northern, Northeastern, Central, and Southern. Simple correlations among all stations within each region

were calculated both from the raw 16-year time series, and from the time series with the linear time trends for each first removed. Those provinces whose "observed yield" time series showed a significant correlation with the yield time series of the nearest site were assigned to be represented by the model output for that site. Both sets of correlation coefficients were used in this determination, although in the few cases of disagreement the set obtained from the series containing the technological trend was the deciding factor.

- The seven selected subsets totalling 54 of the 73 provinces in the nation were found by the above method to represent 91.16 percent of the major rice grown in Thailand in the 1988/89 season.

It is clear that there are much more sophisticated statistical approaches to solving the site selection/aggregation problem than the simple pragmatic methodology used here. We have used some of these in other studies of a similar nature (e.g., Panturat, 1987), and we plan eventually to incorporate the appropriate methodology into the current research. Nonetheless, the method presented above is adequate for our current illustrative purpose.

Potential National Economic Impact of Climate Change on Major Rice Production

Next we convert our geographically distributed major rice production changes into economic terms, and then aggregate these to the national level. As our monetary reference point we take the 1988/89 farm value

Table 11.1. *Major Varieties of Rice Used at the Sites Chosen*

Site	Varieties planted (selected subset)	Variety chosen for present study
Chiang Mai	RD6, RD15, KDML105	RD6
Phitsanulok	RD6, KDML105	RD6
Sakon Nakhon	RD6, RD15	RD15
Surin	RD6, RD15	RD15
Lop Buri	RD15, KTH17	KTH17
Pathum Thani	LPT123, KTH17, KDML105	LPT123
Nakhon Sri Thammarat	RD13, NPY132	NPY 132

(national total) of major rice to be 73,000 million baht (Chuprakob, 1989). We consider the following three strategies with respect to rice production under the BASE and GISS climate scenarios:

(1) Plant the optimal variety in each area under current conditions.
(2) Plant the current varieties under both climates.
(3) Plant the current variety for the BASE climate, but plant the optimal variety under the GISS climate.

We use Figure 11.8 to answer the questions implied by all three strategies as far as major rice production is concerned, and to address strategies (1) and (3) as far as the economic impact potential is concerned.

From Figure 11.8 it is clear that, even under the present climate regime, there would be a considerable advantage associated with planting varieties whose genetic characteristics (as represented by P1, P20, P5) were more appropriate than those we have chosen, given that the optimal planting and fertilizer application practices were followed. The *economic gain* at the national level following strategy (1) would be about 41 percent (30,000 million baht/year), with most of this being realized in the Northern and Northeastern regions.

Under strategy (2), the change to a GISS climate scenario would result in a considerable loss in major rice production in the Northern and Northeastern regions. Over the remainder of the country not much change would result, based on the results from our rice model input configuration. The national level *economic loss* from following this strategy would be about 16 percent (11,500 million baht/year).

Under strategy (3), if we continued to use the same varieties and cultural practice under the present climate, but conducted a plant breeding (and/or selection) program to permit the planting of optimal varieties (with optimal farm cultural practice) under a future GISS climate, then instead of a loss of 11,500 million baht/year we could anticipate an *economic gain* of about 17,000 million baht/year (23.5 percent). In this case, once again, the largest gains could come from the Northern and Northeastern geographical areas of the country.

The above impacts constitute *direct effects* that could be entered into the appropriate sectors of a suitably disaggregated input-output economic model for the country, in order to obtain multiplier effects and final demand information with respect to the gross domestic product. The consequences of such investigations based on weather-related impacts have been reported, for example, by Grubb (1980) and by Cooter (1985).

Policy Implications

The first inference to be drawn from our results is that considerable national benefits would accrue from the completion of a more comprehensive study concerning the optimization of rice farming practice under the present climate. A policy that would promote more extensive rice breeding in response to climate impacts such as those presented, and would enhance farmer-related extension services to bring about optimal cultural practices, would be cost-beneficial.

The second inference to be drawn is that if no appropriate action is taken, then a change to a GISS-type climate could quite likely decrease the national major rice production significantly.

The third clear implication of our study is that any climate change impacts on rice production in Thailand brought about by a GISS-type scenario can be offset by a policy that would ensure careful advance planning in the areas of rice breeding and extension service farm practice information dissemination.

Thailand is a major rice exporting country. What would be the possible change in the demand for these exports brought about by a GISS-type climate change? We produced the results shown in Table 11.2 as a suggestive "first response" to this question. These figures indicate that an absence of the appropriate policy response in the countries shown would produce an increased demand for Thai rice, should it be available.

Conclusions

The results of our study seem to warrant the following inferences:

•   Decreases in rice production in Southeast

FIGURE 11.8. Major rice model estimates for areas represented by rainfed paddy. Includes 91.6% of national major rice area; area not represented shown as stippled.

A: Yield changes shown in boxes: (1) planting optimal variety under current conditions (41.3% increase in total value); (2) planting current variety under both climates (15.7% decrease in total value); (3) planting current variety for base climate and optimal variety for GISS climate (23.5% increase in total value).

B: Crop value changes shown on map in million baht/year. Above • following A1 strategy; below • following A3 strategy.

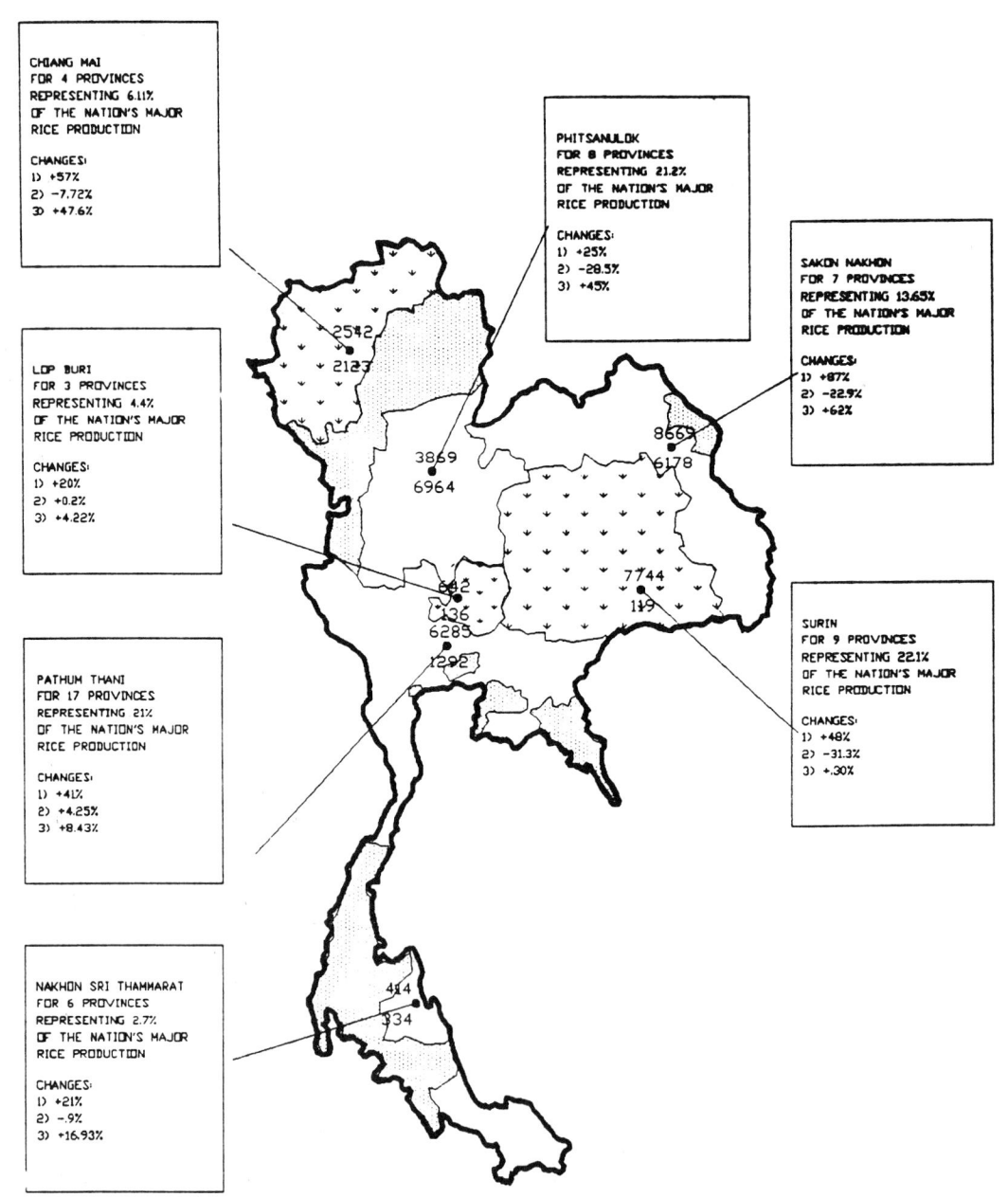

TABLE 11.2. *Potential Impacts of Climate Change on Rice Production (Under Strategy (2)*

Site	Percent Model-Estimated Change in Rice Production From Present
Nakhon Sri Thammarat, Thailand	-0.9 percent
Manila, Philippines	-0.8 percent
Saigon, Vietnam	-5.6 percent
Alor Setar, Malaysia	-7.5 percent
Jakarta, Indonesia	-4.7 percent

Asia, as implied by the GISS GCM $2 \times CO_2$ climate change scenario, can be more than overcome if appropriate and timely actions are taken by the agricultural community.

- A national policy that would enhance such activity in the rice breeding and agricultural extension services would be cost-beneficial.
- Optimization procedures, such as the one implied in the above work, require the incorporation of a policy-relevant value system. This means that if the results are to be of objective and quantitative value to the policymaker's decision making process, then the policymaker and colleagues (economists, resource managers, social scientists, etc.) must be involved in, and provide vital input to, the analysis activity as the project evolves.

References

Chen, R.J., and M.L. Parry (Eds.), 1987. *Climate Impacts and Public Policy*. An IIASA Publication sponsored by UNEP.

Chuprakob, N., 1989. *Agricultural Statistics of Thailand Crop Year 1988/89*. Published by the Center for Agricultural Statistics, OAE, Ministry of Agriculture and Cooperatives, Thailand.

Cooter, E.J., 1985. The assessment of climatalogical impacts on agricultural production and residential energy demand. *OCS Economic Impact of Climate*, **22**.

De Datta, S.K., 1981. *Principles and Practices of Rice Production*. John Wiley, New York.

Eddy, A., and R. J. Sladewski, 1988. *Some Biophysical Implications of Changes in the Water Delivery System of the Atmosphere*. Sixth IWRA World Congress on Water Resources, Ottawa, Canada.

Godwin, D., and U. Singh, 1989. *CERES-RICE V2.00*. International Fertilizer Development Center, Muscle Shoals, Alabama.

Grubb, H.W., 1980. Estimating and using quantitative models to plan and evaluate public sector programs in Texas. *OCS Economic Impact of Climate*, **3**.

Huke, R.E., 1982. *Rice Area by Type and Culture: South, Southeast and East Asia*. IRRI.

IBSNAT, 1990. The CERES RICE Model. *Agro Technology Transfer*, **10**.

Isaranurak, S., and C. Rata, 1988. Effects of climate on agriculture in Thailand. Paper presented at a UNEP meeting on the Societal Economic Impacts of El Nino Event, Bangkok, Thailand.

Janatwat, S., S. Vangnai, P. Duangpatra, *et al.*, 1985. *Soil, Water, Cropping Systems Research Data Bases Relevant to Rainfed Agriculture in Northeast Thailand*. Department of Soils, Kasetsart University, Bangkok, Thailand (USAID supported).

Panturat, S., 1987. *Optimal Sampling to Provide User-Specific Climate Information*. Oklahoma Climatological Survey, Economic Impact of Climate Series, Report No. 23.

12. ASSESSING REGIONAL DAMAGE COSTS FROM GLOBAL WARMING

Robert U. Ayres

Despite obvious difficulties, policymakers in the real world are increasingly forced to quantify and monetize. An example of the usual accounting methodology for estimating economic damages is provided by the recent study by William Nordhaus (Nordhaus 1989a, 1989b). Unfortunately, mainstream economists, including Nordhaus, display a tendency to attach zero valuation to unpriced environmental assets and services. Yet a good or service that is "unpriced" is not necessarily a "free good"—i.e., something that is "valueless" at the margin in the sense that the supply is greater than the demand at any finite price (as might be the case for mongrel kittens or puppies, or old newspapers and magazines, for instance). Nor does "unpriced" mean "priceless," as that term is conventionally used. Things are unpriced, in general, for technical reasons. The usual reason is that there exists no market for the good or service, or for a closely related proxy. The lack of a market may be primarily due to physical characteristics of the good/service—indivisibility[1] being the most common. Lack of separability may be attributed to the nonexistence of appropriate institutions, and therefore may be rectified, in some cases, by institutional innovations.[2] An example of

the latter might be the innovation of "tradeable emissions permits."[3]

This neglect of unpriced values by economists stands in sharp contrast to views expressed by some environmentalists, who argue (implicitly) that environmental values are "priceless" and must, therefore, be protected regardless of cost. However, policymakers in the real world must compromise between the preservation of values that have historically been unpriced (hence, generally undervalued) and mundane short-term considerations of budgetary constraints and competing priorities.

In assessing the likely impacts of global climate change, consideration must be given to many climatic factors, such as the temporal and spatial distribution of temperature, precipitation, evapo-transpiration, clouds and air currents. All of these factors are simulated in so-called global circulation models (GCMs), although the detailed results of the simulations are not as yet regarded as a trustworthy basis for long-range forecasting at the regional level (e.g., Washington and Parkinson, 1986). All the available evidence makes it clear that the impacts of climate warming, assuming warming will occur, will not be uniform among the regions of the earth. There could be some big winners, as well as some big losers. Nor is the ability to respond, either by preventive action or adaptation, equal among affected regions. Thus, regional differentiation is an essential feature of

---

[1] Indivisibility can best be defined in terms of its opposite. A normal good or service is divisible, in the sense that a discrete quantity can be separated physically from the rest for purposes of ownership and exchange. Neither ownership nor exchange is possible without divisibility. A market cannot exist without exchangeability.

[2] Land is a case in point. Markets for land depend on the existence of markets, which depend on the existence of legal titles, which, in turn, depend on the existence of means of recording ownership and boundaries. The latter, in turn, depends on the ability to mark boundaries in a permanent manner, or (better) to redetermine boundaries as needed by surveying.

[3] Tradeability, in this case, depends on confidence on the part of the buyers and sellers that there exists a reliable and trustworthy means of monitoring the use of permits. Thus, if the permit is for the emission of pollutant X, the market depends implicity on confidence that actual emissions of X by each emitter can, and will, be monitored precisely by an independent, incorruptible entity. This, in turn, depends very much on the state of technology.

any analysis, and effective policy responses must reflect this fact. Some, at least, will necessarily be regionally specific.

Of course, the real questions at issue are: (1) What are the consequences for human welfare and how serious are they? (2) What are the most effective policy options to respond to the threat? The second question clearly depends on the answers to the first (although there are some ameliorative policies, such as energy conservation, that would make sense for other reasons).

Assessment of the welfare implications of climate warming are complicated by the fact that most of the important consequences for humans are likely to be indirect (i.e., through human interactions with other aspects of the global environment) rather than direct. These points are considered in some detail later. The chapter then discusses the methodological problems of quantification and monetization of unpriced goods and services in some depth. It reviews Nordhaus's assessment of the economic impact of global warming and offers a critique.[4] It suggests that the major fault of the Nordhaus methodology is the omission (i.e., implicit zero valuation) of unpriced environmental assets and services. Indeed, the omitted items in this case may be of much greater magnitude than the included items. It proposes a practical research plan for developing a coherent, defensible valuation methodology for assessing the regional costs of climate warming in monetary terms. The approach should be generalizable to other cases.

The Framework of Public Debate

Public debate on the climate warming issue is clouded by a profound interdisciplinary and ethical confusion. In attempting to characterize it briefly I risk oversimplification. Nevertheless, it appears that there are two major positions. On one side are those who argue, sometimes with great passion, that the present generation has no right to use up or destroy the environment, which is the common heritage of all humanity, including generations unborn. This view is strongly held

among some intellectual groups, mostly from the rich countries, though it is certainly shared by some traditional cultures.

On the other side are those who argue that (or who act as though) the welfare of future generations may be of some concern to us, but that it is not paramount and should not take precedence over the needs of those alive now. For some religious groups, existence in this world is merely a preparation for the next, and of no great importance in itself. A surprisingly large number, including some Mormons, Jews, Christians and Muslims, expect a "final judgment" or a "second coming" of some sort within the next decade or two. Why then worry about climate warming? For others, such as traditional Hindus and Buddhists, the conditions of worldly existence are beyond man's deliberate control.

Among "modern" Westerners who are oriented to life on this earth, rather than to a future life in heaven, the future is generally discounted on the grounds that our descendants are likely to be richer, or at least in possession of better technology, than we, and thus better able to solve their own problems as well as the problems they will have inherited from us. We may leave them our toxic waste dumps and refuse heaps, but we also endow them with our institutions, our accumulated knowledge and technology, not to mention invested capital. Reasonable people can and do make this argument, even though others challenge the underlying assumptions. Having accepted this framework, the next question that arises is how to choose the "right" discount rate. To answer this question, in effect, means to resolve the question of how to allocate "votes" to future generations, in comparison to those now alive. In resolving this issue, in turn, much depends on whether or not those who come later will have the possibility, in principle, at least, of replacing or repairing any environmental assets that happen to have been destroyed by today's inhabitants (see Page, 1977; Pearce 1990). I return to this issue later.

There is a more fundamental debate, although it is seldom explicitly recognized as such. Public policy in the United States has been based more often on concepts of legal "rights" (or entitlements) than on maximizing benefit/cost ratios. The recognition of such rights is, essentially, a political process, although it has been strongly influenced in

---

[4]Other critiques have been offered, e.g., Hall, 1991.

the past by philosophical and religious elements. In fact, the United States was founded on the basis of recognizing and articulating certain fundamental human rights, such as freedom of assembly, freedom of religion, freedom of speech, and so on. (Many of these notions were derived from the philosophical ferment of the Enlightenment.) In fact, the Bill of Rights eventually became a part of our Constitution. Nobody who follows current events could be unaware of the controversy surrounding the issue of abortion: "right to life" versus the "right to choose" or the "right of privacy." Another growing controversy surrounds the "right to die." Many environmentalists argue that the earth does not belong to humans alone. Other species, even the earth itself, enjoy a fundamental "right to exist," upon which we humans have no moral right to infringe. In fact, in a recent survey taken in Great Britain, respondents strongly agreed with this position (4.4 on a scale of 5).[5] (This right-to-exist is normally taken to apply to species, not to individuals, although animal rights groups go much further; some groups claim rights for dogs and cats that have scarcely yet been secured by the majority of humans living on the planet).

But where policies have been based on absolute rights, and the number of recognized rights keeps growing, there is a possibility, indeed a likelihood, that incompatible "rights" will collide. There is a growing number of examples of political outcomes that are absurd by almost any standard. To mention only one instance, the argument that disabled people have an inherent and absolute "right" to equal access to public transport systems led, by tortuous bureaucratic thinking, to the building of enormously expensive buses with special hydraulic lifts for wheelchairs. (Why not for the bedridden too?) Critics have noted that it would be demonstrably cheaper to provide a free car and driver on call for each handicapped person who actually uses these facilities.

In summary, notwithstanding the strong feelings of some environmentalists, there is no social consensus at present even in the Western nations, still less in the rest of the world, that future generations have any absolute rights that must be protected at the expense of those now alive. In reality, the welfare of the present generation *will* be weighed against the welfare of the next (with those now alive having the only votes). To defend the claims of unborn generations (against those of the present) such claims must be made as specific and quantitative as possible.[6]

Given that public funds are scarce, diversion into unnecessarily expensive protections for the rights of one group will inevitably compromise the ability of the society to help others who may also have rights, and may actually need help more. Entitlements, once awarded, are extremely difficult to take away. Facing the reality of finite uncommitted resources, I agree with the economists who insist (often against strong moralistic opposition) that environmental values cannot be treated as "priceless," either now or in the

---

[5]Data from Green and Tunstall, 1990, cited by Brown, 1990, footnote 6.

[6]A technical issue of importance in comparing future damages or costs with present ones is the choice of "discount rate." It is well known that the use of a discount rate to compare present and future costs or benefits is widely accepted and justified by economists. It is also well known that if the discount rate is high enough, damages (or benefits) expected far in the future are given so little weight in the present that they might as well be neglected. Yet the argument for basing the discount rate on prevailing bank interest rates, GNP growth rates, or the like, is much more appropriate for assessing alternative investments in infrastructure than damages to nature. According to a recent textbook, "A dollar now is worth more than a dollar in the future because it can, in general, be used productively between now and then" (de Neufville, 1990, p. 205)

The problem with this lies in the implicit assumption that a dollar invested "productively" today can be used in the future to repair any damages done to the environment between now and then. In other words, it is assumed that no environmental damage is irreversible, and that future dollars can buy any good or service that exists today. In fact, this assumption is clearly untrue in some cases. Climate warming, species disappearance and the death of 1,000-year old trees are clearly not reversible.

Because of irreversibilities, the interest rate or the GNP growth rate may be quite irrelevant to the choice of a discount rate applicable to evaluating massive long-term losses of an irreplaceable environmental asset.

future. This is one of the themes of this chapter.

I strongly disagree, however, with an unstated assumption that lies behind several of the quantitative policy analyses on the global warming issue (as well as other issues). This unstated assumption is that, if there is no "market price" or generally accepted surrogate for an environmental service, then that service has no "real" monetary value. Few economists would try to defend such a proposition, baldly stated. Yet it is implicitly accepted very widely in practice. The second main purpose of this chapter is to challenge that assumption and suggest some practical means of quantifying environmental losses and policy benefits.

The Problem of Valuation

Assuming quantification, and making due allowances for some overlapping categories (e.g., recreational and environmental benefits of wilderness areas), in principle the above costs can be added together. The unresolved question is how to do it without simply guessing. To place this discussion in context, it is helpful to begin by summarizing and critiquing a well-known recent study by William Nordhaus, a study that has attempted to estimate the economic costs of climate change (Nordhaus 1989a, 1989b).

Nordhaus's Estimates of Damage Costs[7]

The first step in Nordhaus's analysis is a breakdown of the U.S. gross national income or GNI (for 1981) by sector. Next, Nordhaus classified each sector according to its likely sensitivity to climate changes.[8] The most climate sensitive sectors were assumed to be agriculture, forestry and fisheries, which amounted to 3.1 percent of total GNI in 1981. Moderate sensitivity was attributed to sectors such as construction, water transport, utilities, etc. These contributed 10.1 percent of the total. The rest (about 86 percent)

comes from sectors affected negligibly by climate (e.g., mining, finance, manufacturing, etc.). The relevant facts and Nordhaus' conclusions are as follows:

1. *Agricultural Sector*: The expected change in crop yields is a function of location. For instance, field crop yields in the U.S. cornbelt may rise by some 5 percent because of CO_2 fertilization, but may drop by 3 percent to 13 percent, on the other hand, because of a warmer and drier climate (Easterling *et al.,* 1990). Some parts of the globe, including Japan, Finland and Iceland, will enjoy better growing conditions; some parts will be degraded by reduced rainfall. The geographical distribution of crops will change. Solomon has estimated a possible shift of the U.S. cornbelt by several hundred kilometers on a northeast axis (Schneider and Rosenberg, 1990). Costs resulting from such shifts could include:

- Losses due to reduced predictability of weather.
- Costs of additional irrigation infrastructure.
- Environmental degradation—e.g., due to increased pesticide use (needed because a warmer climate with less winter frost will encourage the spread of insect pests and diseases they carry). Agriculture damage costs (offset by the CO_2 fertilization effect) are estimated by Nordhaus as plus or minus $10 billion as an overall impact on all crops.

1a. *Forestry Sector*: Despite management practices, forests are ultimately dominated by climate. The genetic makeup of a forest is normally stable over long periods, and determine(s) its resistance to stress (Sandenburgh *et al.,* 1987). However, trees planted today might grow to maturity in quite different conditions. The earliest and greatest impacts of climate change on forests are expected in the high latitude boreal forest in the Nordic countries. Economic effects might include:

- Loss of timber due to early die-off of overmature trees (Batie and Shugard,

---

[7]This section has been adapted from an earlier publication (Ayers and Walter, 1991).

[8]Gross national income (GNI) differs from GNP by excluding indirect business taxes and capital set aside to replace depreciation.

1990) and increased wildfire frequency and destructiveness.

- Higher production costs due to, for instance, more frequent replanting. On the other hand, CO_2 fertilization might increase growth rates, as noted above.

2. *Sea-Level Rise*: Two kinds of costs can be expected: protection and repair, on the one hand, and irreparable damage (including amenity loss) on the other.

The first category includes:
- the cost of dikes;
- the cost of relocating structures (and people) to inland areas;
- increased water supply costs, due to rising salinity levels in bays, inland waterways, and aquifers (Frederick and Gleick, 1990).

In the second category are:
- social costs associated with relocation of vulnerable rural populations;
- loss of unpriced environmental values of coastal lagoons and wetland ecosystems;
- both financial and opportunity loss of ocean-front recreational amenities and beach tourism.

Impacts of sea level rise (SLR) to coastal regions are potentially massive. Coastlines will move inland up to several hundred meters, in many places, depending on beach slope and characteristics of the beach material (Hekstra, 1990). Salt water will also move upstream via rivers into lowland, freshwater pockets behind coastal dunes, and into groundwater aquifers. The effect will be magnified in some areas where intensive groundwater withdrawal has occurred (e.g., Long Island, New York).

SLR will cause great loss of biologically diverse coastal lowlands and wetland ecosystems (Wilson, 1989). For instance, Indonesia possesses 15 percent of the world's coastline and it is the world's richest nation in terms of the quantity and diversity of wetland ecosystems. Yet as much as 40 percent of Indonesia's productive land surface could be adversely affected by a SLR of only 1

m (Hekstra 1990, p. 58). Worldwide, the land area that would be subject to inundation or made vulnerable by salt water intrusion is about 500 million ha. This is only about 3 percent of all land area, but it constitutes over 30 percent of the most productive cropland area (Hekstra, p. 60). As many as a billion people now live in the most vulnerable areas, including some very large cities. Thus, as has been pointed out, as much as one-fifth of world market valued assets could be adversely affected (Crosson, 1990). Sea level rise damages were estimated by Nordhaus for land loss (15,540 km^2) and protection of high value property and open coasts by levees and dikes. The total market value of the property at risk in the United States is on the order of $100 billion. Nordhaus converted this to an estimated annual equivalent loss of $6.18 billion. (The capital value of property should reflect its continuing flow of benefits, thus reflecting tourism losses implicitly, at least as far as providers such as hotel and motel operators are concerned. What is omitted is the loss of use and option value to users, who may not be able to find equivalent amenities elsewhere.)

3. Other: In addition to the above items, greenhouse warming is expected to increase aggregate demand for airconditioning ($1.65 billion/year) and reduce the demand for space heating ($1.16 billion/year). Assuming average current prices for electricity and fuel, there would be a net annual extra cost to the U.S. economy of $0.46 billion (U.S. EPA, 1988). No quantitative estimates were made by Nordhaus for other goods or services (either market-valued or otherwise). In effect, these were lumped together and included in the uncertainty of the overall estimate (see below). Summarizing the quantified cost items above, the attributable factors can be broken down as follows:

- Agricultural losses: 0 percent
- Changes in energy demand: 8 percent
- Sea-level rise: 92 percent
  (85 percent is for coastal protection costs that involve levees, seawalls, etc., and 7 percent is for loss of land.)

The "bottom line" of Nordhaus's analysis is that the central (most likely) estimate of total annual economic damages for the United States is $6.67 billion (1981 dollars), assuming the damages occurred in 1981. This is equal to 0.28 percent of U.S. gross national income for that year. The error bounds were judged (by Nordhaus) to be quite a bit higher owing to the omitted unquantified items, but still less than 2 percent of national income.

Assuming reasonable economic growth rates for the United States and the rest of the world, gross world income (GWI) in the year 2050 is likely to be more than $26 trillion 1981 dollars (U.S. EPA, 1989; low GNP-case). This is 8.1 times more than U.S. national income in 1981. Nordhaus judged this income scaling factor of 8.1 to be appropriate to extrapolate his estimate of climate-related damages for the U.S. economy (as it was structured in 1981) to estimate global annual damages in 2050. By this method, he estimates that annual world damages due to greenhouse warming are "most likely" to be about $54 billion ($1981), by the middle of the next century, with an upper limit of $520 billion.

Based on expected emissions of 16.9 billion tons of CO_2 equivalents in the year 2050, Nordhaus converted this to marginal shadow damages of emission, namely:

central case: $3.3/ton ($CO_2$-equivalent)

worst case: $36.9/ton ($CO_2$-equivalent)

Critique of Nordhaus Analysis:

There are two possible types of criticism of the Nordhaus assessment: empirical and conceptual. As regards the former, one can quarrel with almost any numerical estimate. This sort of criticism is part of the process of converging on a "best guess." However, the numbers *per se* are not of primary importance for purposes of this chapter.

Conceptual criticisms are another matter. There are several that can be made. One is the use of (gross national) income as a proxy for (national) welfare. While income (and employment) is obviously of concern to policymakers, it is not an appropriate measure of loss or gain from environmental changes. The correct economic measure of welfare change is the change in the sum of producer and consumer surplus. Producer surplus is the difference in the amount of money received by producers for their output and the minimum cost of production (i.e., maximum profit). Consumer surplus is the difference between the maximum amount consumers are willing to pay (WTP) and the actual cost of what they buy.

The difficulty, of course, is that there is no objective or comprehensive source of data on consumer WTP, even for marketed products. Estimates are obtainable only by means of costly (and incomplete) surveys. However, WTP survey methodologies are now fairly sophisticated. Moreover, they are also usable, up to a point, for nonmarket valuation (Cummings *et al.*, 1986).

Another problem is that the structure of the U.S. economy is not typical of the rest of the world. Hence, simple extrapolation is dangerous. Agriculture, forestry and fisheries constitute a much larger percentage of the GNI in virtually all other countries, and they still predominate in the poorest countries. Extrapolating from the U.S. case (where agriculture is only 3 percent of GNP) to the world as a whole therefore grossly underestimates the potential impact of agricultural losses, both on the average, and especially in the poorest countries.[9] Similarly, the pattern of energy use in the United States bears little relation to the rest of the world, although the nature of the bias in this case is unclear. However, it is relatively straightforward to compensate for these biases once they are recognized.

A further conceptual problem is that Nordhaus's calculation of marginal shadow damage coefficients (above) is based on an erroneous assumption with regard to physical damage: It assumes that future damage is simply proportional to greenhouse gas emission rates on a current basis. In actual fact, greenhouse effects—hence damages—are actually proportional to the total accumulation (i.e., stock) of greenhouse gases in the atmo-

[9]The seriousness of the bias in this case is amplified by the fact that the agricultural component of GNI is underestimated in most of the less-developed countries due to the omission of noncash labor services and consumption of home-grown food from private plots (not to mention unreported income from black-market transactions in communist countries).

sphere. (If Nordhaus's model were correct, damages would cease immediately if CO_2-equivalent emissions were miraculously stopped, whereas in reality they would continue for a very long time at the same level.) However, despite the conceptual error, the implied relationship between damages and emission rates may still be defensible as a crude approximation to be used with due caution.

The final and most difficult conceptual problem arises with regard to the omitted damage categories. As noted above, Nordhaus's method starting from national account statistics gives no weight to items not included in those statistics. Climate warming will degrade many existing recreational and touristic facilities without necessarily providing any substitutes. (Of course some coastal areas that were hitherto too cold for swimming may become more tolerable. However if storms become more frequent, SLR will destroy more beaches than it creates.) Major examples will include:

* Reduced opportunities for ski tourism and other winter sports can be expected in the Alps and some other temperate-zone mountain chains owing to reduced snowfall and accelerated melting of glaciers. (In fact, European glaciologists estimate that the volume of ice in Alpine glaciers has declined 60 percent in the past century.)
* Damage to beaches and beach-front property (due to SLR) will result in losses to those whose land is inundated. This problem will be particularly severe for the coastal barrier islands (basically sand dunes) along the Atlantic coast of North America.
* Reduced opportunities can be expected for forest and wilderness based recreation such as fishing, hunting, camping and so on, as a result of accelerated forest dieback.

Conservation systems such as parks, refuges, and preserves that are dedicated to protecting wildlife populations may fail under different climatic conditions. As one observer noted:

> By the mid-21st century, the climate that nurtures Yellowstone National Park could

well be into Canada. The tundra of the Arctic National Wildlife Refuge could be pushed into the sea. (Kerr, 1988, p. 23)

In particular, Nordhaus neglects recreational, ecological and aesthetic assets (except where they are market-valued, as in the case of beach-front property). He makes no attempt to value the national parks and noncommercial forests. He does not even consider other climate-sensitive environmental assets, such as the hydrologic, carbon, nitrogen, sulfur and phosphorus cycles (Turner *et al.*, 1990). Nor does he recognize the value of biological diversity (Wilson, 1989). To make sound public policy it is essential to come to grips with this class of problem. However, before discussing methodology, it is important to say a little more about national account statistics.

Environmental Accounting

The System of National Accounts (SNA) was created for purposes of assessing the economic performance of countries at a time when both natural and environmental resources were still plentiful. In circumstances where such resources are becoming scarce, the SNA is deficient in two major respects. First, SNA does not now include natural resources either mineral or biological as assets, although it does include products of human endeavor such as buildings and equipment. Second, whereas manufactured capital goods are depreciated and their costs of replacement are properly excluded from capital investment (e.g., in arriving at estimates of total invested capital), this is not the case with resource assets or environmental capital. Since resource or environmental assets are not included in the first place, there is no allowance for either depreciation or replacement.

Thus, when a developing country cuts its forests to sell the wood, the sales are recorded as income, but there is no balancing item showing that an asset was depleted (Repetto and Magrath, 1989). Tree-planting activity (if any) would most likely be regarded as a form of "final consumption" by the public sector. (Private forest products firms might do tree planting if encouraged to do so by the tax laws, but the accounting rules governing private firms bear no particular relation to national accounts.)

There is no good rationale for this rather obvious flaw in the SNA. (Even the defenders of the "status quo" in national accounting argue, in effect, that while the problem is real, it is only one of many such flaws and that somehow all of the various inadequacies of the SNA tend to compensate for one another.) It can be argued, perhaps, that the data needed for accurate resource accounts are absent or incomplete. It is perfectly true that estimates of total recoverable reserves of most mineral resources are inaccurate, widely varying, and mostly understated. (In the case of trees or topsoil the total stock can be known with better precision.) But even if the absolute amount of the total stock of oil or gas is not known at all, the annual amount of physical depletion incremental change can be determined relatively accurately. Its monetary value, too, can be assessed to first order by the application of a straightforward price rule. (This is ordinarily done anyhow in government statistics on minerals, metals and fossil fuels.) The details of the price formula matter less than its consistency from year to year.

But the key point is that the output of mines, quarries and oil/gas wells should not be treated by the SNA as a contribution to national income without also being treated as national asset depreciation. Comparisons of GNP or GNI showing increases over time are usually interpreted as evidence of "economic growth." But this interpretation is highly misleading if the country is merely selling off its resource assets without replacing them (as many Third World countries are now doing). For this reason, many environmental economists are now arguing for changes in the SNA and in the methods of calculating GNP to reflect these factors.[10]

Environmental resources such as the nutrient (carbon, nitrogen, sulfur, phosphorus) cycles, the protective stratospheric ozone layer, climatic stability, biological diversity or waste assimilative capacity are also absent from the SNA. Perhaps the *a priori* reasons are better in this case. In the first place, the total stock or the year-to-year changes of the stock of the "resource" is inherently hard to

determine. (In some cases it is even hard to define.) But even if this problem were solved, another difficulty would remain: There is no market price to use in estimating asset depletion. (What is the value of the ozone layer?) Actually, the second difficulty is the most serious. As pointed out above, the absence of exact information on total stocks is not necessarily a critical problem in doing natural resource accounts. What really matters is the incremental change from year to year. The same thing would be true for traditionally unpriced environmental resources. That is to say, it is only the incremental change in the stock that really matters for most purposes.

Unfortunately, it is possible for a country to exhibit a rising GNP that is achieved essentially by selling off or over-exploiting environmental assets (i.e., at the cost of deteriorating environmental resource stocks). As before, failure to account for the environmental depreciation overstates the apparent growth. Indeed, given the importance of primary products as a component of total hard currency exports for many developing countries,[11] it is by no means clear that many countries are actually growing economically at all, since they don't properly measure the depletion of environmental resources. Misleading indicators (such as GNP) have an impact on policy. National governments tend to measure their own performance in terms of GNP growth rates. They tend to select policies (such as extraction and consumption of natural resources) that will have an immediate and positive effect on the measure and reject or ignore policies (such as energy conservation) that will not. Fixing the measure of economic output won't by itself fix the policies, but it would unquestionably help.

Repetto gives an interesting illustration of this point (Repetto, 1990). He notes that electric utilities have severely criticized environmental regulation on the grounds that the costs of producing electricity have been

---

[10]See, for instance, Ahmad *et al.* 1990; Bartelmus *et al.* 1989; Hueting 1980; Peskin 1989; Peskin and Lutz 1990; Stahmer 1990.

[11]Primary products (both renewable and nonrenewable) as a percentage of total exports in 1988 were 100% for Uganda, 99% for Ethiopia, 98% for Nigeria, 93% for Zaire, 93% for Somalia, 97% for Ghana and Bolivia, 89% for Burma, 83% for Kenya, 81% for Tanzania, 75% for Columbia and 71% for Indonesia (World Bank, 1990, Table 16).

increased without increasing the value of the product (electric power). In effect (it is argued) the productivity growth of the electric power sector has been adversely affected by environmental regulation. However, this conclusion is based on the conventional definition of productivity, which is based on the conventional measures. Taking into account not only the electricity produced but also the pollutants emitted, and using EPA's "best" estimates of the marginal damage costs of those pollutants (SO_2, NO_x, and particulates) Repetto shows that the "conventional" productivity growth rate of 0.5 percent per annum (1971-1979) and 0.4 percent (1980-1987) for the sector would be revised upward to 1.2 percent in the first period and 0.8 percent in the second period.[12] If utility executives and regulators took the quantifiable environmental benefits into account in their calculations, they would be much less reluctant to invest in emission controls.

It is not my present intention to concentrate on the problem of fixing the flaws in the SNA. This is an important topic, but somewhat aside from the theme of this chapter. However, it has one thing in common with the narrower problem of global warming being addressed here: the need to quantify expected environmental resource losses, especially in monetary terms.

Monetization Methodologies

As noted in the opening of this chapter, the existence of a market price depends on various factors having no relation to intrinsic value. Exchangeability depends, in general, on separability. Separability, as applied to many environmental goods or services that are essentially "flows, rather than "stocks," depends also on the existence of certain legal institutions (such as recognized title), and on a technical ability to monitor the service flows and ascertain their sources as well as their "sinks." As it happens, decision-makers can sometimes avoid the problem of monetization. There are two cases, especially, where this is feasible: (1) where cost is truly not an issue, as perhaps in the case of national defense, and (2) where "no regret" policies can be identified.[13] However, that is not possible when the decision to be made is essentially a constrained resource allocation (budgeting) problem: How much money should the nation and the world be prepared to pay to avoid or repair damages from greenhouse climate warming, bearing in mind the simultaneous need to deal with other problems? There is no generally accepted theory for computing exact values of nonmarket resources directly from objective economic data (e.g., shadow prices of traded goods or services). Nevertheless, there are two potentially useful tools for filling in the gaps (e.g, Sinden and Worrell, 1979).

One interesting approach is the hedonic price (HP) method, which seeks to infer "shadow" values for unpriced environmental services from observed price data on other (exchangeable) commodities or services of a more aggregated nature, such as land. For instance, land (or house) prices in various locations reflect a wide variety of measurable attributes, including potentially measurable environmental ones such as sootfall, odors or noise. By statistically regressing land prices against all of the measurable attributes, the "explanatory power" of each of them can be determined, in principle. The result is a model in which the price of land is given as a function of all of the identified variables, including the environmental ones. One of the first applications of this method was an estimation of the value loss due to air pollution in St. Louis neighborhoods (Ridker and Henning, 1967; Ridker, 1967). Ridker and Henning estimated that the presence of SO_2 and H_2S in the air reduced the average value of a house by $245, other factors remaining equal. Other valuation studies of this kind include those by Jaksch and Stoevener (1970), Anderson and Crocker (1971, 1972) and Freeman (1974). Most of these studies obtained significant results. However, Freeman reported two studies in which no significant relationship was found.

Unfortunately, the number of attributes potentially contributing to the price of a particular parcel of land is very large. Income

---

[12]This estimate assumes damage estimates increase in proportion to GNP.

[13]A "no regret" policy is one that has positive net benefits for all scenarios. For a detailed discussion of the conditions under which this is possible, see Sinden and Worrell, 1979.

differentials, for instance, may have more effect on local land prices than pollution. Moreover, while many attributes are potentially measurable, actual measurements of variables such as visibility, SO_2, NO_x, H_2S, particulates or ozone concentration are rarely available as a function of location except on a regional rather than local scale. Further, land sales transactions in any given local area and time period are very limited in number and are therefore not particularly representative. On the other hand, regional averages are not very helpful for putting shadow prices on such "disservices" as SO_2, NO_x, etc., because of the wide variability of both pollutant concentrations and land prices within each region. Despite these generic difficulties, there are instances where the HP method may be applicable. The problem of evaluating the impacts of climatic (and life-zone change) effects on a regional basis may be such an instance, as will be seen.

The other important economist's tool is the contingent valuation (CV) method (Brookshire *et al.*, 1976; Cummings *et al.*, 1986). Superficially, the method is based on surveying a population's willingness to pay (WTP) for a specified environmental improvement or to avoid some specified degradation. At a more fundamental level, the CV method can be regarded as a specialized application of the theory of multi-attribute utility theory (MAUT). In this regard, it shares several underlying assumptions with the HP method sketched above. The CV method has already been mentioned in connection with the possibility of estimating consumer surplus. It has been applied, in practice, by the EPA to estimate the economic value of eliminating low level ozone-related health effects, such as headaches and respiratory problems.

The CV method has also been applied by Chestnut and Rowe in connection with the National Acid Precipitation Assessment Program (NAPAP, 1990). The method was used in an attempt to measure the monetary value of changes in visibility due to atmospheric haze caused by sulfate particles. In brief, Chestnut and Rowe used a "consensus function" of the following form:

$$HWTP = B \times \ln(VRB/VRA)$$

where HWTP is average "household willingness to pay" to increase visibility, B is a co-

efficient to be determined by survey, VRA is the actual visual range, and VRB is the background visual range in the absence of sulfates. VRA is estimated to be 25 km in the presence of sulfate particles, and VRB is estimated to be 76 km in their absence (NAPAP, 1990). Based on surveys in several parts of the country, B values were derived by Chestnut and Rowe, ranging from $78 to $1,061. The average for eastern studies was $270, and the average for all parts of the country was $270.

Assuming 50 million U.S. households, the Chestnut-Rowe model suggests aggregate annual losses due to haze from sulfate particles alone ranging from a low of $3.9 billion to a high of over $50 billion. Using the national average value of B would yield an estimated loss of around $10 billion per annum. Critics suggest these figures may be too high because they may (improperly) include health effects with pure visibility effects. However, the order of magnitude is probably about right and, as remarked earlier, even a rough estimate is better than none.

Despite its adoption by EPA and NAPAP, the CV method is still far from being a standard tool. In practice, it often tends to yield results that appear inconsistent and are hard to interpret at best.[14] Its underlying

---

[14]For instance, in the EPA-sponsored survey of low-level ozone effects, people who had experienced headaches estimated that they would be willing to pay $180, on average, to eliminate their last headache(!). Figures were similarly obtained for each of the other symptoms. They were then asked to estimate the number of occurrences each month so that the interviewers could obtain a total value. Respondents were then asked to confirm that they would pay this amount per month. Faced with this amount, most respondents recanted and sharply reduced their estimates of WTP. (In fact, the revised result averaged out to $2 per headache.)

It is not clear why the discrepancy between the initial and revised estimates is so great. One possibility is that the two questions are fundamentally different in some way that is not obvious. A second possibility is that people really don't know what they would be willing to pay until the situation actually arises. Another possibility is that interviewees understand very well that the WTP question is hypothetical and that they will not have to pay any real money. Given the circumstances, respondents may assume that there is no budgetary constraint to

behaviorist assumptions are controversial, and there are critics who question, on fundamental grounds, whether the method can ever yield useful results, even in principle (e.g., Fischhoff, 1991). However, the more extreme critics are behavioral psychologists who are also generally skeptical about the behavioral assumptions underlying microeconomic theory. Thus, while these critics have raised very important questions, their skepticism is not generally shared by the economics profession.

Valuation of Trees and Forests

For the sake of concreteness, I now focus on one specific category of environmental resource that is likely to be particularly vulnerable to climate change: undisturbed noncommercial forests and "wild lands." As noted above, this is one of the major omissions from Nordhaus's evaluation and, indeed, from the SNA. How to proceed?

One cannot approach this question naively by attaching values to the remaining undisturbed forest or other wild lands in terms of the market price of lumber or whatever mineral resources may be found there. The fact is that old-growth forests can be logged and "converted" to lumber more or less overnight. Or they can be burned and "converted" to pasture. Similarly, wetlands can be drained and plowed. But the reverse transformation is not nearly so fast, if indeed it is possible at all.[15] Commercial forest, crop or pasture land cannot be turned into "old growth" forest, except by the passage of many decades or centuries.[16]

Ever since the colonization of this continent, and well into the present century, forests were simply in the way of agricultural development. The forests of New England were cut for fuel and to clear land for crops. Most of the central Appalachians have been cut a number of times for land clearing and charcoal production. The primeval forests of Tennessee, Kentucky, Ohio, Michigan and other states west of the Appalachians and east of the Mississippi were mostly burned to clear the land for farming. Since the rich farmlands of the Midwest became accessible to the markets of Europe (via the Mississippi) eastern farmlands have gradually been reverting to recreational or commercial forest, but except for a few state or national parks the eastern forest is almost entirely "second growth." It is slow to regrow and still far from maturity. Meanwhile the exhausted former cotton lands of the Old South have been converted to commercial wood pulp and lumber production. The southeastern commercial softwood forests have little or no recreational value.

The last of the great virgin forests of the United States are in the Rockies, the Pacific Northwest, and Alaska.[17] Some are already locked up in national parks or protected "wilderness" areas. The privately owned remainder is under growing pressure. Conservation and environmentalist groups are arguing for a radical increase in the area of virgin forestland to be protected from log-

---

consider and that they are being asked to "send a message" to government about the perceived importance of the ozone problem. When faced with the true monthly financial implications of their "message" and reminded of their budgetary constraints, however, a more realistic estimate is put forward.

[15]In the humid tropics land that is cleared for pasture may not be reforestable. If the soil is too mineralized, exposure to the sun may cause it to harden into a kind of rock (similar to adobe). Major areas of Indonesia have been degraded in this way. Amazon Basin soils contain very little humus and, once cleared, major areas may suffer this fate.

[16]Commercial forests in Europe are typically

logged in three stages, at 20-30 years, 40-50 years and finally 80-100 years. In warmer climates, such as the southern United States, the cycle can be shorter. However, a mature oak-beech forest will have large trees up to 300 years old. Redwoods and sequoias, and some tropical species, do not reach maturity for a millennium or longer.

[17]There is almost no virgin old-growth forest in Western Europe. (It was all cut down centuries ago.) The prevailing pattern is "mixed use," meaning a combination of recreational uses, private woodcutting for fuel and commercial logging. The value of the standing crop of wood, alone, ranges from $44 to $67 per hectare, depending on the country, with an average value of about $55. However, the average value of all uses taken together is just over $200 per hectare—a multiplier of 3.7. Recreational and noncommercial values clearly dominate by a large factor, at least in Europe.

ging. Meanwhile, the logging companies seem to be in a race to remove as many redwoods and Douglas firs as possible before they are forced to stop owing to increasing public pressure. The federal deficit of recent years is probably the only factor that has held back the pressure to increase the amount of forestland permanently set aside for wilderness and parks. This represents a major shift in social priorities and political response since the nineteenth century (or even a few decades ago). The rising uproar over the destruction of the tropical forests of the Amazon Basin is but another example of the shift of attitudes.

It is certainly *not* true that the political process allocates public goods (and services) as efficiently, or even in the same way, that a free competitive market would do if the resources in question were amenable to pricing and trading. The political process is far less impersonal and more haphazard in its functioning. Nevertheless, it does allocate resources, which is also the main function of markets. The fact that the political process is less efficient is largely due to the fact that it deals mainly with indivisibles or intangibles that are not readily exchangeable in principle. Thus, if one tried to infer the social value ("shadow price") of public goods from the way they are politically allocated, the resulting estimates would often be seriously in error. (That is to say, if the "real" social value could be determined by objective means, it would usually be significantly different from the "apparent" social value, as inferred from the political allocation.)

Having said all this, however, it is also probably true that reversals are important indicators. Gross overvaluations or undervaluations of public goods eventually correct themselves. Overvaluations lead to a public perception that too much public money is being spent on certain programs; conversely, undervaluations correspond to a perception that too little is being spent. Although the correction process is quite slow (due to bureaucratic inertia and the power of vested interests), long swings of this sort do occur. (For instance, military preparedness was probably undervalued in the two decades before Pearl Harbor. It may have been overvalued somewhat during the last four decades.) At any rate, it is quite plausible to suggest that public "goods" whose average long-term

share of discretionary public spending is decreasing are most likely to be overvalued, while those whose average share is increasing are most probably undervalued.

Environmental goods and services are currently undervalued by the political process, according to this test. However, this fact, in most cases, does not help in terms of quantification and monetization. This is because most environmental goods and services do not have any direct market-priced counterparts. However, there are a few that do. One case in point is the one being considered here, namely the ecological and recreational services provided by undisturbed, virgin forests and other wild lands. There is a market value for such land: It is what lumber companies (or others) are willing to pay to acquire it.[18] But the argument of the last few paragraphs suggests that the real social value of the virgin forests that are left is now significantly higher than the private (i.e. market) value of the land. By "significantly higher" I mean several times higher, not just a few percent higher. This is because the difference has to be fairly large to overcome the inertia of an historical trend of long-standing duration.

The argument so far is only semi-quantitative. It merely places a lower bound under the value of remaining wild lands. Except for the redwood and Douglas fir forests, the commercial value of most wild land is not even very high. On the other hand, urban shade trees are very valuable indeed, as indicated by the prices people are willing to pay tree service companies for their maintenance. Another indication is the fact that the penalties imposed by municipal shade tree commissions for illicit cutting (arboricide) are ris-

---

[18]There is an important clarification that needs to be made here. Suppose there is a parcel of virgin forest, and that real estate developers are willing to pay a very high price for a piece cut out of the middle of it. Is this not an instance of the free market allocating resources properly? The answer must take into account the following two points: (1) much of the value of the land to the developer is likely to be attributable to the amenity value of the surrounding undeveloped land; but (2), on the other hand, the environmental and ecological value of the virgin forest would be severely damaged by the intrusion. In short, there are positive spillover effects from the undeveloped land to the developed area and negative spillovers in the other direction.

ing sharply. New York City now allows for a 90 day jail term and up to $1,000 fine per tree. New legislation in New Jersey would permit fines of up to $20,000 for cutting a very fine old tree, based on a replacement cost formula once approved by the Internal Revenue Service for calculating losses from damage to trees.

However, I think the argument could be made much more quantitative by the application of either the HP or the CV methodology, or both. As noted above, the HP methodology normally depends upon the existence of voluminous data on prices, as well as other variables with which those prices can be correlated. The trick is to identify underlying variables (or combinations) that can account for the price differences among actual transactions. From this point it is relatively straightforward to determine the marginal prices of the underlying variables themselves.[19] The methodology as described above is limited in its applicability because of the data requirements. With respect to land prices, in particular, there is a further difficulty because transaction prices depend on a host of local factors, many of which are themselves intangible.

However, on a regional scale, many of these difficulties would actually be reduced or eliminated. Sensitivity of vegetation to climate has been investigated on a regional basis by Emanuel and colleagues (1985) and Leemans (1990) using the so-called Holdridge classification of "life zones" (see Figure 12.1). It is based on three climatic variables, namely bio-temperature (a function of the average temperature and length of the growing season, in degree-days), mean annual precipitation, and potential evapotranspiration. The third variable is a measure of the humidity of

the region. Since the broad scale distribution of terrestrial ecosystems is determined to a large part by the regional climate, one can relate the character of natural vegetation to climatic variables such as average annual temperature and precipitation.

The proposed methodology would then be as follows: Instead of using actual transaction prices for individual parcels of land, it might actually be preferable to construct price indices for land of several different types (functioning farms, building lots, wooded, pasture, or other) in a number of sample counties and by the Holdridge life-zone system. It would then be possible to do a cross-sectional analysis, regressing land price indices, by land-use type, against standard geographic-climatic variables such as average rainfall, bio-temperature, potential evapotranspiration and others. These variables determine the Holdridge life zone, as noted above. Thus, in effect, one could derive relationships between climate and land price, *ceteris paribus*. The output of such a project would be to determine the implicit prices—a costly task, but well worth the attempt, even if the results are incomplete and imperfect.

Given the availability of time series data on land price indices one could also explore the correlation between prices and year-to-year climatic variations at a given location, holding other factors constant. While either the cross-sectional or the longitudinal analysis would be useful, a combination of the two would be optimal.

Notwithstanding its controversial behavioral foundations, the contingent valuation (CV) approach also appears applicable to the problem of evaluating impacts of climate change. Here the basic scheme would be to carry out surveys in a number of communities and elicit willingness-to-pay (WTP) for "insurance" against various climate-related contingencies.

The climate-related contingencies would include, for instance, more very hot days in the summer, increased probability of drought and lower water tables, increased probability of insect pest outbreaks, accelerated loss of large shade trees in parks, yards and on local streets, possible loss of redwoods, etc. All of these possibilities can be framed in probabilistic language. For instance, such a survey might ask questions like this: "Suppose you were told by the Department of

---

[19]To be precise, the object is to determine the partial derivative of the total price with respect to each of these underlying variables. For instance, we would like to know how land prices are influenced by climate variables, other factors remaining the same. The derivative of land price with respect to a climate variable (e.g., bio-temperature) is defined as the marginal price of the land with respect to that variable. It describes the effect of a unit change in that variable on the price. From this sort of data, one can, in principle, carry out an integration to determine the total aggregated impact of climate change on the value of all land.

FIGURE 12.1. The Holdridge life-zone classification system.

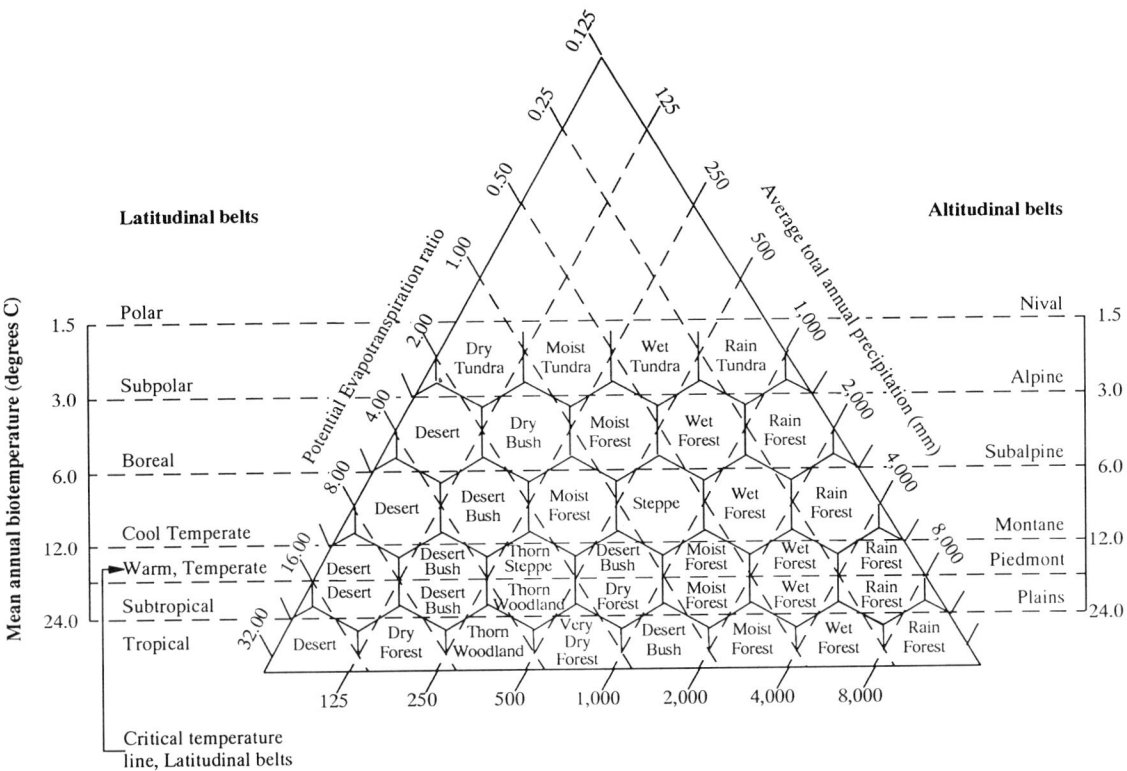

Agriculture that 50 percent of the shade trees in your neighborhood are likely to die from the effects of drought or insect pests (or both) over the next 10 years, unless you take action. How much would you personally be willing to pay each month to cut the expected loss from 50 percent to 10 percent?"

Obviously, it is important to guard against the sort of internal inconsistency noted in some previous applications of the CV method (see footnote 16). There are several ways of minimizing such problems. One is to emphasize budgetary constraints. For example, in connection with the sample question above, it might be helpful to ask: "Assuming your monthly income continues at its present level, what expenditures would you be willing to cut in order to find this money?" Another approach is to include questions in the survey about a variety of other contingencies and risks, some of which are independently analyzable by more objective means. Examples include earthquake, flood and tornado damage; airplane crashes, and ma-

jor medical problems. In all of these cases theoretical WTP can be compared with actual, quantitatively verifiable premium payments to insurance companies.

Of course, both types of study would be costly and time-consuming. However, the costs of research would be insignificant in comparison with the ecological and other damage costs being quantified, and the potential savings that could accrue through the timely application of appropriate ameliorative policies. It may seem unscientific to speculate at this point on the results of research that has not been done. Yet, almost no large-scale research is done in the absence of some plausible speculation, called "hypothesis formation." For whatever it is worth, I suspect that if visibility losses from atmospheric haze in the United States alone have an annual value in the ballpark of $10 billion ($4 billion to $50 billion), the full range of potential climate-related losses to forests and ecosystems, as sketched above, would be valued on a WTP basis at least 10 times higher than visibility alone, on an annual basis.[20]

In short, my hypothesis is that the American public would be prepared to pay something like $100 billion per year and possibly much more than that to protect these precious environmental assets, assuming only that the protection follows clearly and directly from the payment. In other words, it does not necessarily follow that the public would be willing to pay $100 billion more annually in federal taxes on the basis of political campaign promises to pass more laws to protect the environment. People have become understandably cynical about the capability of the Congress and the bureaucracy to translate political rhetoric into effective action. Moreover, I postulate (that is, I would bet) that the implicit valuation of environmental assets will rise over the next few decades at a rate considerably faster than GNP. (This means, incidentally, that application of conventional discounting methods would not make the problem vanish simply because the major impacts are fairly far in the future.)[21]

I cannot help noting that the sum in question ($100 billion) is far less than we currently pay for military protection against vaguely defined "threats to national security." It is far less than we currently pay for medical and health services (12% of GNP and growing). It is comparable to what we pay for life insurance or accident insurance (7% of GNP)—which, of course, do not actually protect life or prevent accidents. On the basis of admittedly superficial comparisons like these, I do not think the magnitudes I have suggested could be dismissed out of hand as absurd. Of course, that is by no means evidence that the number I have suggested is correct. A considerable amount of research would have to be done to develop credible evidence, one way or the other.

However, I am persuaded that there is a very strong argument for funding and carrying out the necessary research, with particular application to the global warming problem. If the magnitude of the potential environmental damage (or the public's willingness-to-pay to prevent it) is in the range suggested by

Nordhaus' analysis, certain implications for policy follow. In particular, the obvious conclusion from Nordhaus's numbers is that the problem of greenhouse warming is much less serious than many other problems and therefore does not warrant extraordinary responses or expenditures by government.

Many environmentalists (and others) feel that Nordhaus's analysis is much too optimistic and that it essentially misses the point. But they lack convincing arguments to attack and counter his detailed quantitative arguments. In the world of constrained budgets and fiercely competing government priorities, however, feelings like this no matter how strongly held don't count for much. The only effective arguments for defending any policy proposal with tax or fiscal implications against the conservative defenders of the status quo are those that can be backed up by a scientifically defensible quantitative systems analytic methodology.

References

Ahmad, Y.J., S. el-Sarafy and E. Lutz (Eds.), 1990. *Environmental Accounting for Sustainable Development*, World Bank, Washington, D.C.

Anderson, R.J., Jr., and T.D. Crocker, 1971. Air pollution and residential property values. *Urban Studies*, **8**:171-80.

―――― and T.D. Crocker, 1972. Air pollution and property values: A reply. *Review of Economics and Statistics*, **54**:470-73.

Ayers, R.U., and J. Walter, 1991. The greenhouse effect: Damage, cost and abatement. *Environmental and Resource Economics*, **1**:237-70.

Bartelmus, P., C. Stahmer and J. V. Tongeren, 1989. SNA framework for integrating environmental and economic accounting, 21st General Conference, IARIW.

Batie, S.S. and H.H. Shugard, 1990. The biological consequences of climate changes: An ecological and economic assessment. In: *Greenhouse Warming: Abatement and Adaptation*. N.J. Rosenberg, W.E. Easterling, P.R. Crosson, *et al.* (Eds.). Resources for the Future, Washington D.C., pp. 121-32

Brookshire, D.S., B.C. Ives, and W.D.

---

[20]Obviously these numbers are far higher than the estimates given by Nordhaus.

[21]See footnote 8.

Schultze, 1976. The valuation of aesthetic preferences. *Journal of Environmental Economics and Management*, 3:325-46.

Brown, P.G., 1990. Fiduciary responsibilities and the greenhouse effect. Working Paper, School of Public Affairs, University of Maryland, College Park (Working Group on Equity and the Greenhouse Effect).

Crosson, P.R., 1990. Climate change: Problems of limits and policy responses. In: *Greenhouse Warming: Abatement & Adaptation*. N.J. Rosenberg, W.E. Easterling, P.R. Crosson, *et al.* (Eds.). Resources for the Future, Washington D.C., pp. 69-82.

Cummings, R.D., D.S. Brookshire and William D. Schultze (Eds.), 1986. *Valuing Environmental Goods: An Assessment of the Contingent Valuation Method*. Rowman and Allenheld, Totowa, N.J.

de Neufville, R., 1990. *Systems Analysis*. MIT Press, Cambridge, Mass.

Easterling, W.E. III, M.L. Parry and P.R. Crosson, 1990. Adapting future agriculture to changes in climate. In: *Greenhouse Warming: Abatement and Adaptation*. N.J. Rosenberg, W.E. Easterling, P.R. Crosson, *et al.* (Eds.). Resources for the Future, Washington D.C., pp. 91-104.

Emanuel, W.R., H.H. Shugard and M.L. Stevenson, 1985. Climate change and the broad scale distribution of terrestrial ecosystem complexes. *Climatic Change*, 7:29-43.

Fischhoff, B., 1991. Value elicitation: Is there anything in there? *American Psychologist*, 46:835-47.

Frederick, K.D., and P.H. Gleick, 1990. Water resources and climate change. In: *Greenhouse Warming: Abatement and Adaptation*. N.J. Rosenberg, W.E. Easterling, P.R. Crosson, *et al.* (Eds.). Resources for the Future, Washington D.C., pp. 133-46.

Freeman, A. Myrick III, 1974. On estimating air pollution control benefits from land value studies. *Journal of Environmental Economics and Management*, 1:74-83.

Green, C.H., and S.M. Tunstall, 1990. Oil and water? Environmental economics and environmental ethics. Conference on Ethics and Environmental Policies, Fondazione Lanza, Milan, August 30.

Hall, D.C., 1991. Social costs of CO_2 abatement from energy efficiency and solar power in the United States. 66th Annual Conference, Western Economic Association International, Seattle, Wash., June 30-July 3, 1991.

Hekstra, G.P., 1990. Sea-level rise: Regional consequences and responses. In: *Greenhouse Warming: Abatement and Adaptation*. N.J. Rosenberg, W.E. Easterling, P.R. Crosson, *et al.* (Eds.). Resources for the Future, Washington D.C., pp. 53-68.

Hueting, R., 1980. *New Scarcity and Economic Growth: More Welfare through Less Production?* North Holland, Amsterdam.

Jaksch, J.A., and H.H. Stoevener, 1970. Effects of air pollution on residential property values in Toledo, Oregon. *Special Report (304)*. Oregon State University Agricultural Experiment Station, Covallis.

Kerr, R., 1988. Report urges greenhouse action now. *Science,* 241:23-24.

Leemans, R., 1990. *IIASA Biosphere Project*. International Institute for Applied Systems Analysis, Laxenburg, Austria.

National Acid Precipitation Assessment Program, 1990. *Integrated Assessment; Questions 1 and 2, Draft final report (1)*. National Acid Precipitation Assessment Program, Washington D.C., August 1990.

Nordhaus, W.D., 1989a. To slow or not to slow: The economics of the greenhouse effect. Annual Meeting, American Association for the Advancement of Science, New Orleans, January 1989.

Nordhaus, W.D., 1989b. The economics of the greenhouse effect. In: *International Energy Workshop*. International Institute for Applied Systems Analysis, Laxenburg, Austria, June 1989.

Page, T., 1977. *Conservation and Economic Efficiency*. Johns Hopkins University Press, Baltimore, Md.

Pearce, D.W., 1990. Economics and the global environmental challenge. *Millennium: Journal of International Studies*, 19:365-87.

Peskin, H.M., 1989. A proposed environmental accounts framework. In: *Environmental Accounting for Sustainable Development*. A. el-Sarafy and E. Lutz (Eds.). World Bank, Washington D.C.
———— with Ernst Lutz, 1990. A survey of resource and environmental accounting. In: *Industrialized Countries, Environmental Working Paper (37)*. World Bank, Washington D.C.

Repetto, R., 1990. The concept and measurement of environmental productivity: An exploratory study of the electric power industry. Towards 2000: Environment, technology and the new century (symposium). World Resources Institute and the OECD, Annapolis Md., June 1990 (background paper).

Repetto, R., and W.D. Magrath, 1989. *Wasting Assets: Natural Resources in the National Income Accounts*. World Resources Institute, Washington D.C.

Ridker, R.G., 1967. *Economic Costs of Air Pollution*. Praeger, New York.
———— and J.A. Henning, 1967. The determinants of residential property values with special reference to air pollution. *Review of Economics and Statistics*, **49**:246-57.

Sandenburgh, R. C Taylor and J.S. Hoffman, 1987. Rising carbon dioxide, climate change, and forest management: An overview. In: *The Greenhouse Effect, Climate Change, and U.S. Forests*. W.E. Shands and J.S. Hoffman (Eds.), The Conservation Foundation, Washington D.C., p. 113 et seq.

Schneider, S.H., and N.J. Rosenberg, 1990. The greenhouse effect: Its causes, possible impacts, and associated uncertainties. In: *Greenhouse Warming: Abatement and Adaptation*. N.J. Rosenberg, W.E. Easterling, P.R. Crosson, *et al.* (Eds.). Resources for the Future, Washington D.C., pp. 7-34.

Sinden, J.A., and A.C. Worrell, 1979. *Unpriced Values: Decisions Without Market Price*. John Wiley, New York.

Stahmer, C., 1990. Towards integrated environmental and economic accounting. In: *SNA Handbook (Draft)*. UNSO, New York, chapter 1.

Turner, B.L., W.C. Clark, R.W. Kates *et al.*(Ed.), 1990. *The Earth as Transformed by Human Action*. Cambridge University Press, Cambridge.

United States Environmental Protection Agency, 1988. *The Potential Effects of Global Climate Change on the United States*. Draft Report to Congress, United States Environmental Protection Agency, Washington D.C., October 1988.
————, 1989. *Policy Options for Stabilizing Global Climate, Draft Report to Congress*. USEPA, Washington D.C., February 1989.

Washington, W.M., and C.L. Parkinson, 1986. *An Introduction to 3-Dimensional Climate Modelling*. University Science.

Wilson, E.O., 1989. Threats to bio-diversity. *Scientific American*, **261(3)**:60-66, September 1989.

World Bank, 1990. Poverty. *World Development Report 1990*. Oxford University Press, Oxford.

Part III

Developing Regional Policies

for Climate Change

Climate change will affect every region of the world in different ways and to different degrees. Thus, although a global effort would be the most effective for controlling the rate of warming, regional entities are uniquely qualified to develop adaptive strategies.

Previous chapters in this volume have identified some of the most vulnerable regions of the world, in particular those that are already subject to flooding or are drought-prone. Many of these areas are some of the most impoverished. Addressing an additional problem like global warming, to which so much uncertainty can be attached, is beyond their capabilities; all of their resources are already committed to dealing with current problems of a very basic nature. Thus, the theme of many of the chapters of this book is to focus on problems that not only will be exacerbated by climate change, but that will also have an immediate payoff.

In the first chapter of Part III, the author's focus is on identifying current vulnerabilities and solving present problems that are linked to the impacts of future climate change in the regional context, without planning for "uncertain" impacts. Although this is not a substitute for direct action against global warming, the complexities of the problem, the degree of uncertainty associated with predicted changes, and the general pattern of inaction tend to reinforce each other. Thus, N. S. Jodha argues, this approach can help insofar as the problem is accentuated by cumulative types of changes such as deforestation and desertification.

The next two chapters of Part III focus on Brazil. Goldemberg stresses the role that developing countries have to play in reducing greenhouse gas emissions. Assuming that an increase in the standard of living of the population is inevitable, he claims that adopting new energy-efficient technologies immediately, instead of retrofitting later,

would result in a much slower rate of growth in energy consumption than is usually predicted for the developing world. It is proposed that industrialized nations help in achieving this goal, possibly through the establishment of a carbon tax.

In Chapter 15 the focus is on the vulnerability of northeast Brazil to climate change and the need for sustainable development. Magalhaes describes how policy responses have been aimed at alleviating the short-term impacts of droughts, but have failed to provide long-term solutions that would increase the regional capacity to cope with them. In fact, the prevailing social, environmental and economic conditions tend to intensify the effects of drought, causing an even greater transfer of wealth to the rich. The importance of implementing a strategy for sustainable development is emphasized.

The need for the development of coordinated regional policies by political entities is the theme of Chapters 16-18. Burton proposes that, having identified the potential impacts of climate change, regions should seek to enhance their resilience and simultaneously reduce their vulnerability. He elaborates on the importance of sustainable development, but also outlines the opportunities presented by a regional economic analysis that identifies the strengths and weaknesses of a region with respect to climate change. In an example of the Great Lakes region, he suggests that climate change could increase the need for energy, while reducing the hydropower-generating potential. Thus, it would be prudent to start developing less centralized systems for generating electricity and avoid increased reliance on coal-fired plants, which contribute to global warming.

In Chapter 17, Moomaw focuses on the institutional aspects of developing regional policies. He argues that the northeastern United States and eastern Canada will be better positioned to deal with global warming if they develop coordinated policies for transportation, land-use planning, utility regulation, energy development and taxation. He proposes the establishment of a Northeast Regional Council for Sustainable Development that would provide a forum for developing compatible policies, while leaving the decision-making authority with the individual political jurisdictions.

In the chapter by Saha on the Southwest Indian Islands, the main thesis is that these islands are too small to be taken seriously by the international community unless they join together. Because of their limited resources, they need to pool their efforts to try to increase their knowledge base and develop a multilevel institutional structure to stimulate both political and public awareness. Island communities are some of the most at-risk populations, and they need to develop effective policies and contribute to the international debate.

In the concluding chapter, Ferenc Toth presents an overview of, and summarizes, the general conclusions from a UNEP project conducted in Southeast Asia to identify socioeconomic impacts of, and policy responses to, climate change. Because climate change assessments are not very helpful in attracting the attention of policymakers, the policy exercises were designed to link the potential regional impacts of climate change to long-term and complex social and economic problems, to ongoing and planned large-scale development programs, and to long-term objectives for overall socio-economic development.

The chapters in this section outline a number of the approaches that can be taken to address the potential impacts of global climate change. They emphasize three major themes:

- the need for sustainable development;
- the need for cooperative action among neighboring political entities;
- the need to identify issues that have immediate significance and that address current problems.

Any international or regional forums that can be established to facilitate realization of these goals should be encouraged. We emphasize again that, because international treaties are hard to negotiate, a more regional approach for developing preventative and adaptive strategies holds the most promise for action.

13. UNDERSTANDING AND RESPONDING TO GLOBAL CLIMATE CHANGE IN FRAGILE RESOURCE ZONES

N. S. Jodha

Judging by the amount of scientific discourse, the number of political resolutions, and the interest generated in the mass media, the projected global climate change, resulting from increased emissions of CO_2 and other trace gases, is one of the major concerns in the world today. This is not without cause because, if these changes continue unabated, they will have the potential to erode all of the gains made by our current civilization. The phenomenon has elicited widespread concern, considerable resource allocation, serious scientific work and numerous exercises in policy advocacy. Despite these efforts, the progress, in terms of sufficient understanding of the phenomena and concrete steps to abate and adapt to the changes they produce, is extremely slow and insignificant.

In fact, persistent uncertainties concerning the precise extent of the predicted changes—as well as their temporal and spatial dimensions, the generation of panic rather than the promotion of concrete remedial measures by ongoing debates, and the frequent recycling of the same information and similar recommendations at global warming symposia—tend to suggest a state of stagnation or crisis in the field. The purpose of this somewhat provocative statement is neither to belittle the ongoing serious research in this area, nor to underrate the complexity and enormity of the problem. Rather the intent is to draw attention to the constraining framework that obstructs any real breakthrough in terms of precise predictions, as well as concrete steps by policymakers, to handle the problem.

In such a given situation, it is useful to diagnose the "crisis syndrome" (if the term is acceptable) and look for new perspectives and approaches to resolve the problems involved. This chapter argues for the identification of options, the adoption of which will not be unduly obstructed by "uncertainties" and related problems. It indicates an approach that might facilitate the evolution of adaptation strategies to global change in a regional context, despite the persisting uncertainties of the change "scenarios." The focus of discussion is on:

1. the identification of certainty components within the complex of uncertainties associated with global change processes and scenarios;
2. the identification of responses with reference to (1) above; and
3. the possibility that the impacts of such responses or adaptation strategies (i.e., item 2) would extend to the uncertainty components of the predicted change in some cases.

The approach is illustrated by references to some areas in South Asia. In a more practical context, the approach relates the current problems and their remedial measures to the possible negative impacts of future climate change in the regional context. It is possible that options identified through such an approach, besides handling immediate development problems, may help to strengthen the regional capability to withstand the impacts of climate change, despite the unknown factors involved in such a change.

Outline of the Chapter

In the section that follows, we briefly discuss the factors obstructing development of abatement and adaptation strategies to global climate change and their implications. Different dimensions of the crisis situation characterizing the "global warming debate" are commented upon.

In the section titled "Acting Despite Uncertainties," we comment on the current perspectives on global environmental change (or global climate change) that reflect imbalances in the approach to the problem, leading to an underemphasis on the "certainty components" of the change process. Taking a lead from the altered perspectives on the problem, the section titled "Regional Adaption Strategies" outlines an approach to facilitating adaptation measures, despite the uncertainties of climate change scenarios.

The approach is illustrated by references to the prevailing situation in the fragile resource zones of South Asia (i.e., the dry tropical region of India, and to a limited extent, the mountain region of the Himalayas). This chapter uses the terms *global climate change* and *global warming* interchangeably. In some instances *global environmental change* is used in place of global climate change although the former may have broader connotations.

Global Warming Debate: "Crisis Syndrome"

The ongoing debate on global warming, along with its consequences and remedial measures, reflects some type of crisis. By this, one implies that a situation exists in which one knows about the problem and yet does not know enough about it to develop an effective remedy In the context of global warming, the complexities of the problem, the degree of uncertainty associated with predicted changes, and the general pattern of inaction (in terms of anticipatory strategies) tend to reinforce each other. The crisis is, therefore, related to both the science behind the debate and the action that should follow the debate.

Science Behind the Global Warming

The science-related elements of the global warming "crisis" are reflected in the limitations of current models in terms of capturing the total reality of the change and its processes (IPCC, 1990; Jager, 1988; Schneider and Rosenberg, 1989; Wuebbles and Edmonds, 1988). For instance, although General Circulation Models (GCMs) are able to predict the increase in mean levels of global temperature following the accumulation of trace gases in the atmosphere, the regional dimensions of the extent of warming are still vague and uncertain. Similarly, several knowledge gaps and uncertainties exist regarding the time dimension of the change and the critical values of doses of trace gases that the system can absorb without reacting in terms of global climate change. The ability of such models to determine the role of other factors, such as cloud cover and global sinks (such as oceans and forests), is also questionable (Abelson, 1990; Rogers and Fiering, 1989; Spencer and Christy, 1990; White 1990). Scientists dealing with the problem of global climate change are becoming increasingly aware of this problem. Everyone advocates more research on the subject so as to reduce the range of uncertainties. However, the problem lies in pushing for action (which may involve huge dislocation costs) without a precise understanding of the issues involved.

Confusion and Conflicts on the Action Front

Another important dimension of the "crisis syndrome" characterizing the global warming debate is the general failure to take any effective steps to solve the problem. Without belittling the progress made on the Ozone Treaty, the steps undertaken by individual countries in Europe and North America, and the useful spadework done by IPCC, WMO, UNEP and others on various aspects of the problem (Flavin, 1989; IPCC, 1990; UNEP and the Beijer Institute, 1989), it can be stated that, when seen in the context of the enormity of the problem, the concrete measures necessary to control global warming or to facilitate the adaption to such changes are not yet visible. This lack of serious and concrete effort is reflected in the open or concealed indifference of policymakers to global warming issues; the persistent differences in the perspectives of both developed and developing countries on the various dimensions of the problem (Agarwal and Narain, 1991; Parikh, 1991), the confusions and conflicts on perceived losses and responses (Glantz *et al.*, 1988; Gleick, 1989), the priority given to debate over action and, most importantly, the choice of a "wait and see" approach by developed and developing nations, albeit for different reasons (UNU and IFIAS, 1989). Moreover, the risks involved in initiating action without concrete contexts (due to uncertainties of climate change scenarios) are often used as an argument for inaction.

Pressures for Action

However, notwithstanding the uncertainties of global warming and their implications, the context for policy decisions is rapidly changing. Because of factors such as the strong concern-generating potential of the global warming problem (i.e., doomsday scenarios), the percolation of scientific discourses into the mass media, the increased environmental awareness of an "informed" public, the growing pressure of environmental nongovernmental organizations (NGOs), and the impacts of recent unusual weather events, such as severe droughts and hot summers (especially in developed countries), the noise level of the debate on global warming is much higher today than ever before. There may not be full awareness of the limitations of the scientific evidence on the subject, but the debate has acquired sufficient strength to force the policymakers into action.

Nevertheless, these changes have not radically altered collective anticipatory planning on global warming. Consequently, apart from agreements such as the Ozone Treaty and some initiatives by individual countries (UNEP and the Beijer Institute, 1989), most public responses are still confined to increasing resource allocation for scientific research on global warming; progressive "globalization" with an increased number of global warming meetings; the gradual upgrading of the political status of such meetings; and efforts to disentangle the involved dilemmas (through official and nonofficial working groups). Substantive action on the problem, in the form of focused and collective strategies, is still nonexistent, notwithstanding the World Bank, UNEP, and WMO initiatives leading to environmental safety tags on project files.

In fact, the crux of the problem lies in the fact that the choice for concrete action in unknown or uncertain contexts goes against the general risk-averse nature of human beings, unless they happen to have strong gambling instincts. In addition, the problem involves the "global commons." Thus, such decisions involve the concurrence of the entire community of nations, and they all have diverse perspectives, resource capacities, and perceptions vis-a-vis the issues involved (Dorfman, 1991). Moreover, the perceived potential costs of economic dislocation, changes in consumption, and investment decisions associated with possible action strategies against global warming are too high, and they further discourage the actions, even in a "speculative mode." Consequently, "wait and see" is the preferred option of most countries under the existing circumstances.

Is "Wait and See" an Option?

A product of the uncertainty of global change scenarios and the risk-averse (and calculating) nature of policymakers, the wait-and-see approach to global warming seriously conflicts with the cumulative nature of the "warming" problem. Despite other uncertainties, scientific understanding is more certain on the heat trapping potential of trace gases and their increasing accumulation in the atmosphere. Consequently, waiting for more precise predictions on global change before embarking on abatement/adaptation strategies would mean committing Earth to additional warming and its consequences, that is, increased risks and costs (Flavin, 1989; IPCC, 1990; Jager, 1988; Pearce *et al.*, 1990).

A detailed look at the factors affecting the choice (in terms of waiting or acting), summarized by Pearce *et al.* (1990), suggests that there are more factors favoring anticipatory environmental policies. However, the time preference or discounting procedures, the usual tools in investment decisions analysis, would favor waiting. Similarly, the cost-effectiveness of responses that are directly linked to an improvement in information on global warming would favor waiting. But other factors, such as the cumulative nature of the warming and associated escalation in costs; the possibility of irreversible changes in some cases; the uncertainty of impacts, which may take the form of shocks and catch the society unprepared; and the imperatives of incorporating environmental concerns into narrowly focused conventional development strategies (Dorfman, 1991), are compelling enough to support a choice for concrete policies and action strategies regarding global warming.

Moreover, other factors do not favor a wait-and-see option. First, waiting (as against action) tends to dilute the concern, commitment, and enthusiasm for the problem and, hence, slackens the preparedness to act at the appropriate time. In the global warming context, this may mean the slackening of support for the conduct of scientific research necessary for reducing the range of uncertainties. The past history of resource allocation for different research activities, fluctuating according to the changing whims and biases of decision-makers, would bear this out.

Second, it is difficult to imagine how long a waiting period is involved in the wait-and-see approach. Despite remarkable improvements in the human capacity and the means to document and simulate the behavior of nature, the latter is too complicated to be fully understood within a short enough period to facilitate concrete policy decisions. Hence, the time lag involved cannot be overstated. The pressure for even tentative action strategies and provision of a more concrete focus to scientific work may probably reduce the time lag somewhat. In the past, there have been cases where pressure for concrete output (implied by certain policy decisions) has changed the pace and perspective of scientific research.

The third issue again relates to human nature as it is reflected at the negotiating table. The history of international treaties shows that the preparation and finalization of major agreements (unless they are between victors and vanquished) is a long and drawn-out process. Depending upon the gains and losses perceived, and rigidities associated with the respective positions of involved parties, progress on international treaties is slow and incremental. Treaties on global warming, owing both to its complexities and the involvement of nations with diverse perceptions, may take a very long time and add further to the waiting period (even after scientific understanding of the problem has improved). The Ozone Treaty (in terms of the relatively short time it took) is of limited value as a precedent, because the issues and potential adjustment costs involved in the control of CFCs are far lower than those involved in controlling global warming.

To sum up, the cumulative nature of the global warming problem and its associated risks do not permit the luxury of a wait-and-see option; if such an option is chosen, the waiting period will be far too prolonged and the situation might acquire irreversible proportions during this period, especially if the change takes the form of major "shocks" or "surprises." Hence, unless the community preference is for inferior options, implied by forced adaptations to global warming, there is no choice but anticipatory actions to address the situation, even if its precise details are not yet available (Flavin, 1989; Gleick, 1989; IPCC, 1990; Jager, 1988; Pearce *et al.*, 1990).

Acting Despite Uncertainties

The risks associated with the wait-and-see approach on the one hand, and the difficulties and costs of developing and implementing abatement/adaptation measures without concrete contexts on the other, are, in a way, the core of the crisis on the global warming action front. If society needs anticipatory planning for global warming, for this to be relevant and effective, such planning needs contexts with reasonable degrees of certainty and definitiveness. To resolve this dilemma, there must be a focus on the possible options and approaches where problems of uncertainty, time preference (or discounting), and information gaps are somehow overcome. The process of searching for such options to facilitate "action despite uncertainties" can start with a search for certainty components within the complex of uncertainties associated with the pace and pattern of global environmental change and its impacts. The response strategies, then, could build upon the certainty components of the problem.

Decisions about and management of certainty components (not obstructed by the conjectural nature of change scenarios) may also have a slow and small impact on components of the problem with strong uncertainty attributes. Thus, under circumstances that discourage concrete action on global warming, such an approach can form a basis for evolving "dual purpose" strategies to handle the problems of global warming (Cooper, 1989). The important components of such strategies would involve the following steps:

1. Identifying the certainty components within the overall complex of causes, processes, and consequences of global warming.
2. Focusing on measures directed to certainty components that have the potential to extend to uncertainty components.
3. In order to make (1) and (2) above operational, (a) identifying linkages among current problems and potential climate change impacts; (b) understanding the extent of convergence between measures directed against current problems and potential responses to regional impacts of climate change; and, finally, based on steps (a) and (b) above, (c) identifying short-term response strategies (which are less affected by the uncertainty of change scenarios and associated problems) that have the potential to mitigate the risks of long-term climate change.

Thus, a search for certainty and its consequences in a complex of uncertainties is the essence of the approach suggested by this chapter. This will be illustrated by reference to specific areas in South Asia. However, the whole approach outlined above begs a significant question: Are there any certainty components in the entire process of global warming and its impacts that may serve as a basis for concrete action strategies? This may require a close look at the current overall perspectives on the problem.

Global Environmental Change: Skewed Perspectives

Global environmental change has several dimensions in terms of the multiplicity of causative factors and involved processes, the manifestations of consequences, and the diversity of potential response approaches. However, owing to certain historical and institutional reasons (such as the involvement of specific scientific disciplines in the initial work on the subject), the newness of the problem, as well as the "noise potential" of specific aspects of the problem, only some dimensions of the global environmental change phenomenon have received the major

focus of scientific work, debate in the media, and policy advocacy exercises. These are also the dimensions (involving complex bio-geochemical variables and their interactions) that have a greater degree of uncertainty associated with them. Thus, uncertainty, in the context of global change scenarios, could be partly a product of imbalances in the current perspectives on the subject.

To elaborate on these issues, we may take the lead from some recent conceptual work on the subject. Turner and colleagues (1990) discuss two dimensions of global environmental change: (1) "systemic" changes and (2) "cumulative" changes. Broadly speaking, a systemic change is one that takes place in one locale with implications for the system elsewhere. The underlying activity need not be widespread or global in scale, but its potential impact is global in the sense that it influences the operation and functioning of the entire system, as manifested through the subsequent adjustments in the system. Carbon dioxide emissions from limited activities that have impacts on the great geosphere-biosphere systems of Earth and cause global warming are a prime example. Cumulative types of changes refer to localized, but widely replicated, activities where a change in one place does not cause change in other places. When accumulated over time, they may acquire a scale and potential to influence the total global situation in specific ways. Widespread deforestation and extractive land use systems with their potential impacts on global environment serve as examples. Both types of changes are the products of nature/human interactions, and they are linked to each other in several ways.

Other conceptual leads, which facilitate a critical look at the current approaches and thinking on global changes, relate to (1) geocentric perspectives and (2) anthropocentric perspectives on the change, its causes, and consequences (Chen et al., 1983; Clark, 1988; Price, 1990; Turner et al., 1990). The key focus of the former is on Earth and its geobiophysical variables—with all of their complexities and unknowns, as emphasized by natural scientists. Anthropocentric perspectives, on the other hand, emphasize nature-society interactions, as well as the processes of change that have a primary focus on their importance for society.

The "cumulative type of changes" and

"anthropocentric perspectives" present a situation that is much simpler and involves fewer uncertainties and unknowns. It is much easier to relate these perspectives to the human approach to constraints and opportunities presented by the environment. Yet, since they are not sufficiently emphasized in global-change-related work, their potential for guiding response strategies remains underutilized (Kasperson *et al.*, 1989). Table 13.1 presents some details indicating the skewed nature of perspectives on global change and their implications. Many more details and concrete examples can be pooled under the major categories presented in the table.

Besides referring to the above-mentioned imbalance in perspectives, the primary purpose of alluding to the above concepts is to highlight the possibility of evolving approaches based on the "cumulative" type of changes and of using them as a basis for exploring "action strategies, despite the uncertainties of the change scenarios" discussed earlier. As Table 13.1 shows, cumulative changes are easy to understand; it is also easy to evolve response strategies to manage them and adapt to their impacts.

Furthermore, the impacts of response strategies on cumulative changes may also extend to systemic changes and their consequences. This is because of the linkages between cumulative and systemic changes, especially when the former acquire large-scale and widespread dimensions (Turner *et al.*, 1990). Table 13.2 illustrates the above linkages. Accordingly, large-scale deforestation (especially of tropical rain forests), various forms of desertification, and extractive land use systems that are extended particularly throughout the developing countries (Glantz 1987), in their respective ways, can either contribute to the components of systemic change (e.g., growth of greenhouse gases) or reduce the capacity of natural sinks (e.g., forests) to absorb CO_2.

The same factors may indirectly contribute to systemic changes by creating circumstances that generate pressure on socioeconomic systems to engage in activities (e.g., technologies, industries) that create more employment and income, but also contribute to the emission of trace gases. Seen from this perspective, some of the phenomena (directly unrelated to use/misuse of the biophysical resource base), such as Third World poverty, indebtedness, and unequal trade/exchange arrangements, also indirectly contribute to global "systemic" changes. The understanding of such linkages can help in developing an approach that can integrate the concerns about global environmental change and current problems of Third World poverty and hunger (Dorfman, 1991; Kates, 1990).

The potential links between the impacts of "cumulative changes" (listed in Table 13.2) and the impacts of "systemic changes" (i.e., global climate change through trace gases) are still stronger and relatively easy to understand. To illustrate this, we can refer to the likely impacts of global climate change on agriculture—the dominant sector in the developing world. The impacts of changes in precipitation patterns, temperature, humidity, etc., following or during global warming, would influence the resource base and production environment of agriculture (called first-order impacts); the structure and functioning of farming systems (called second-order impacts); and macro-level agricultural policies, infrastructure, and support systems (called third-order impacts).

The cumulative changes (e.g., deforestation, desertification, etc.) also have impacts that fit into these first-, second- and third-order impacts. For example, deforestation, by influencing the hydrology of the region, influences the resource base and production environment of agriculture. It affects the farming system by weakening the "forestry-farming linkages" and has an impact on macro-level fuel and energy policies because it constrains biomass supplies. Similar examples can be given for other variables. Hence, these cumulative changes have the potential to accentuate the impacts of global climate change and vice versa. This perspective can help link the issues of current and local (or regional) concerns with those of future and global concerns. This, in turn, can facilitate the resolution of the problem of "inaction" on global warming issues by approaching them through current problem perspectives, especially in developing countries. On the basis of the aforementioned linkages, one can identify strategies to manage the emergence and impacts of cumulative changes, and this may extend to systemic changes in some cases. In the next section this will be elaborated in the context of a concrete regional example.

TABLE 13.1. *Indicators of the Skewed Perspectives on Global Environmental Change*[a]

Elements prominently focused	Elements under emphasised
Systemic type of change	Cumulative type of change
Focus on biogeochemical variables and their interaction processes relating to the functions and operation of geosphere/biosphere systems of Earth.	Localized and widely replicated changes in different variables and process of resource use (when accumulated) influence the global systems.
Geocentric perspective	Anthropocentric perspective
Focus on physical dimensions, typically in the natural science framework; concentration on geobiological variables and their complex interaction patterns, with little direct incorporation of human dimension of changes and change processes.	Primacy of nature/society interactions with focus on their importance to the society; potential mechanism for understanding and handling "cumulative changes" (with some possibility of influencing impacts of "systemic changes").
Other associated aspects	Other associated aspects
Emphasis on long-time horizon (decades/ centuries) and inter-generational issues; focus on terminal impacts involving selected variables (e.g. sea-level and temperatures rise, shift of climatic zones, etc.) affecting fundamental equilibrium of world system and atmosphere; analytical methods and material used involve high degree of complexity and sophistication, information on several unknowns, limited transparency (for uninitiated ones), multiple uncertainties, and conjectural nature of predictions.	Sensitivity to both intragenerational and intergenerational issues; analytical approaches simpler and oriented to integration of change processes in current problem-solving mode; predictions, action/advocacy focus on short or medium planning horizon, greater ease and possibility of associating causes, consequences of and responses to change; greater possibility of integrating geocentric and anthropocentric perspectives.
Advocacy and action	Advocacy and action
High "scare and noise" potential of issues covered, (e.g, doomsday predictions); approaches to abate/adapt to changes: obstructed by uncertainty of change scenarios, induce higher discounting of the potential options, inject vagueness about gains and sacrifices, and create more panic and debate than concrete action.	Possibility of evolving options within the received (and modified) framework of handling current crisis situations in local contexts; greater scope for clearly associating cost and benefits, greater certainty of potential options and their easy acceptability to decision makers; possibility of dual purpose options to handle current and future "impacts."

a. For various issues and examples that could fit into the following grouping of perspectives see Chen *et al.* (1983), Clark (1985), Flavin (1989), Garcia (1981), Glantz *et al.* (1988), Jodha (1989), Kates (1990), Price (1990), Turner *et al.* (1990), Wuebbles and Edmond (1988).

TABLE 13.2. *Global Environmental Change: Linkage Between Cumulative and Systemic Types of Changes*

Cumulative type of changes (Examples)	Cumulative[a] change supporting systemic[b] change		Cumulative change accentuating impacts of systemic change		
	Directly[c]	Indirectly[d]	Types of Impacts[e]		
			1st order[e]	2nd order[f]	3rd order[g]
Deforestation, overgrazing; depletion of biomass potential; reduced biodiversity; etc.	x	x	x	x	x
Desertification, soil erosion; soil/water salinisation; ground water depletion; frequency/ intensity of droughts/impacts; disturbance to hydrology; etc.		x	x	x	x
Land use changes, reduced agricultural diversity/flexibi- lity, and resource regeneration processes; resource extractive crops/technologies; increased external dependency; etc.	x	x			x
Third World poverty; indebted- ness; trade barriers, unequal exchange; etc.		x			x

a) Localized and widely replicated changes with potential for impact at global level.
b) Even a small change in one place affects changes in other places and has potential for directly influencing the operation and functions of the fundamental systems of Earth (e.g. CO_2 emission; see text).
c) By adding to emission of greenhouse gases and/or reducing their "sinks."
d) By generating pressure on the economic/social system for additional resource generation through activities contributing to global warming.
e) Impact on physical resource base and production environment of agriculture (e.g., moisture regime, soil erosion hazards, seasonality, length of growing season).
f) Impact on components of farming systems (e.g., adapted cultivars, crop combination, soil-moisture/agronomic management).
g) Impact on macro level agricultural systems/policies (e.g., irrigation, R&D, relief, etc.).
   (For sources to find examples on the above, see note under Table 13.1.)

However, before moving on to the regional context to indicate the possibility of making the above framework operational, it can be added that the anthropocentric perspective, which in a practical context focuses on human survival and welfare considerations while dealing with different phenomena (including global change), may prove instrumental in generating concerns and components for evolving strategies to manage cumulative changes. Thus, cumulative types of changes and anthropocentric perspectives have some form of complementarity and this can be harnessed while evolving strategies to "act despite uncertainty" on global warming scenarios.

Regional Adaptation Strategies

The approach outlined above can be illustrated by reference to any region for which details on current crises with long term implications are available. For the following reasons, the fragile resource regions of the world are eminently suitable cases for this purpose. Because of the fragility of their biophysical resource bases, they are quite vulnerable to rapid decline from even a small disturbance. The current pressures of population and market forces have already initiated resource degradation processes in such areas. One can quite easily identify a degree of convergence between the impacts of these processes and their remedies on the one hand, and the impacts of future climatic change and adaptations to them on the other.

Accordingly, we examined the situation in the fragile resource zones in southern Asia, namely arid and semiarid tropical region of India and the lower and middle hill areas of the Himalayan region. Despite being very different ecosystems, the two regions share a number of commonalities, such as inaccessibility, marginality, internal diversity, specific niche and human adaptation experiences in generally high risk, low productivity environments. Moreover, in both the cases, their increased integration with mainstream market systems and rapid population growth have similar impacts in terms of resource degradation and emerging indicators of unsustainability. Moreover, the key operational elements in the proposed approach to adaptation strategies to climate change (based on convergence between the impacts of [1] current crises and [2] regional climate change, and convergence between remedial measures against the two), consist of those indicators of unsustainability that have been studied by the author in the case of both regions (Jodha, 1991; Jodha *et al.*, 1992). Hence, despite various ecosystemic differences, the framework presented in this chapter can be illustrated by references to both or any of the regions covered. The following discussion uses the example of the dry tropical region only; a similar discussion for mountain areas is given in Jodha, 1992.

Regional Climate Change

The scenarios for future climate change in the dry tropical areas share the general limitations of change scenarios for other regions. Accordingly, there are no concrete predictions. Yet, on the basis of projections by various general circulation models (GCMs) on the arid and semiarid tropical regions (35°S to 35°N), temperature increases in the order of 0.5 °C to 4.0°C are predicted. Increases in precipitation are predicted for most, but not all, of the semiarid tropics. This increase is expected to take the form of convective rainfall, which could imply a higher intensity, but not necessarily an increased frequency, of precipitation. Thus, the seasonally dry tropics would have potentially high rainfall, high runoff (high soil erosion in plantless areas with undulating topography) and high evaporation, without necessarily having a lengthened or improved growing season. In view of the existing high variability of seasonal and annual rainfall, any trend toward decreased precipitation would have significant negative impacts in terms of prolonged and recurrent droughts (Jager, 1988).

Beyond the above conjectural information, not many details on future climatic changes are available from the GCMs. However, lack of certainty about change scenarios, as mentioned earlier, is the key reason for suggesting the approach to adaptation strategies put forward by this chapter. In keeping with the idea of cumulative changes as a focal point in evolving regional strategies to tackle future global climate change, we refer to certain dimensions of current realities faced by the dry regions (and mountain regions). The dimensions representing negative processes are, however, not unique to these two regions alone, but can be observed, and have been documented, in several similar ecosystems in the developing countries (ACTS, 1990; Allan *et al.*, 1988; Blaikie and Brookfield, 1987; Dregne, 1983; Garcia, 1981; Glantz, 1987; Grainger, 1982; Jodha *et al.*, 1992; Price, 1981; Rieger, 1981). Being so pervasive, when pooled over space, they have significant direct or indirect impacts at the international level. Internationally, the growing magnitude of the problem of "climatic refugees," the vanishing resilience of traditional farming systems and sustenance strategies of the people, the increasing need for large-scale and perpetual subsidization (in biochemical, physical, and economic terms) of agriculture in fragile resource zones, the rapidly widening gaps in food supplies, and worsening ecological conditions in fragile areas are some of the manifestations of localized cumulative changes that have begun to acquire a global dimension (Jodha, 1990; Jodha *et al.*, 1992). Both the potential contribution of these processes to "systemic" changes and the accentuation of the latter's impacts in the global context (Table 13.2) are not difficult to imagine.

Indicators of Unsustainability

As part of our inquiry into conditions for sustainable agriculture in fragile resource zones (Jodha 1991), we were alerted to over two dozen variables showing verifiable or measurable negative trends during a period as short as four to five decades. Some of them, such as the increased extent of landslides in the mountains and the deepening of water tables in dry tropical plains, were related to the resource base. Others were related to production flows, such as the decline in crop yield and biomass availability from grazing lands in both the mountains and the dry plains. Yet other negative trends were related to the infeasibility of resource-regenerative agricultural management practices (e.g., specific rotations and crop diversification) in both areas. Some of these changes are more visible (e.g., the extent of landslides and the decline in crop yields) than others (e.g. the substitution of shallow-rooted crops for deep-rooted crops owing to the erosion of topsoil). Again, some represented the consequence of negative processes (e.g., reduced diversity), while others formed part of the processes themselves (e.g., encouragement of mono-cropping through new technology). In totality, they manifest reduced production prospects for the present generation compared to past generations. When viewed in the context of the central element of the "sustainability" phenomenon (i.e. inter-generational equity) the above changes represent the emergence of inter-generational inequity and, hence, we call them indicators of unsustainability (Jodha, 1991).

Inquiry into the factors and processes associated with the above negative changes indicated the following:

- Disregard for the specific resource and environmental characteristics of these areas (fragility, marginality, diversity and so on) by private and public intervention resulted in the emergence of unsustainability indicators.
- Pressure generated on resources by population growth, market forces, and public interventions encouraged the disregard of regional resource specificities.
- Redundancy (due to infeasibility or ineffectiveness) of traditional technologies and the failure to develop new technological options, which could enhance the use-intensity and productivity of land resources without degrading them, was another factor responsible for the negative changes.
- Climate is part of the regional resource endowment, but, in the overall context of the functioning of the regional agricultural system, the role of climate-related factors in some of the negative changes (e.g., those related to groundwater recharge and drought) could be separately identified (Table 13.3).

Search for Sustainability

Restoration of the "sustainability" of agriculture in both the regions implies arresting and reversing the negative trends mentioned above. For this, one has to understand the factors and processes associated with the indicators of unsustainability. Thus, the problem of sustainability is approached through unsustainability. Looking at the nature and extent of unsustainability indicators, one can devise technological and institutional options to manage the current situation and plan for sustainability. Such options may be focused on resource protection, upgrading, and scientific management; growth and stability of crop production; and diversity and flexibility of resource use practices.

Such measures, besides strengthening the resource regeneration processes and resilience of farming systems in the regions, can help satisfy the conditions associated with sustainability. The latter means the ability of the agricultural system to maintain or enhance its production performance without damaging

TABLE 13.3. *Some Measures Directed Against Current Problems with Potential to Mitigate the Negative Impacts of Regional Climate Change in Dry Tropical Region of India*

Current problem[a]	Possible technological/institutional areas of intervention	Negative climate impacts likely to be affected by interventions		
		1st[c]	2nd	3rd
Deepening/salinization of groundwater (C, T, N, D)[b]	R&D on water conservation, management; irrigation development, water use regulations	x	x	x
Top soil/fertility erosion, plantlessness, sand movement (T, D)	Land rehabilitation programs, technologies for marginal lands, silvi-pastoral programs/technologies	x	x	
Land depleting usage, cropping patterns, technologies (T, N, D)	Resource/crop technologies for soil building/binding/productivity, institutional support systems	x	x	x
Reduced crop/animal yields, per capita land, etc (T, N)	Crop/resource-centred technologies, for high yields; population/ employment programs		x	x
Increased drought frequency/severity, relief-dependency (C)	Drought-resistant crop technology, resource conservation programs; alternative employment; linkage with stable areas/markets		x	x
Infeasibility of resource stabilizing, regenerative practices (T, N)	Reorientation of agricultural R&D and support systems with focus on sustainability		x	x

a   These problems represent "cumulative type of changes" (see text).
b   Capital letters indicate the causes: C = climate-related factors; T = technological failures; N = institutional (population, market, state intervention related factors); D = disregard of use capabilities of resources, and nonavailability of relevant technologies.
c   First-order impacts related to the resourcebase and production environment of agriculture. Second-order impacts related to components of the farming system. Third-order impacts related to macro-level agricultural and related policies and programs (see text for details).

"the essential ecological integrity of the system." More importantly, because of the aforementioned potential complementarity between the indicators of the current crisis and the negative impacts of regional climate change, it is possible to use remedial measures on current problems as a part of the strategy to mitigate the impacts of regional climate change. Thus, measures directed toward restoring the sustainability of fragile-resource agriculture can also help strengthen its capacity to withstand the negative impact of regional climate change. Such possibilities are broadly illustrated in Table 13.3. However, to understand this fully, it

will be useful to elaborate upon the climatic impacts with which we are dealing.

Climate-Impacts

The potential climatic impacts, listed along with their sequential interactions (Jodha, 1989), are called first-, second- and third-order impacts (see Table 13.3). Among the important climatic variables subject to change as a result of global warming are temperature, solar radiation, precipitation, humidity, evapo-transpiration, soil moisture and runoff. Their likely immediate impact will be on the major components of the physical resource base and the production environment of agriculture (in our illustration). They are called first-order impacts and they would cover variables such as moisture regimes, growing seasons, micro-climatic stress, seasonality and stability of weather, disease and pest complexes, biomass potential, photosynthesis, plant-input interactions, soil chemistry, and erosion hazard.

In turn, the changes in the above variables will influence the components and features of farming systems. Covered under second order impacts, they will relate to adapted plant and animal species, combinations and linkages of agricultural activities, moisture management and water security measures, farm activity calendars, agronomy and input practices, risk-management mechanisms, production flows, and yields and returns.

Changes in the structure and functioning of farming systems would directly or indirectly induce changes in macro-level agricultural systems, policies, and strategies. Designated as likely third order impacts of climatic change, they include irrigation systems and strategies, relief policies and programs, agricultural infrastructure, input supply systems, research and development (R&D) priorities and strategies, marketing and trade systems, intersectoral linkages, agricultural planning strategies, and employment and income distribution.

With the help of the above details on climatic impacts, it is not difficult to understand that the negative orientation of any of the above impacts can further aggravate the crisis situations manifested by indicators of unsustainability in the region. Similarly, the severity of the climatic impacts will be much greater if the current negative trends in the regions cannot be reversed. The crucial point in the above context is that, despite potential links between the two, the present crisis is more definite, while the projected negative climatic impacts are uncertain; consequently, measures against the former are more easily acceptable by policymakers. Besides, any improvements, such as land resource upgrading, technologies against yield reducers (such as salinity or drought), increased diversity and flexibility of agriculture and promotion of off-farm employment and income generation,

while solving the current problems, may equip the regions to withstand future climatic impacts better, without explicitly planning for them. Finally, since there is no exclusive effort involved in developing adaptation strategies to climatic change, there are no costs and risks in terms of redundancy and wastefulness of the effort, if the predicted change scenario does not materialize.

We illustrate this for the dry tropical region in Table 13.3, which focuses on some of the components of the current crisis (indicators of unsustainability). It also lists the broad areas of technological and institutional interventions necessary to manage the current problems. These interventions have the potential to influence or mitigate the likely negative impacts of regional climate change. For instance, measures directed at dealing with emerging salinity of groundwater and declining of water tables in terms of R&D on water management, irrigation development and water use regulation can also provide protection against the relevant impacts of climate change. The latter include disturbance in the length of the growing season, increased frequency of weather aberrations (first order impacts); disturbance to moisture management systems and agronomy (second order impacts); and irrigation systems and water pricing policies (third order impacts). Similarly, interventions in terms of alternative crop technologies, land rehabilitation programs, and public relief programs directed to current problems can have mitigating effects on the different negative impacts of climatic change.

The details presented in Table 13.3 can be presented differently. As indicated by Table 13.4, the areas of convergence between current and (climatic impact-based) future problems can be seen through the orientation of agricultural R&D in the dry tropical area. Table 13.4 presents areas of convergence for R&D and institutional support that have direct links with current problems and likely climate impacts. The promotion of technologies with this focus can act as a major component of adaptation strategies to regional climate change, despite the uncertainties associated with it.

The above discussion presents a broad picture of the situation in dry tropical areas of India where, by using available information

TABLE 13.4. *Areas of Convergence Between Regional Climate Change Impacts and Current Agricultural Problems Focused through Orientation of R&D in Dry Tropical Region, India*

Selected areas of potential negative impacts of regional climate change	Selected current problems					
	Ground water depletion	Soil erosion	Crop yield decline	Forest, pasture health	Protection against drought	Income employ- ment
First order impacts						
Weather aberrations	R ᵃ	R	H	H	R	N
Length of growing season			S		R	
Plant-input interaction		R	S	H		
Soil erosion hazard		R	S	H		
Second Order Impacts						
Adapted cultivars	R		R		R	
Agronomic practices		S	S	S	S	
Risk adjustment				R	S	N
Yields/returns		R	R			R
Third Order Impacts						
Irrigation systems	R				R	N
Relief policies				H	S	R
Agricultural support systems			N		N	N
Income/employment policies			N		N	N

a   Capital letters indicate the broad areas of R&D. More than one R&D area could be relevant for specific situations indicated by the Table. However, we have indicated the most relevant one in each case (see Jodha, 1991, for details).

R =   Resource-centered R&D: physical/biological measures to manage slope, drainage, soil, moisture; soil-binding, building plants/crops; fast growing, high yielding annual/perennials, local resource renewability.

S =   Systems-oriented R&D: for combining conservation and productivity, crop-resource centered management options, diversified interlinked resource use.

H =   R&D focused to harness "niche," comparative advantage, and resource diversity; sensitive to resource capabilities; wider adaptation of crops, stability and growth of biomass (especially perennials).

N =   Institutional measures - policies, programs, projects (related to technological and other measures).

on current problems and currently contemplated (or even adopted) remedial measures, one can relate the current problems to future climate-change impacts. In addition, one can identify measures that are primarily directed toward handling the current problems, but which can have some mitigative influence on the problems created by future climate-change impacts. Thus, the approach can serve a dual purpose.

Summary

To sum up, material in this chapter has attempted to illustrate one approach for developing adaptation strategies to global climate change in the regional context, despite the uncertainties of predicted change scenarios. The approach tries to identify certainty components of the problem by using the lead offered by concepts such as "cumulative types of change" and "anthropocentric perspectives" on the problem. This approach, illustrated by reference to the dry tropical areas of India and, to a limited extent, the mountain region of the Himalayas, can help link the current problems in a regional context to problems associated with future global climate change. Remedial measures, conceived for current problems, can also offer a by-product in the form of the increased ability of a region to withstand the impacts of climate changes.

One can identify a number of current measures in sectors other than agriculture (e.g., restriction on exhaust emissions from motor vehicles as a part of better living conditions in urban areas, or a greater emphasis on nonconventional energy sources due to the high cost of crude oil) that will fit into the framework presented in this chapter. The only key requirement is the under-standing of the linkages between the current problem and its remedial measures and the aspects/impacts of global warming.

With the help of more specific situations from different sectors, the focus of the approach can be further sharpened. Apart from the diminished role of "uncertainties" in obstructing action on the current problem, the approach has a number of other advantages. These include:

- Options are easy to conceive and acceptable to decision makers, particularly in the developing world where efforts are concentrated upon current problems.
- The problems caused by intercountry differences in perspective and externalities (in terms of the inability to restrict the gains of remedial measures to oneself) would not obstruct action under this approach.
- Under this approach there are no risks of redundancy of options and associated resources if predicted change scenarios do not materialize, because the remedial measures are not designed to handle unknown and unidentified factors.

However, it should be noted that this approach cannot substitute for the measures and approaches required to deal directly with the global warming problem. This approach is helpful only to the extent that "cumu-lative" change plays a role in global warming and the accentuation of its impacts. Its strong point is that it helps in integrating the concerns of current problems with those of the future impacts of global warming and advocates dual purpose strategies to treat the two, without being unduly obstructed by the uncertainties of change scenarios.

Acknowledgements

The author is grateful to Greta Rana for her valuable comments on the first draft of this chapter.

References

Abelson, P.H., 1990. Uncertainties about global warming. *Science*, **247**(4950): 1529.

ACTS, 1990. *The Nairobi Declaration on Climatic Change*. African Centre for Technology Studies Press, Nairobi.

Agarwal, A., and S. Narain, 1991. *Global Warming in an Unequal World*. Centre for Science and Environment, New Delhi.

Allan, N.J.R., G.W. Knapp and C. Stadel (Eds.), 1988. *Human Impacts on Mountains*. Rowman and Littlefield, Totowa, N.J.

Blaikie, P.M. and H. Brookfield, 1987. *Land Degradation and Society*. Methuen, London.

Chen, R.S., E. Boulding and S.H. Schneider (Eds.), 1983. *Social Science Research and Climate Change: An Interdisciplinary Appraisal*. D. Reidel Publishing, Dordrech.

Clark, W.C., 1985. *On the Practical Implications of the Carbon Dioxide Question*. WP-85-43. International Institute of Applied Systems Analysis, Laxenburg, Austria.

———, 1988. The human dimensions of global environment change. In: *Committee on Global Change, Towards an Understanding of Global Change*. National Academy Press, Washington D.C.

Cooper, C.L., 1989. Epilogue. In: *Greenhouse Warming: Abatement and Adaptation*. N.J. Rosenberg, W.E. Easterling, P.R. Crosson, *et al.* (Eds.).

Resources for the Future, Washington D.C.

Dorfman, R., 1991. Protecting the global environment: An immodest proposal. *World Development*, **19**(1):103-110.

Dregne. H.E., 1983. Desertification of arid lands. *Advances in Arid Land Technology and Development*. Haswood, New York.

Flavin, C., 1989. Slowing global warming: A worldwide strategy. *World Watch Paper 91*. The World Watch Institute, Washington D.C.

Garcia, R. 1981. *Drought and Man. Vol. I: Nature Pleads Not Guilty*. Pergamon Press, Oxford, U.K.

Glantz, M.H. (Ed.), 1987. *Desertification: Environmental Degradation in and Around Arid Lands*. Westview Press, Boulder, Colo..

———, B.G. Brown, and M.E. Krenz, 1988. *Societal Responses to Regional Climate Change: Forecasting by Analogy*. National Center for Atmospheric Research, Boulder, Colo.

Gleick, P.H., 1989. Climate change and international politics: Problems facing developing countries. *Ambio*, **18**(6).

Grainger, A. 1982. *How People Can Make Deserts: How People Can Stop and Why They Don't*. Earthscan, London.

IPCC (Intergovernmental Panel on Climate Change), 1990. IPCC First Assessment Report: Overview. WMO and UNEP, Geneva.

Jager, J, 1988. Developing policies for responding to climate change [Summary of discussions and recommendations of the workshops held in Villach and Bellagio, September-November, 1987] WCIP-1, WMO/TD-No. 225. World Meteorological Organization, United Nations Environment Programme, Geneva.

Jodha, N.S., 1989. Potential strategies for adapting to greenhouse warming: Perspectives from the developing world. In: *Greenhouse Warming: Abatement and Adaptation*. N.J. Rosenberg, W.E. Easterling, P.R. Crosson, *et al.* (Eds.). Resources for the Future, Washington D.C.

———, 1990. Rural common property resources: Contributions and crisis. SPWD Foundation Day Lecture May 16, New Delhi: Society for Promotion of Wasteland Development. Reproduced in *Economic and Political Weekly (Quarterly Review of Agriculture)*, **25**(26):A.65-A.78.

———, 1991. Sustainable agriculture in fragile resource zones: The technological imperatives. *Economic and Political Weekly (Quarterly Review of Agriculture)*, **26**(13):A.15-A.25.

———, 1992. Global change and environment risks in mountain ecosystems. In: *Global Environmental Risk* (UNU Volume). R.E. Kasperson and J.X. Kasperson (Eds.), (in press).

———, S.M. Virmani, S. Gadgil, *et al.* 1988. The effects of climate variations on agriculture in dry tropical regions of India. In: *The Impact of Climate Variations on Agriculture: Assessment in Semi-Arid Regions. Vol. 2.* M.L. Parry, T.R. Carter and N.T. Konjin (Eds.), Kluwer Academic Publishers, Dordrecht, The Netherlands.

———, M. Banskota and T. Pratap (Eds.), 1992. *Sustainable Development of Mountain Agriculture* (Vol. 1 and 2). Oxford and IBH Publishing Company, New Dehli.

Kasperson R.E., K. Dow, D. Golding, *et al.* (Eds.), 1989. *Understanding Global Environmental Change: The Contributions of Risk Analysis and Management*. A Report of an International Workshop, Clark University, Worcester, Mass.

Kates, R.W., 1990. Hunger, poverty, and human environment. Distinguished Speaker Series, Center for Advanced Study of International Development, Michigan State University, East Lansing.

Parikh, J. K., 1991. *Billion Dollar Misunderstanding: A Long Way to Go for Fair Environmental Negotiation*. Indira Gandhi Institute of Development Research, Bombay.

Pearce, D., A. Markandya, and E.B. Barbier, 1990. *Blueprint for a Green Economy*. Earthscan, London.

Price, L.W., 1981. *Mountain and Man: A Study of Process and Environment*. University of California Press, Berkeley.

Price, M.F., 1990. *The Human Aspects of Global Change, Final Report*. Environment and Societal Impacts Group, National Centre for Atmospheric Research, Boulder, Colo.

Rieger, H.C., 1981. Man versus Mountain: The Destruction of Himalayan Ecosystem. In: *The Himalaya: Aspects of Change.* J.S. Lall and A.D. Moodie (Eds.). Oxford University Press, Delhi, p. 27.

Rogers, P., and M. Fiering, 1989. Climate change: Do we know enough to act? *Forum for Applied Research and Public Policy* (Winter Issue), pp. 6-13.

Schneider, S.H. and N.J. Rosenberg, 1989. The greenhouse effect: Its causes, possible impacts, and associated uncertainties. In: *Greenhouse Warming: Abatement and Adaptation.* N.J. Rosenberg, W.E. Easterling, P.R. Crosson, and J. Darmstadter (Eds.). Resources for the Future, Washington D.C.

Spencer, R.W., and J.R. Christy, 1990. Precise monitoring of global temperature trends from satellites. *Science*, **247** (4950):1558-62.

Turner B.L., R.E. Kasperson, W.B.Meyer, *et al.*, 1990. Two types of global environmental change: Definitional and spatial-scale issues in their human dimension. *Global Environment Change*, I(1):14-22.

UNEP and the Beijer Institute, 1989. *The Full Range of Responses to Anticipated Climate Change.* United Nations Environment Programme, Nairobi.

UNU and IFIAS, 1989. Industrial metabolism: Restructuring for sustainable development. Report of workshop held at Maastricht, The Netherlands.

White, R.M., 1990. The great climate debate. *Scientific American*, **263**(1):36-43.

Wuebbles, D.J., and J. Edmonds, 1988. *A Primer on Greenhouse Gases.* National Technical Information Service, U.S. Department of Commerce, Springfield, Va.

14. GLOBAL CLIMATE CHANGE: THE ROLE OF DEVELOPING COUNTRIES

José Goldemberg

Less developed countries (LDCs) contribute to CO_2 emissions in two distinct ways: fossil fuel burning and deforestation. Carbon dioxide emissions from both developed and developing countries are responsible for 55 percent of the greenhouse effect and the resulting potential climate changes.

No significant differences exist in the way energy is produced and used in LDCs or in the industrialized nations. The structure of societies in developing countries is dual in character, with large and prosperous urban centers surrounded by a "sea of poverty" in the form of slums or rural areas. In these areas few of the comforts of everyday life, such as electricity, running water and sewage disposal, exist. Patterns of consumption in the prosperous areas of urban centers are approximately the same as in industrialized countries, based on the burning of fossil fuels. However, except in China, the majority of LDCs are located in the tropics and do not use large amounts of energy for heating.

In the rural areas of large parts of Africa, Asia and Latin America noncommercial energy sources are used mainly for cooking, either in the form of fuelwood, charcoal or agricultural wastes. Deforestation is a significant result of such use. Significant deforestation also occurs as a result of land clearing for crop production and pastures.

Emissions of CO_2 from fossil fuel burning are known quite accurately. Figure 14.1 shows the evolution of commercial energy consumption from the mid-1960s to 1988 in a number of areas (BP 1989). Figure 14.2 shows the tonnage of carbon emitted as a result of such consumption around the world (Marland *et al.*, 1988). LDCS (including China) contributed 1.33 Gton/year to CO_2 emissions in 1985, 19 percent of the total worldwide tonnage of carbon emitted as a result of fossil fuel consumption.

On the other hand, deforestation—usually accompanied by slash-and-burn practices (after some hardwoods are taken out for commercial use such as furniture making)—is responsible for another 1.66 Gton/year (24 percent of the total). This figure is the best estimate based on a number of authors whose values range from 0.9 to 2.5 Gton. Releases of CO_2 from tropical forests dwarf the estimated additional 0.1 Gton released by harvesting the trees of temperate and boreal forests (Houghton, 1990).

The contribution of LDCs to total CO_2 emissions, which is already high (43 percent), is likely to increase because the industrialized countries have made significant progress in halting energy growth (and consequently CO_2 emissions), as shown in Figure 14.1. In the LDCs, however, commercial energy consumption has been growing for a long time and shows no indication of saturation.

Such unavoidable growth in energy consumption in LDCs—in addition to the continued deforestation of tropical forests—could jeopardize all of the efforts of industrialized countries to stabilize the composition of the atmosphere. Stabilization of the emissions of greenhouse gases will be a difficult task, unless the industrialized nations reduce their emissions to offset the increase in emissions of the developing countries. A model outlining in detail how this could be done has been proposed by Goldemberg and others in a book with the suggestive title *Energy for a Sustainable World* (Goldemberg *et al.* 1987).

Stabilizing emissions, however, is not the same as stabilizing the composition of the atmosphere. This will require a reduction in global emissions. According to a study conducted by the U.S. Environmental Protection Agency (Lashof and Tirpak, 1989), emissions of CO_2 would have to be reduced by 50 percent in the next 15 to 20

FIGURE 14.1. World commercial energy consumption between 1969 and 1988, expressed as million tons of oil equivalent.

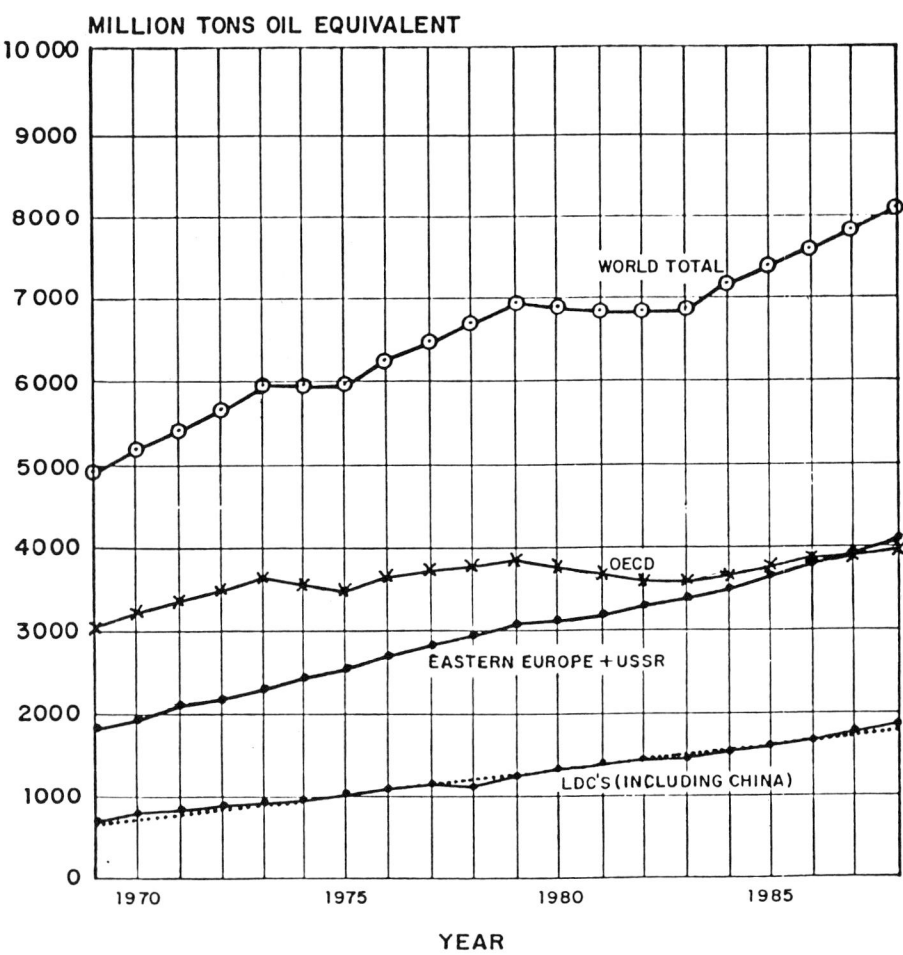

years, if a stable atmosphere is to be achieved early next century. In addition, realization of this goal requires the phase out of CFC's and reductions in the emission of other gases. We will discuss in this chapter how LDCs can contribute to such goals through the more efficient use of energy, a larger share of renewables, and reforestation.

Energy Consumption in LDCs

The growth in energy consumption in LDCs can be explained by a combination of factors:

1. "horizontal growth" as the rural areas modernize and switch to commercial energy use, e.g., replacing fuelwood with LPG (liquefied petroleum gas) or kerosene;

2. "vertical growth" as industrialization and urbanization proceed, increasing the consumption of coal, oil derivatives, electricity and other modern energy carriers;

3. population increase: the growth in energy consumption and corresponding affluence that comes with it will, in due time, reduce the rate of population growth, but not in the time periods considered here, i.e., the next 20 to 30 years.

The expansion of commercial energy supplies grew in LDCs at an average annual

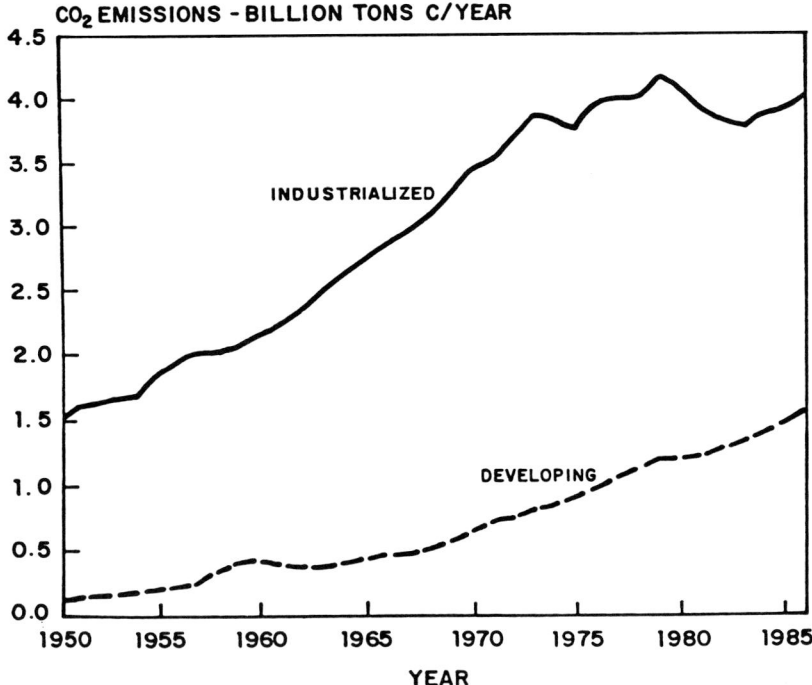

CO$_2$ EMISSIONS - BILLION TONS C/YEAR

INDUSTRIALIZED

DEVELOPING

YEAR

FIGURE 14.2. Total annual fossil CO$_2$ emissions between 1956 and 1986 for industrialized and developing countries.

rate of 3.6 percent per capita in the 1970s. This growth has been costly, especially for oil-importing developing countries, whose oil imports grew at an average annual rate of 6.3 percent during the 1970s. By 1981, low- and medium-income developing countries, respectively, spent 61 percent and 37 percent of their export earnings on oil imports (The World Bank, 1983).

If the pattern of the 1970s were to persist, the average per capita rate of primary commercial energy use in developing countries would increase from 0.55 kW in 1980 to 2.3 kW in 2020 (Table 14.1). This implies that aggregate commercial energy use in these countries would increase from 1.8 Terawatts (TW) in 1980 to 15 TW in 2020, in light of the expected near doubling of the population. This increase in total energy use, predicted using "historical trends," is equivalent to 1.5 times the total world energy use in 1980 (10.3 TW). Increasing energy supplies to meet such energy requirements at reasonable costs and without major environmental or security problems would be exceedingly difficult.

Even somewhat less ambitious energy growth scenarios pose formidable economic challenges. The World Bank estimated in

1983 that a 2.5 percent annual growth in per capita commercial energy use between 1980 and 1995 would entail investments in new energy supplies for developing countries averaging some $130 billion annually (in 1982 dollars) between 1982 and 1992 (The World Bank, 1983). Half of this investment would have to come from foreign exchange earnings, requiring an average annual increase of 15 percent in real foreign exchange allocations to energy supply expansion. In this less ambitious expansion in energy production, oil imports by oil-importing developing countries would still increase by almost one third, to nearly 8 million barrels of oil per day by 1995.

More recently, Schramm (1990) discussed the magnitude of the capital mobilization problem based on a detailed compilation of the electric power expansion plans for more than 70 developing nations carried out by the World Bank. The total installed generating capacity in the developing countries is projected to increase from 471 to 855 GW in the 1990s. The corresponding projected annual electric power growth rate in the 1990's is about 6.6 percent, substantially lower than the historical rates of 10 percent in the 1970s and 7 percent in the 1980s.

TABLE 14.1. *Projection of the Energy Consumption in the LDCs in the Year 2020*

	Total consumption (TW-year/year)	Consumption per capita (kW)
World Energy Conference (low)[a]	6.8	1.1
International Institute for Applied Systems Analysis (low)[b]	7.0	1.2
Energy for a Sustainable World[c]	7.4	1.2
Projection based on the 1970s' trend (3.6% per year)	15.0	2.3
Consumption in 1980 in LDCs	1.8[d]	0.55
Consumption in 1980 in industrialised countries	7.0	6.3

a. WEC (1978).
b. Haefele *et al*. (1981).
c. ESW in Goldemberg *et al*. (1987).
d. Excludes 1.5 TW of noncommercial energy.

Capital requirements needed to sustain this projected growth over the 1990s total about $760 billion in 1989 U.S. dollars, or more than one trillion nominal dollars, of which about 60 percent is for generation, and the remainder is for transmission and distribution. According to Schramm (1990), these needs are approximately one order of magnitude higher than the funds available from the World Bank. The staggering costs of providing such increases in energy supply lead many to believe—if rarely to state publicly—that living standards cannot be substantially improved in developing countries.

Most of the LDCs are comparatively bereft of fossil fuels. The principal remaining reserves of coal are in the former Soviet Union, the United States and the Peoples' Republic of China. The largest remaining reserves of oil are in the Persian Gulf region, and the reserves in the United States are dwindling. Ten years from now, if OPEC (Organization of Petroleum Exporting Countries) were to gain full control of oil prices, availability of oil at affordable prices might be limited.

Less developed countries have, therefore, good reasons to worry about their future access to energy sources, even disregarding any environmental considerations. By and large, it is recognized by many in the LDCs that GDP has to grow fast, and that energy growth has to grow even faster, in order to build the necessary infrastructure of roads, heavy industry and housing. No politician in the LDCs would dare defend any other policies. However, these goals are in conflict

with the available financial resources, as pointed out above. In addition, such growth poses a serious threat to the global environment.

Aspirations of impoverished people in these countries for betterment of their living conditions most probably will not be fulfilled, thus leading to serious political and economic instabilities. Internal and external conflicts are bound to increases, the former because large segments of the population might try to impose a redistribution of income (and therefore of access to energy) through force; the latter because regional distribution of energy resources is quite heterogeneous around the world, and the stronger countries will try to guarantee access to oil and other natural resources.

The Conventional Energy Consumption Scenario for LDCs

Such gloomy projections are translated into numbers in some of the leading studies conducted to predict future energy consumption in LDCs. These studies, conducted by the World Energy Conference (1978) and the International Institute for Applied Systems Analysis (Haefele *et al.*, 1981), take into account the rising cost of oil and a number of other factors.

The WEC study was carried out by the Energy Research Group at the Cavendish Laboratory, Cambridge, England, for the Conservation Commission of the World Energy Conference. It is a global analysis that looks to the year 2020. An important feature of WEC analysis for developing countries is that it takes noncommercial energy into account.

The IIASA study (carried out between 1973 and 1979 by analysts at the International Institute for Applied Systems Analysis in Vienna) makes global projections to the year 2030, by aggregating projections for several regions of the world. The primary emphasis in this study is on commercial energy. The basic approach in these two studies consists of two steps:

- estimating future energy demand on the basis of assumptions about future demographic and economic trends together with historical correlations between such trends and energy demand;
- matching this demand to a mix of energy supplies.

This mix is chosen so that it is compatible with estimates of the energy resources base and, in the case of new energy supply technologies, with judgments about how much energy can be produced by these supply technologies at various future dates.

The WEC and IIASA studies have one long-run objective in common: to shift from oil to more abundant energy resources, such as natural gas, coal and nuclear energy, while preserving a significant growth of energy. According to these studies, consumption per capita in the LDCs could reach 1.1 or 1.2 kW by 2020, as compared to 0.5 kW/capita in 1980, with a 10- to 15-fold increase in nuclear output (Table 14.1).

By comparison, energy consumption in industrialized countries, excluding energy used for warming in the winter (which represents approximately 30 percent of total energy consumption), was 6.3 kW/capita in 1980. These numbers indicate a level of amenities three times lower—on the average—in developing countries than in industrialized countries. Since some energy efficiency improvements have been incorporated into this scenario, the situation in the LDCs might be a little better than that. It is clear, however, that the IIASA scenario will limit the potential for development of LDCs.

Are there any better prospects for developing countries compatible with a better level of life and due regard to environmental considerations? The issue was addressed in a book by Goldemberg and others (1987) entitled *Energy for a Sustainable World*, published in 1988, which gave rise to the ESW scenario.

The Energy for a Sustainable World Scenario

The Energy for a Sustainable World (ESW) scenario was constructed for a hypothetical developing country whose needs for energy services resemble those of Western Europe in the 1970s (excluding space heating, which few developing countries require, with the notable exception of China). Table 14.2 shows that one can do much better than IIASA's projection by introducing energy-saving technologies. It is obvious in this table that the ESW study incorporates a higher degree of energy-saving technologies and provides a level of amenities for the population higher than the IIASA projection. The per capita energy use for each activity is the product of the activity level and an energy intensity corresponding to either the most energy-efficient technology on the market today or an advanced technology that could be commercialized within about a decade.

The resulting total energy use per capita in LDCs, about 1.2 kW/capita according to the ESW, is only 20 percent higher than the average consumption in 1980, if one includes commercial and noncommercial energy sources, as indicated in Table 14.1. If this goal is achieved by the year 2020, total energy consumption in the LDCs would have to rise from the present 3.3 TW to 7.4 TW as a result of per capita energy and population increase. Large improvements in living standards characterizing the ESW scenario can be achieved with such small increase in per capita energy use, for two reasons:

- The shift from traditional, inefficient noncommercial fuels, to modern energy sources (electricity, liquid and gaseous fuels, processed solid fuels, etc.). The importance of this change is plain: In Western Europe, where noncommercial fuel use is small, per capita final energy use for purposes other than space heating was only 2.3 kW/capita in 1975, about 2.5 times the total per capita consump-

TABLE 14.2. *Per Capita Activity Levels for Selected Activities in Developing Countries*

Activity	Activity Index	1975 activity levels for developing	Activity levels for the ESW scenario	Activity levels for the IIASA[a] low scenario
Domestic hot water	Liters of hot water per capita per day	9[b]	50	20[b]
Service sector development	Square meters of commercial floor space per capita	1.1	5.4	2.6
Car use	Number of cars per capita	0.0107	0.19	0.047
Air travel	Passenger-kilometer per capita per year	14	345	82
Truck freight	Tonne-kilometers per capita per year	545	1,495	1,378
Rail freight	Tonne-kilometers per capita per year	189	814	625
Steel production	Kilograms per capita per year	21[c]	320	64
Cement production	Kilograms per capita per year	77[c]	479	NA[d]
Nitrogen fertilizer production	Kg of contained N per capita per year	3[c]	26	NA[d]

a. These are population-weighted average values for all developing countries except the centrally planned Asian economies. Unless otherwise indicated, the parameters shown are from A.M. Kahn and A. Holzl, *Evolution of Future Energy Demands till 2030 in Different World Regions: An Assessment Made for the Two IIASA Scenarios* (Laxemburg Austria: International Institute for Applied Systems Analysis, April 1982). End-Use analysis was not carried out for the centrally planned Asian economies in Kahn and Holzl.
b. The parameters presented in the International Institute for Applied Systems Analysis (IIASA) study are converted from kcal per capita to liters of hot water per capita, assuming the water is heated by 30°C.
c. From United Nations, *1979/80 Statistical Year Book.*
d. Steel is the only basic material for which demand levels are explicitly indicated in the IIASA scenarios.

NA = not available.

tion of developing countries, although per capita GDP was 10 times as large.

• The adoption of new and more energy-efficient technologies. Some of the assumed technologies for the residential sector illustrate how large increases in amenities can be achieved without increasing energy consumption to Western Europe levels. For example, the average household is assumed to have a refrigerator-freezer that is as energy efficient as the most efficient two-door unit available in Europe in 1982, a 315-liter unit requiring 475 kWh per capita year (or less than one-third of the electricity required by the average U.S. refrigerator-freezer). Similarly, the level of lighting services is assumed to be the equivalent of five 75 -watt incandescent bulbs operating four hours a day, though this light is delivered by compact fluorescent bulbs that draw only one-fourth as much electricity.

"Leapfrogging" as a Development Policy in LDCs

From the previous discussion, it is clear that one way for developing nations to address the dilemmas posed by environmental and economic concerns is to "leapfrog" the technological path followed by industrialized countries in the past. This means incorporating energy-efficient technologies into their process of development. Long-term studies of the evolution of "energy intensity" (i.e., energy consumption by unit of gross domestic product [E/GDP]) for a number of countries indicate that such a ratio increases in the initial phases of development when the heavy industrial infrastructure was put in place, goes through a peak and then decreases steadily (Martin, 1988). This point is illustrated in Figure 14.3, where per capita oil consumption is plotted against U.S.$1,000 of GDP for countries at various stages of economic development.

The data indicate that latecomers in the development process follow this same pattern with much less accentuated peaks; they don't have to reach high values of the E/GDP ratio, even in their initial stages of industrialization, because they can benefit from the modern methods of manufacturing and more efficient systems of transportation already developed. In other words, the "decoupling" of energy and GDP growth—considered inseparable in the past—is not a general feature of modern economies. As pointed out by Drucker (1986), Strout (1985) and Williams *et al.* (1987), two primary reasons account for this trend. First, there has been a saturation in the consumption of consumer goods, and in industrialized societies economic activity has been moving in the direction of services and not heavy industry. Second, there has been a revolution in materials, with a clear shift toward those of less energy-intensity.

Such trends were evident in the industrialized countries before the oil crisis of 1973, with the increase in oil prices only accelerating the pace of structural change. This is evident from the study conducted by Strout (1985), which showed that consumption of

FIGURE 14.3. Evolution of the energy intensity in different countries, expressed as tons equivalent of petroleum per U.S.$1,000 of gross domestic product.

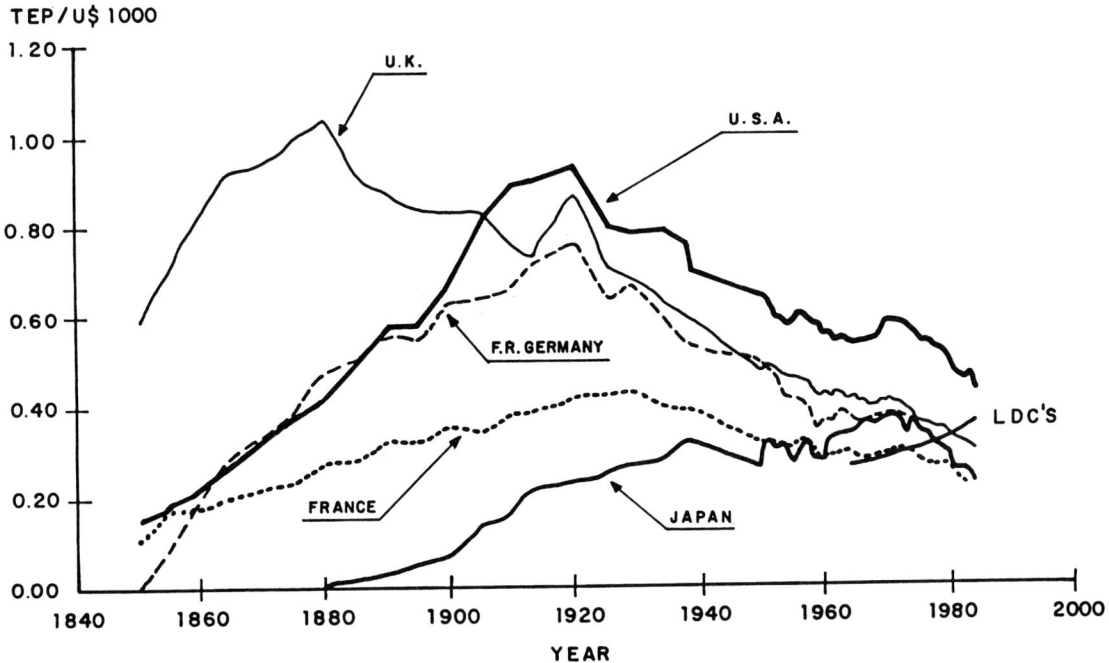

ten energy intensive materials in 52 countries (wood pulp, paper and paper-board, chemical fertilizers, hydraulic cement, steel products, copper, lead, zinc, aluminum and tin) closely correlated with GDP up to per capita incomes of U.S.$2,000 (in 1970 dollars); above that, the strong coupling between these two indicators disappears (Figure 14.4). Thus, in more affluent societies, where most of the infrastructure is already in place, the structural shifts mentioned above dominate the picture.

If we concentrate our attention on the industrial sector of the U.S. economy, the energy intensity of this sector decreased 3.5 percent per year in the period 1973-1986, faster than the overall energy intensity (Williams *et al.* 1987). This decrease was due to two equally important reasons.

1. There was a shift from energy-intensive manufacture of materials to lighter materials, leading to a reduction in energy consumption of 1.6 percent/year. There

is an increasing tendency for developed countries to import energy-intensive materials from developing countries. In 1986 the net imports of embodied energy in goods and services traded by the United States reached about 3 percent of total energy use in the United States. It is estimated that 25 percent of the primary metals industry of the United States will be shut down in 10-20 years (Roop, 1986).

2. There occurred a reduction in the energy intensity of many industrial sectors, accounting for 1.9 percent/year.

It can be concluded from this that the decline in energy intensity in the developed countries is taking place for a variety of reasons and that the fast pace of technological advances in goods and materials, coupled with greater attention to efficient energy use, has led to an accelerated decline in energy intensity over the years.

FIGURE 14.4. Energy use in materials consumption, expressed as barrels of oil equivalent (BOE) per capita, as a function of gross domestic product (GDP).

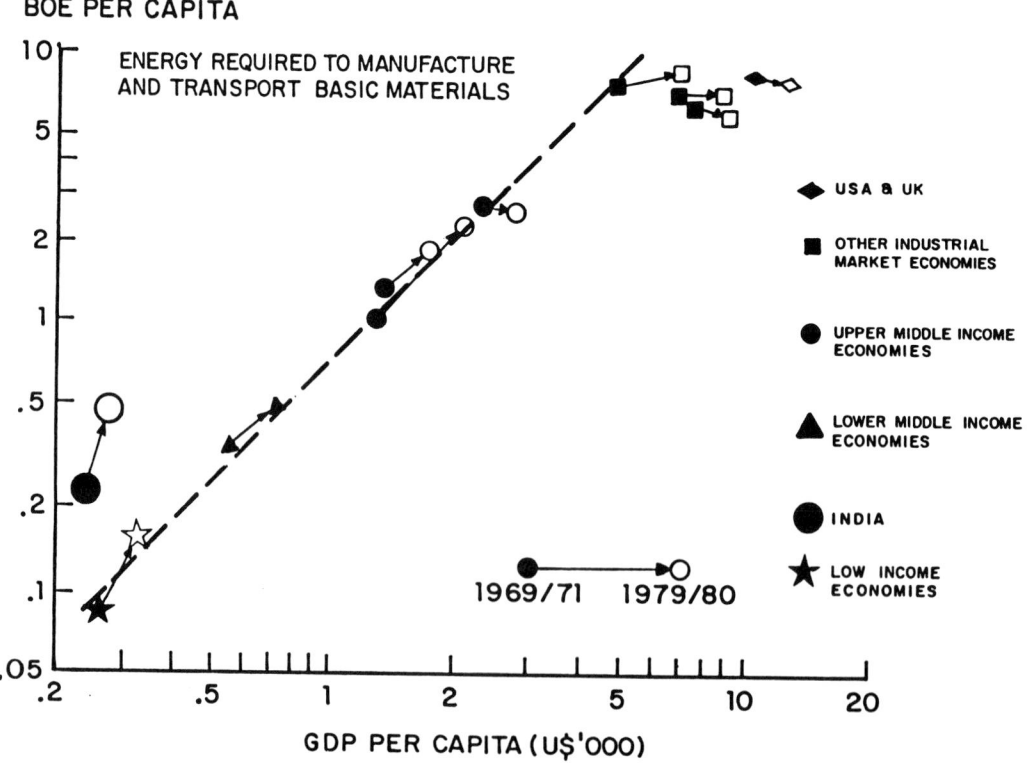

Environmental considerations are bound to accelerate such trends, and "energy conservation" (a synonym for efficient energy use) is likely to play a very important role in the future. The political leadership in the OECD (Organization for Economic Cooperation and Development) countries—pressured by public opinion—will probably adopt public policies designed to cut CO_2 emissions by imposing new taxes on emissions or introducing new mandatory regulations. One example of such regulations is the CAFE efficiency standards imposed by the U.S. Congress in 1975, resulting in an increase in the overall fuel performance of American cars by more that 50 percent.

In contrast, as shown in Figure 14.3, the energy intensity in LDCs is increasing and could jeopardize, in terms of greenhouse gaseous emissions, all efforts of industrialized countries to reduce them. What are the possibilities for introducing modern energy efficient technologies in such countries?

Opportunities for Leapfrogging in Developing Countries

We will discuss here a number of examples of leapfrogging possibilities, relying heavily on data from Brazil, which is representative of the conditions prevailing in a number of developing countries. These examples are gasoline consumption in automobiles, ethanol as an automotive fuel and electricity consumption in refrigerators.

Gasoline Consumption in Automobiles

Gasoline consumption in automobiles is a major contributor to total energy consumption in developed countries and, as industrialization and urbanization proceeds, it is also becoming important in developing countries. Fuel efficiency is improving gradually in the developed countries through a number of improvements and technological advances that include the following:

- the optimization of motor tuning and air-fuel mixtures through the use of electronic devices;
- other engine improvements (e.g., four

valves per cylinder);
- improvements in the aerodynamics and tires;
- the use of sophisticated gear boxes;
- weight reduction as a result of smaller engines, front-wheel drive, and better materials.

The fuel efficiency of vehicles in the United States has improved from an average of 13.3 mpg in 1973 to 19.2 mpg in 1987. It is evident that considerable improvements have taken place in the last 15 years in the United States and there are hopes that further gains can be obtained between now and the year 2000. Gains in terms of fuel consumption can be enormous: If vehicles in the U.S. reach the level of 45 mpg in the year 2000, this would mean savings of 2 quads (about 1.0 million barrels/day).

It is interesting to inquire whether fuel-efficiency improvements are also taking place in the developing countries where the large cars manufacturers such as Ford, Fiat, Volkswagen and General Motors are transferring their factories. Table 14.3 compares the gasoline consumption in France and Brazil in 1988 for a number of car models. As can be seen in this table, gasoline consumption for similar models is approximately 25 percent to 30 percent higher in Brazil than in France at 90 km/h (road driving). In urban traffic conditions, consumption in Brazil is slightly higher; for the Fiat models it is appreciably higher. It is clear, therefore, that considerable savings in fuel consumption could be gained in Brazil by adopting the up-to-date technology currently used in Europe or in the United States for the manufacture of motor vehicles in Brazil. The same applies to other developing countries.

The Alcohol Program in Brazil

The alcohol program in Brazil, an innovative response to the oil crisis of the 1970s, is another example of technological "leapfrogging." Faced with an increasing deficit in its trade balance caused by an enormous jump in petroleum prices, Brazil decided to substitute pure ethanol and gasohol (mixture of ethanol and gasoline) for fuel in automobile engines. Aside from economic considera-

TABLE 14.3. *Gasoline Consumption of Automobiles in France and Brazil* (liters/100 km)

Model	At 90 km/h	At 120 km/h	Urban traffic	Avg. Power (kW)
France[a] Fiat (avg. of 15 models)	4.9	7.0	7.0	25-50
Brazil[b] Fiat (avg. of 6 models)	7.6[c]	—	10.0	43-61
France[a] Volkswagen (avg. of 20 models)	6.0	7.7	10.4	66-102
Brazil[b] Volkswagen (avg. of 7 models)	7.8[c]	—	10.6	57-82
France[a] Ford (avg. of 11 models)	6.0	7.8	9.9	70
Brazil[b] Ford (avg. of 7 models)	7.5[c]	—	10.5	69

a. "Consommations conventionelles de carbvurant des voitures particulieres Octobre 1988," 21 *eme* edition—Agance Francaise pour la Maitrise de l'Energie.
b. *Quatro Rodas* (a monthly trade magazine in Brazil—1988 and 1989 issues).
c. Extrapolated from 80 km/h (data from *Quatro Rodas*).

tions, ethanol has a higher octane ratio and other technical advantages over gasoline (Reddy and Goldemberg, 1990).

Production of ethanol from fermenting sugarcane juice rose from 900 million liters in 1973 to 4.08 billion liters in 1981, of which 1.88 billion liters were turned into hydrated ethanol (92 percent to 94 percent ethanol by weight). The remaining 2.2 billion liters became anhydrous ethanol mixed with 20 percent gasoline. In 1989 some 12 billion liters of ethanol replaced almost 200,000 barrels of gasoline a day in approximately 5 million Brazilian automobiles. The alcohol industry created 700,000 jobs. The excellent performance of ethanol-fueled automobiles significantly improved the quality of air in polluted megalopolises such as São Paulo and Rio de Janeiro. Above all, Brazil, a developing country, established an entire fuel cycle—involving an energy source (sugarcane) for end-use devices (alcohol-fueled automobiles) that provided an energy service (transportation)—that does not exist in industrialized nations.

The average cost of ethanol produced in the southern region of Brazil is 18.5 cents per liter. At this price, ethanol could compete successfully with imported oil when the international price of oil was $24.00 per barrel. When oil prices fell in the mid-1980s, however, the Brazilian ethanol program faced a serious economic crisis and had to be subsidized by the government. The resulting strong pressure to remove subsidies led to major efforts to improve the productivity and economics of sugarcane agriculture and ethanol production. As a result, ethanol costs have fallen 4 percent a year over a span of ten years. Brazilian ethanol distilleries have become the best in the world and compete strongly in the international market.

The effective cost of ethanol can be decreased even more if *bagasse*, the residue left over after sugarcane is crushed and drained of its juice, is burned efficiently to make steam. Today, low-pressure steam-turbine systems for electricity generation can produce about 20 kWh per ton of sugarcane. Higher pressure steam turbines used for co-generation of steam and electricity and the use of modern new technologies—as discussed below—could make alcohol distilleries net energy-exporting enterprises.

Electricity Consumption in Refrigerators

Refrigerators and freezers represent 12 percent of all energy consumed in the residential sector in the United States and approximately 20 percent of residential electricity consumption. In Brazil electricity represents 44 percent of all energy used in the residential sector, one-third of which (32 percent) is in refrigerators and freezers. The dominant type of refrigerator in use in the United States, and to a lesser extent in Europe and Brazil, is a two-storage compartment: one with a temperature about $+5°C$ (the refrigerator) and another one at approximately $-18°C$ (the freezer). Usually the volume of the freezer is about 25 percent of the total volume.

The performance of refrigerators has improved by a factor of two from 1973 to 1988 in the United States, and further gains are expected by 1993. The maximum energy consumption for a typical unit produced in 1990 will be 1.85 kWh/year/liter as compared to 2.8 kWh/year/liter for the average units in use in 1988 (and 4.1 kWh/year/liter for 1972). In Brazil levels of consumption are much higher, about 2.8 kWh/year/liter even for the new models, but there is a trend toward a reduction in consumption. Clearly there is a lot of room for improvement.

Gains of a factor of two in Brazil, where approximately 2 million refrigerators are sold every year, could mean savings of 600 million kWh. This would yield $30 million in electricity consumption saved annually (at 5 cents/kWh) or savings in investments in new electricity-generating plants of approximately 160 MW (approximately $320 million annually). A factor of two improvement in efficiency in Brazil is consistent with complying with the 1993 minimum efficiency standard adopted in the United States. The cost of more efficient refrigerators would probably not exceed $10 to $20 per unit.

The Enhanced Use of Renewables in LDCs

In the ESW study, LDCs were not treated separately from the industrialized countries because it was assumed that, by the year 2020, they would reach a level of consumption equal to that of Western European nations in the mid 1970s, with essentially the same commercial mix of fuels.

The "modernization" of energy consumption in LDCs will still include, however, a large consumption of biomass in a very inefficient fashion (fuelwood cooking and conversion to charcoal among others); some of it will be replaced by fossil fuels. In addition, modern methods of conversion of biomass into electricity, through gasification and gas turbine generators, are becoming a distinct new possibility (Williams and Larson, 1989). Furthermore, the conversion of sugarcane into ethanol is also gaining ground, and the enormous quantities of bagasse produced in this process could be gasified and used for electricity generation. Figure 14.5 indicates the enormous gains in efficiency one could gain using gas turbines.

The use of such turbines with natural gas (which can be found in small and medium size deposits even in many non-oil producing developing countries) will increase the share of natural gas (and biomass) in the future energy mix of LDCs. Burning gas has several environmental advantages when compared to oil and coal, the most conspicuous one being a reduction of 30 percent to 50 percent in CO_2 released into the atmosphere per unit of energy produced.

New developments in wind, solar thermal and photovoltaics as electricity sources make these sources a potentially larger contributor in the next few decades in the LDCs (Williams 1989). Such a contribution, which was not significant in the ESW scenario, could represent more than the expected contribution of nuclear power.

A reexamination of all such possibilities for LDCs up to the year 2025 led to some new conclusions, including an expanded role for hydroelectricity in the developing nations (Goldemberg *et al.*, 1990). Although in 1985 hydroelectricity represented only 2.3 percent of the world primary energy consumption, the contribution of hydroelectricity could reach 10 percent in developing countries in the year 2025. Hydroelectricity is an attractive energy source, being renewable and not a producer of CO_2. This does not mean, however, that there will be no environmental impacts. To produce 10 percent of the primary energy of LDCs would require flooding extensive areas—roughly 100 million ha. Because many of the hydrostations will be located in

FIGURE 14.5. Electricity generated by sugarcane-based co-generation systems to show the potential for increasing the yield as compared with typical existing systems. CEST = Condensing Extraction Steam Turbine; BIG/STIG = Biomass Gasification/Steam Injected Gas Turbine; BIG/ISTIG = Biomass Gasification/Intercooled Steam Injected Gas Turbine.

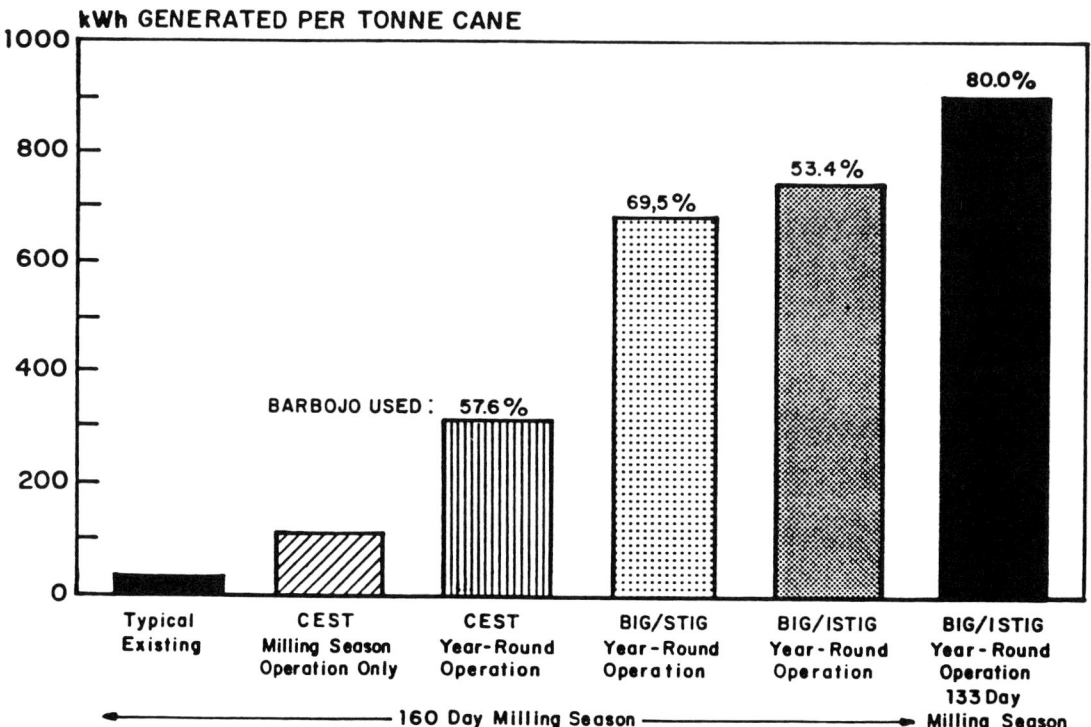

forested areas, biodiversity will be lost and submersed biomass could produce methane, an effective greenhouse gas. If the biomass is cut and burned before flooding, appreciable amounts of CO_2 will be emitted into the atmosphere.

If some of these more innovative approaches for meeting future energy needs are not adopted, present total emissions of 1.33 Gtons of carbon per year from LDCs will grow to 2.2 Gtons of carbon per year (a growth of 45 percent), while primary energy consumption will go from 3.5 exajoules to 6.8 exajoules. This represents a growth of 77 percent in noncommercial energy use or approximately 200 percent in commercial energy use.

The Role of Nuclear Energy

It has been argued recently that nuclear generated electricity could play a significant role in supplying the energy needs of industrialized and developing countries alike, thus avoiding the problems of CO_2 emissions. Although nuclear reactors effectively do not emit CO_2, their widespread use in the future faces very serious hurdles, ranging from the negative reactions of the public to unacceptably high costs.

Such problems are reflected clearly in data for annual global nuclear grid connections as shown in Figure 14.6. These connections reached a peak of almost 32 gigawatts per year in 1985 and decreased dramatically in recent years. According to present projections, they will drop to zero in 1996 in the occidental world, with the possible exception of France, and in the year 2000 in the communist and former communist world.

Hopes have been raised that a new generation of "intrinsically safe reactors" could turn the tide of public opinion in favor of nuclear energy, which would go through a revival in the next decade. In reality, the nuclear industry is in a quandary; because it is so committed to the present type of reactor in operation (mainly pressurized light water reactors) the switch to new types of reactors—safer as they might be—could undermine confidence in the hundreds of reactors already operating.

It is therefore unlikely that, in the next few decades, nuclear generated electricity will play a very significant role in the world energy picture in general and in developing countries in particular.

Actions Needed

As we have already shown, developing countries can avoid the ruinous path followed by industrialized nations during the course of their development, and a very reasonable level of living can be reached by the year 2025 with a supply mix that would not seriously increase greenhouse gas emissions. How-

ever, two steps would have to be taken in the immediate future to make it happen. These are discussed below.

Technology Transfer to LDCs

Usually the technologies transferred to LDCs are inefficient from an energy viewpoint. As old-fashioned factories are replaced by modern ones in the industrialized nations, some of them are disassembled and shipped to LDCs where they are still accepted. Such replacement is done not only for commercial reasons and to extend the useful life of factories, but also as a result of environmental concerns in the industrialized countries. An example of the first case is a Volkswagen assembly plant that was transferred to Brazil in the 1950s and remained in operation until recently.

A typical example of the second alternative is given by the refusal of some LDCs to

FIGURE 14.6. Annual global nuclear grid connections. WOCA = World Outside Communist Area. (Source: IAEA "Nuclear Power Reactors in the World, 1989," and estimate for plants under construction).

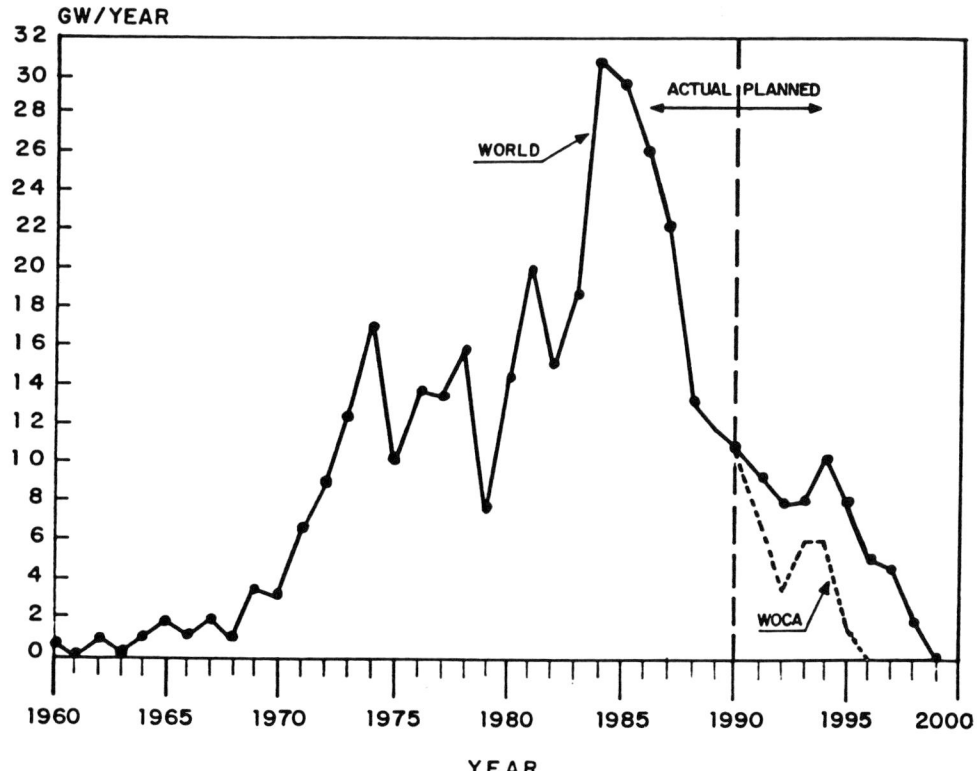

join the Montreal Protocol and reduce the use of CFCs (and eventually their phaseout). It is argued by some LDCs—mainly India—that substitutes for CFCs are expensive and that the adoption of costly replacements will put them at an economic disadvantage. The only solution to this problem is to transfer to LDCs, under favorable commercial terms, the technologies needed to reduce the use of CFCs because this is in the self-interest of industrialized countries. Internal mechanisms to compensate industries for their costs would have to be established by governments in the industrialized nations. An important step in this direction was made in the London Conference on Ozone (held July 1990) where a fund was set up by the industrialized countries to assist LDCs in phasing out CFC production.

Reforestation

Again, from the viewpoint of LDCs it is hard to see the economic benefits of engaging in very large reforestation programs in order to recapture CO_2 from the atmosphere. To be effective in reducing the concentration of CO_2, the areas to be reforested are of the order of 10 million to 15 million ha (capable of capturing 0.1-0.2 Gtons of carbon annually); it would be difficult to obtain an economic return from such large areas. Consequently, most of it will have to be done as an ecologically oriented reforestation and, therefore, financed on a concessional basis.

In the particular case of the African countries where deforestation is directly linked to energy use, such as cooking, the adoption of modern cooking methods using kerosene and gas stoves is necessary. In our opinion, the effort to replace fuelwood cooking stoves has not proved to be successful, and the sooner that rural populations are introduced to modern cooking methods based on the use of kerosene or liquefied petroleum gas, the better this will be, both from their viewpoint and from the viewpoint of environmental conservation.

Figure 14.7 presents a qualitative overview of the role of different actions that could reduce CO_2 emissions (Grubb, 1990): energy conservation, fuel switching (due to the increased use of gas) and the expansion of CO_2 sinks (reforestation). The historical trend in the growth of CO_2 emissions can be reversed through a combination of these actions.

Financial Needs

To obtain the financial resources needed for the actions outlined above in the developing countries, the creation of a general carbon tax, designed to constitute a fund to prevent climate change, was proposed. Estimates were made of the costs of a preventive strategy in a background report prepared for the Noordwijk Conference by McKinsey and Associates (McKinsey and Company, 1989). These estimates were made for the following actions:

- *Continued funding for a CFC phaseout:* of the order of $150 million to $200 million per year initially, and $600 million to $700 million per year should a 100 percent phaseout of CFCs be pursued.
- *Expanded forest management funding:* up to a maximum of $10 billion to $15 billion per year (12 million ha annually of reforested areas) at an approximate cost of U.S.$1,000/ha.
- *Funding of fossil fuel energy conservation:* of the order of $10 billion to $15 billion per year; this number is based on a preliminary cost assessment for the introduction of efficient energy conservation measures in the developing world and interfuel substitution.

The total amount of money needed for the operation of the fund (of the order of $30 billion annually) could easily be amassed by setting a levy of U.S.$1 per barrel equivalent of petroleum consumed. Because many forms of energy conservation have a positive return, once the hurdle of incremental investments has been overcome, or a marginal subsidy has been provided, funds initially committed could conceivable be lower than those cited in the McKinsey report.

Although both the technical feasibility and favorable economics of energy conservation actions—on a life cycle cost basis—are well established, they represent realistic op

FIGURE 14.7. Measures for reducing CO_2 emissions. The "trend" line shows the expected increase in emissions under current conditions. Potential savings from energy conservation, fuel switching and reforestation are projected over time.

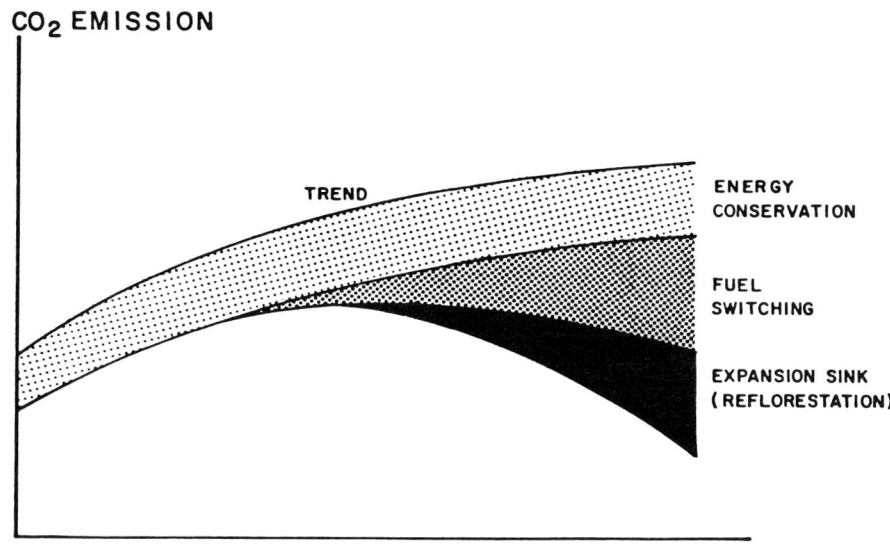

tions only to the extent that institutional mechanisms can be found to reduce significantly the very high discount rates (far higher than the market rate of interest) implicit in most energy-user decisions relating to investments in energy efficiency. Various strategies for coping with this problem have been tried. Some have been failures—most notably the U.S. residential energy conservation tax credit, the single most expensive federal energy conservation program, for which there is little evidence of energy savings. But there have also been notable successes. For example, Greene (1989) has shown that in the United States the CAFE standards for new cars, required by the Energy Policy and Conservation Act of 1975, were more than twice as important as price increases in achieving the light vehicle fuel economy gains that have been made since the mid-1970s (reflected in a 4.7 quad decrease in U.S. energy use in 1986).

To achieve the high energy savings indicated as feasible in the above studies, it may be necessary to supplement present policy approaches that promote efficiency improvements for specific technologies (e.g., efficiency standards and various utility programs) with new generic approaches (in addition to getting the right prices) that can have a much

wider impact. High priority needs to be given to changing utility regulatory and general financial policies to motivate utilities, and private industry generally, to pursue the opportunities for more efficient use of energy. Particularly promising are some recent proposals that, in essence, convert first costs into operating costs for consumers—notably proposals that would enable electric utilities to be transformed into energy service companies (e.g., by leasing energy-efficient equipment to their customers) or enable independent energy service companies to sell, via auctions, "saved electricity" to utilities (Cichetti and Hogan, 1988). These utilities would, in turn, resell the associated energy services to the customers receiving the energy-saving equipment or materials.

In contrast with such optimistic visions, several groups have been claiming that costs to achieve a "CO_2 freeze" would be much higher. Kaya et al. (1989) estimate costs at $200 billion per year. Manne and Richels (1990) estimate costs as high as several trillion dollars over the next few decades, assuming there will be little additional technical innovation. This assumption has been severely criticized by Williams (1990).

The work of Kaya and Manne and Richels hinges heavily on GDP as an indicator of well-being. This does not, however, adequately take into account environmental costs that are bound to become more and more important in the near future. Clearly more work is needed in this field, but the role of technological innovation will determine the success of any policy adopted.

International and Equity Issues

The establishment of a carbon tax or an "insurance fund" of the size proposed above (U.S.$1 per barrel of oil equivalent) raises several issues. First, it would not serve as a market mechanism in industrialized countries to discourage inefficient use of fossil fuels. For that purpose a much greater tax would be needed—such as a doubling of the current price of coal or oil—which is very unlikely to happen.

As far as developing nations are concerned, two points should be stressed: First, these countries would have to be exempted from contributing to the fund until such time as the transferral of adequate technologies enables them to minimize emissions of greenhouse gases. Second, the application of the tax or fund will have to be made in concessional terms. Transferral of technologies to replace CFCs, to introduce energy-efficient devices in LDCs, and to modernize the industrial sector that manufactures them would have to be made in such a way as to induce the LDCs to carry out the necessary industrial and institutional adjustments. Otherwise, the LDCs will argue that environmental considerations are overriding their vital development considerations.

It might also be appropriate to point out that concerns for the environment became very strong in developed countries before the problem of poverty was completely solved in these societies. Even at present, in the United States increasingly large amounts of funds are being spent on cleaning the environment, while the poverty problems of a fraction of the population remain unsolved. This point is made by some in order to stress the need to separate the two problems, even at the risk of having environmental concerns qualified as "concerns of the rich." In reality, policies to protect the environment cut across social categories and are clearly of interest to all.

Funds for official development aid (ODA) of the OECD countries (including multilateral banks such as the World Bank) presently amount to $50 billion and represent 0.2-1.0 percent of the GDP of OECD members countries. This money enables governments in LDCs to finance power plants, roads, railways, heavy industry, sanitation and education. World Bank loans, which have amounted to approximately $14 billion annually in recent years, have long payback periods and attractive rates of interest. Bilateral aid, in general, is given in the form of concessional grants to the poorest nations.

Although a great effort is made every year to increase the funds set aside for such activities, this task has not proved to be easy. In large countries, tight fiscal policies and a growing reliance on market mechanisms have reduced the availability of public funds for development aid. As a consequence, it is essential that the new carbon tax should come from additional resources (i.e., that the environmental actions structured to reduce greenhouse gas emissions do not come from presently existing official development funds).

If adopted, a carbon tax or insurance fund could be proportional to the contribution of different countries to total CO_2 emissions. The internal mechanism needed to collect the revenue could be left to them: in same cases it could come from the government budget (i.e., from general taxes) and in others through additional taxes that signal to the population the purpose of the new action taken. Once this is done, there could be a mechanism to increase the taxes gradually, should emissions of greenhouse gases not respond to the actions taken.

The new carbon fund would correspond to less than 0.5 percent of the total GDP of the richer countries, including the former Soviet Union (and other centrally planned economies), where energy is used rather inefficiently. The absolute amount of money involved is, however, significant, and many nations will resist its adoption. Even if approved by international treaty, some countries might not sign it. The only precedent for an action of this size was the Marshall Plan, devised for the reconstruction of Europe after World War II, in which the United States invested 1 percent of its GDP annually in the period between 1948 and

1952.

Some of the money spent on reducing CO_2 emissions undoubtedly will result in economic development. Reforestation will probably be the first example of that, because it avoids land degradation and generates employment. More important than this is the introduction of energy efficient technologies as a fundamental ingredient in the process of economic development. To allow the LDCs to make costly mistakes in their individual development, paralleling those that we see—with the benefit of hindsight—in the industrialized countries, is not only irrational but will require, later on, extensive retrofitting and modernization. It is in their interest—and in the interest of industrialized countries to avoid a continuation of these policies.

Conclusions

1. If properly conducted, growth and development of the LDCs is not necessarily a threat to the stabilization of the atmospheric composition.
2. Industrialized countries can help in this process by providing technical and financial resources.
3. Although large sums of money are needed, these funds are not larger than those spent today in official development aid (ODA) by the industrialized countries: $50 billion or 0.4 percent of their total gross domestic product.
4. To increase such spending not for philanthropic actions—which characterizes ODA in general—but to stabilize the composition of the atmosphere, which is in the interest of all nations, is within our grasp.

References

BP Statistical Review of World Energy, 1989. British Petroleum Company.

Cichetti, C., and W. Hogan, 1988. *Including Unbundled Demand Side Options in Electric Utility Bidding Programs.* Energy and Environmental Policy Center Report No. E-88-07, Kennedy School of Government, Harvard University, Cambridge, Mass.

Drucker, P.F., 1986. The changed world economy. *Foreign Affairs,* **65**:768-91.

Goldemberg, J., T.B. Johansson, A.K.N. Reddy, *et al.*, 1987. *Energy for a Sustainable World.* Wiley Eastern Ltd., New Delhi, 1988; and World Resources Institute, Washington D.C., 1987.

———, 1990. *Energy for a Sustainable World* (an update with emphasis on developing countries). Belaggio Seminar on Energy Efficiency for a Sustainable World, June 25-30, 1990.

Greene, D., 1989. *CAFE or Price? An Analysis of the Effects of Federal Fuel Economy Regulations and Gasoline Price on New Car MPG, 1978-89,* Oak Ridge National Laboratory, Oak Ridge, Tenn.

Grubb, M.J., 1990. Issues and options in implementing greenhouse gas control agreements. Paper for IPCC Energy and Industries Subgroup on greenhouse gas emissions. London.

Haefele, W. *et. al.*, 1989. Energy in a finite world—A global systems analysis. International Institute for Applied Systems Analysis, Ballinger, Cambridge, Mass.

Houghton, R., 1990. Tropical deforestation and atmospheric carbon dioxide. Climate Change, **19**:99-118.

Kaya, Y., K. Yamaji, and R. Matsuhashi, 1989. A grand strategy for global warming. Tokyo Conference on the Global Environment and Human Response Toward Sustainable Development. September 11-13, 1989.

Lashof, D.A., and D.A. Tirpak (Eds.), 1989. *Policy Options for Stabilizing Global Climate.* USEPA, Washington D.C.

Manne, A., and R. Richels, 1990. CO_2 emission limits: an economic cost analysis for the USA. *Energy Policy* (in press).

Marland, G., T.A. Boden, R.C. Griffin, *et al.*, 1989. *Estimates of CO_2 Emissions from Fossil Fuel Burning and Cement Manufacturing, Based on the United Nations Energy Statistics and the U.S. Bureau of Mines Cement Manufacturing Data.* Oak Ridge National Laboratory, Oak Ridge, Tenn.

Martin, J.M., 1988. L'intensité energetique de l'activité economique: Les evolutions de trés long periodes livrent-elles des enseignements utiles? *Eco. Soc.* **49**:27.

McKinsey and Company, 1989. Background paper on funding mechanisms prepared for the Ministerial Conference on Atmospheric Pollution and Climate Change. Noordwijk, The Netherlands, November 6-7, 1989.

Reddy, A.K.N., and J. Goldemberg, 1990. Energy for the developing world. *Scientific American*, **263**(3):63-72.

Roop, J.M., 1986. The trade effects on energy use in the U.S. economy: An input-output analysis. In: *North American International Association of Energy Economists Conference: Changing World Energy Economy*. Cambridge, Mass., Nov. 19, 1986.

Schramm, G., 1990. Electric power in developing countries: Status, problems, projects. *Annual Review of Energy*, **15**:307-33

Strout, A.M., 1985. Energy-intensive materials and the developing countries. *Materials and Society*. **9**:281-330.

WEC (World Energy Conference), 1978. *World Energy Resources 1985-2020*. IPC Science and Technology Press, Guildford, U.K.

Williams, R.H., 1989. *Low-Cost Strategies for Coping with CO_2 Emission Limits*. Center for Energy and Environmental Studies, Princeton University, Princeton, N.J.

Williams, R.H., E.D. Larson, and M.H. Ross, 1987. Materials, affluence and industrial energy Use. *Annual Review of Energy*, **12**:99-144.

Williams, R., and E. Larson, 1989. Expanding roles for gas turbines on power generation. In: *Electricity: Efficient End-Use and New Generation Technologies and Their Planning Implications*. T.B. Johansson, B. Bodlund, and R.H. Williams (Eds.). Lund University Press, Lund, Sweden.

Williams, R.H., 1990. Will constraining fossil fuel carbon dioxide emissions really cost so much? (Princeton University, draft April 13).

World Bank, 1983. *The Energy Transition in Developing Countries*. World Bank, Washington D.C.

15. UNDERSTANDING THE IMPLICATIONS OF GLOBAL WARMING IN DEVELOPING REGIONS: THE CASE OF NORTHEAST BRAZIL

Antonio R. Magalhaes

Characteristics of Northeast Brazil

Northeast Brazil (NEB) is a large developing region. Its 1,548.6 thousand km^2 (18.2 percent of the country) make it larger than most countries in the world. With 43 million inhabitants in 1990 (28.5 percent of the Brazilian population), it is second in America only to the United States, Brazil and Mexico in population size.

NEB is one of the five macro regions of Brazil and the least developed one. If the NEB were developed, Brazil as a whole would be developed. Indeed, most of the present underdevelopment problems of the country are rooted in NEB; the peripheries of the big cities and the slums of Rio de Janeiro and São Paulo are populated with poor rural families that migrated from this region.

Two dimensions identify NEB as a target region. One is its underdeveloped condition. The other is the semiaridity that covers 60 percent of its total area.

Geographically, the region is rather heterogeneous, comprised of five great ecosystems (Figure 15.1):

- The Coastal Zone (98,300 km^2) is a humid area, with average precipitation of 2,000 mm/year. The main activities are sugarcane and cocoa production in large landholdings. Major cities are located in this area, where the manufacturing sector and one third of the regional population are concentrated.
- The Cerrado area (436,900 km^2) comprises the scarcely populated western part of the NEB (parts of Bahia, Piaui and Maranhao). The main economic activity is soybean production in large landholdings. Most of the inhabitants are recent settlers who migrated from the Southern Region of Brazil. Annual precipitation exceeds 1,000 mm.
- The Pre-Amazonia, in the northern part of Maranhao State, is a transition zone between the Semiarid and the super-humid Amazonia. Precipitation is high. In the last 30 years, human and agricultural settlements have encroached on this area, where 10 percent of the NEB population is presently living. Rice production and cattle raising constitute the main activities.
- The Agreste is a transition zone between the Coastal Forest Zone and the Semiarid. Rainfall is about 800 mm/year and variability is high. Severe drought events in the Semiarid usually reach this area. For this reason, it is often included in the Semiarid statistics. Population density is high. Agriculture (food production) and cattle raising are the main activities.
- The Semiarid Sertao is the biggest and most problematic area. In a sense, it constitutes the main source of the NEB underdevelopment. With relatively few natural resources, shallow soils, scattered caatinga vegetation and high climatic variability, this area is subject to frequent droughts. Precipitation ranges between 300 and 800 mm/year. The Semiarid is highly populated: 20 million people live in the Sertao and the Agreste. The natural resource and economic base is insufficient to support the population; poverty is widespread throughout the area. Agricultural activity predominates and consists of food crops (beans and maize) and cotton, together with extensive cattle raising. But climatic variability brings a high risk to agriculture. Droughts cause crop failures and unemployment, as well as social problems such as malnutrition and migration.

FIGURE 15.1. Map showing the five subregions of Northeast Brazil described in the text.

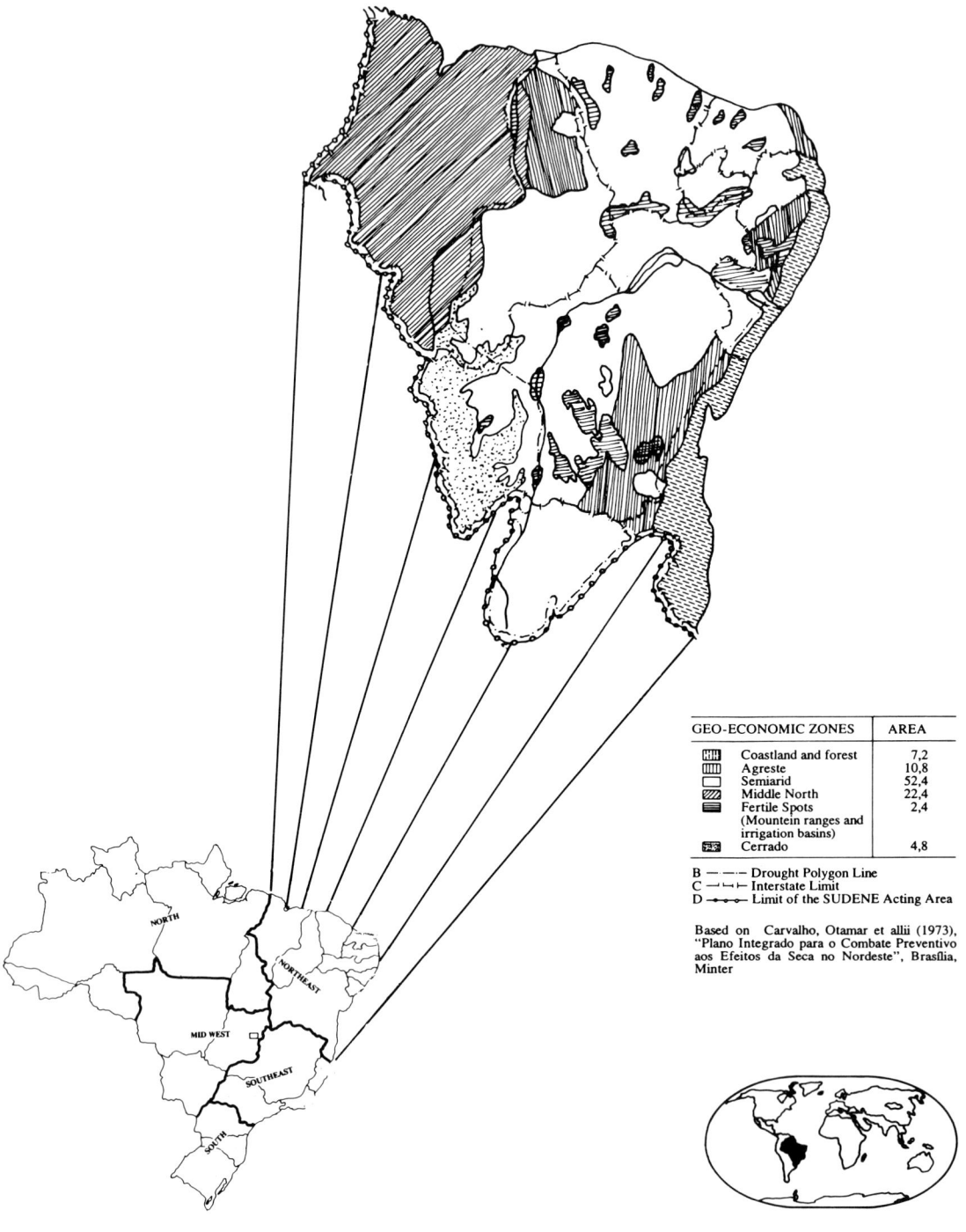

GEO-ECONOMIC ZONES		AREA
	Coastland and forest	7,2
	Agreste	10,8
	Semiarid	52,4
	Middle North	22,4
	Fertile Spots	2,4
	(Mountein ranges and	
	irrigation basins)	
	Cerrado	4,8

B — — — Drought Polygon Line
C —⊢⊢ Interstate Limit
D —•—•— Limit of the SUDENE Acting Area

Based on Carvalho, Otamar et allii (1973),
"Plano Integrado para o Combate Preventivo
aos Efeitos da Seca no Nordeste", Brasília,
Minter

Economic and Social Characteristics

Though poverty conditions are extreme in the Semiarid, they pervade the other areas as well. Land ownership is highly concentrated; 3.4 percent of the larger landholdings occupy 40 percent of the total land area, while the 29 percent smallest ones occupy 1.6 percent. The majority of the small producers have no land at all and, thus, have to work as share-croppers on large landholdings. They are the most vulnerable population and constitute the majority of the poor of the NEB.

Water resources are scarce in the Semiarid. There is only one permanent river crossing it, the São Francisco. It supplies the area with hydroelectricity (all of which is already committed), water for irrigation and transportation. Water trade-offs between energy production and irrigation are an important issue in the São Francisco. The Parnaiba River is on the north border between the Semiarid and the Pre-Amazonia. Smaller water flows occur in the Atlantic Forest Zone.

As to infrastructure, there is a large network of roads, both within the NEB and linking it to other regions. All of the NEB states are on the Atlantic Coast and have access to maritime transportation through existing harbors. All state capitals but one are on the coast and each has airports appropriate for cargo and passenger airplanes. Long distance communications are available in almost all cities, even the small ones. Energy is supplied from hydroelectric plants in the São Francisco River and from the Tocantins River, the latter in the Amazonian Basin. However, the supply is limited, and by the middle of the present decade, the demand for energy may exceed supply.

The economy of the NEB has experienced a rapid growth in the last few decades, dynamic sources of growth coming from industrialization. The industrial sector represents 26 percent to 30 percent of the GDP. Since Northeast Brazil is an open region in the country, its industry needs to be competitive by national standards. The main subsectors include petrochemicals, textiles, clothing and food production. However, the size of the manufacturing sector is still very small, representing only 10 percent of total Brazilian manufacturing GDP. Agriculture's share in regional GDP represents 17 percent and is based on sugar and alcohol production, cocoa, and fruit production in the Coastal Forest Zone; cotton, beans and cattle in the Semiarid; soybeans in the Cerrado; and rice, soybeans and cattle in the Pre-Amazonia. There is a huge unemployment problem, amounting to 60 percent of the labor force. It is largely concealed in the low productivity agriculture and services sectors.

The effects of population growth and human activities in the NEB have been historically catastrophic for the environment. The coastal forests have been replaced by sugarcane, and the original Semiarid caatinga vegetation has almost disappeared. The same is now occurring in the Cerrado and Pre-Amazonia areas. Most of the rich fauna of Northeast Brazil has disappeared. Agricultural practices and excessive wood use, such as for firewood and fence building, have been conducive to overall deforestation and, in some places, desertification. Soil erosion is a major environmental and economic problem, resulting in a continuous fall in agricultural productivity.

Regional Policies

During the last 40 years, Northeast Brazil has been the object of regional development policies, with the establishment of planning and financial agencies and the implementation of numerous programs and projects. Though such policies have suffered from insufficient funding, inadequate political interference and discontinuities, a diversified infrastructure of transportation, communications and energy production and distribution has been created. A large quantity of water is now stored in thousands of reservoirs all over the area. Manufacturing has been modernized and irrigation and other agricultural projects have been introduced. Notwithstanding the fact that the environment has been destroyed, the living conditions of the majority of the population have not improved; the economic base is insufficient and wealth has become more concentrated. Under such circumstances, the challenge of sustainable development represents a formidable problem to be faced by policymakers, scientific advisers and society as a whole.

Chapter Objectives

The aim of this chapter is to discuss the effects of, and responses to, global warming on Northeast Brazil. The present climatic variability in the Semiarid region provides an analogy for the study of future regional climatic change. Though the chapter refers to NEB as a whole, the analysis is mostly centered on the Semiarid subregion because of its specific vulnerability to climatic change and its role in determining the social and developmental status of NEB. The section that follows deals with impacts of, and responses to, past and present climatic variability. Subsequent sections focus on a possible scenario of climatic change and possible future climatic change impacts on the region. Potential policy responses to regional climatic change are also discussed with emphasis on the implementation of a sustainable development strategy. The final section presents some conclusions and recommendations.

Impacts of Climatic Variability and Policy Responses

Climatic Variability and Droughts

Current climatic conditions in NEB are highly variable. In the Semiarid subregion, the rainfall interannual variability coefficient may be as high as 60 percent. Both intra-annual and spatial variability are equally high. Floods and droughts are frequent. Out of every ten years, some kind of drought is to be expected in eight of these years. Also, there may be floods in drought years owing to the concentration of heavy rainfall in a few days.

During the winter of 1990 the Semiarid underwent a partial drought that affected more than 1 million people. Its effects lasted until the beginning of the 1991 rainy season, with the result that hundreds of thousands of rural workers became unemployed. At the same time, in August 1990, the city of Recife suffered heavy floods that caused the death of 40 people and destroyed houses and public infrastructure.

Droughts may be partial or total, according to its geographic extension, duration and intensity. An extreme drought usually covers all the Semiarid and Agreste and extends to the Coastal Forest and the Pre-Amazonia. Some places are more drought-prone than others. If there is a drought, there is an 80 percent chance that it will hit the states of Ceara (93 percent of which is semiarid) and the interior of Rio Grande do Norte, Paraiba and Pernambuco.

The impacts of droughts on the society and the economy of Northeast Brazil, as well as governmental responses to them, provide good examples of likely impacts of, and possible responses to, climatic change.

Impacts on the Population

One characteristic of climatic anomalies in overpopulated, developing areas is their heavy impact on local inhabitants. While in developed areas the impacts are mainly of an economic and environmental nature, in developing areas they are mostly social.

The NEB population is most vulnerable to droughts. However, while some groups are especially vulnerable and numerous, others profit from droughts. An analysis of the literature on the impacts of climatic variability in the NEB (see Magalhaes *et al.*, 1988; Magalhaes and Bezerra Neto, 1989) shows that in general :

- The rural population is more vulnerable than the urban population. Hence, rural-urban migration is increased during drought periods.
- Owners of small farms are more vulnerable than owners of large farms.
- Landless agricultural producers, such as rural workers and sharecroppers, are more vulnerable than landowners.
- Food agriculture is more vulnerable than raw materials agriculture, and both are more vulnerable than cattle raising.
- Large landowners, urban traders and politicians may be net gainers by buying cheap land and cattle or organizing government relief actions. This type of gain is becoming less frequent with an improvement in government responses, although retroactive action is still more frequent than desirable.

The impact of droughts on the NEB population occurs in three complementary ways. The main impact comes from the sudden interruption of the income stream of the poor population (rural workers, sharecroppers and small farmers). Second, there is in certain areas scarcity of water, even for drinking. Third, there is a drop in food supply that leads to food scarcity and price increases.

Once a drought is established, agricultural producers decide not to plant. Large landowners avoid dedicating land for sharecropping and allocate all of it for forage production so as to protect their cattle-raising activity. Within a few days, hundreds of thousands of workers become unemployed. In the absence of relief aid, the stricken population has a choice between migrating or starving to death. Indeed, there are estimates that, during the extreme droughts of the past (end of nineteenth century), about 100,000 to 200,000 people died of hunger and thirst and a great part of the remaining population migrated to the Amazon region (Furtado, 1964, pp. 159-160).

Recently the situation has changed because of increased availability of infrastructure, stored water resources and government-sponsored relief actions. But the degree of vulnerability of the rural population is still the same. The poor are now 30 times more numerous than a century ago, and the environment is much less productive than before because of devegetation, soil degradation and cultivation of marginal lands.

With drought, water becomes more scarce. Notwithstanding the water storage behind dams, in drought years many reservoirs become dry and so do other water supplies. The distance between the people's homes and their water supply becomes longer, sometimes about 100 km. Water quality also becomes a major concern.

Food scarcity is two-faced. On the one hand, subsistence agriculture is bankrupted. Families who used to produce their own basic food are faced with crop losses. At the same time, they cannot buy food in the market because of loss of income.

Without governmental assistance, the situation deteriorates rapidly. Thousands of rural workers invade cities and loot private and government storehouses (Barreira,1989). Hunger becomes a reality; living conditions deteriorate; malnutrition and infant and general mortality increase. Fortunately, in recent years government action has been able to avoid some of the worst consequences.

Impacts on the Economy

Agriculture is the sector of the economy most affected by drought, with dryland farming suffering the most serious consequences. Table 15.1 provides data on the impacts of climate on agriculture during the five-year drought of 1979-83 (including the extreme drought of 1983), as compared to the nondrought years of 1976-78.

Typically, irrigated areas are not affected, or may even benefit from drought, because it becomes easier to control the quantity of water needed by the plants and to profit from increased prices. However, small irrigation projects that depend on water sources that are reduced or depleted may be affected. In the case of extreme droughts, especially in back-to-back drought situations, even large irrigation projects may suffer.

Cattle raising is the last activity to be affected by drought. The aggregated data show that only in extreme drought events,

TABLE 15.1. *Northeast Brazil, Impacts of Droughts on Agriculture* (1976-1978 = 100)

	1979	1980	1981	1982	1983
Cattle	104.6	111.6	112.9	112.8	105.1
Cotton	72.6	61.8	55.3	67.7	28.9
Rice	96.7	106.8	66.9	142.3	44.9
Beans	113.7	83.4	74.9	121.5	37.9
Corn	90.3	59.7	36.4	81.6	22.1

Source: Magalhaes and Reboucas 1988, p. 288.

like 1958 and 1983, is there some reduction in the cattle herd. However, small farmers may have to sell their herds at depressed prices, if they don't have the means to keep them. Large farmers, who own forage reserves or have the means to transfer the cattle to nondrought areas, may buy the cattle and increase their herds. In this case, there is a perverse wealth redistribution caused by the drought.

Industry and services are much less affected, or not affected at all. Some industries, like the agroindustries, which use inputs from agriculture, are affected by drought. In Northreast Brazil, however, the majority of those industries have been able to substitute inputs from other regions.

A similar situation may occur within the service sector. On the one hand, those services that are dependent on agriculture suffer a reduction. On the other hand, some of the workers and farmers unemployed by drought look for other activities within the service sector, which swells during drought years.

On the whole, the gross regional product is affected. A loss of income in the agriculture sector implies a reduction in aggregate demand and this depresses the regional GDP. Government relief programs, however, have been able to avoid a drop in the aggregate level of economic activity.

Impact on the Environment

There are no specific studies on the effects of drought on the environment in the NEB. Indeed, the NEB Semiarid ecosystem is adapted to the prevailing pattern of climatic variability. The caatinga vegetation is well equipped to overcome long periods of droughts and proliferates with the eventual return of rains. It has been the rapidly growing human settlements, especially those associated with land clearing and agriculture, that have caused the environment to become more sensitive to droughts.

On the one hand, droughts sharply reduce the productivity of natural resources and cause an increase in environmentally destructive activities, such as collecting wild fruits and roots. In some places, this may be the cause of desertification. On the other hand, the abandonment of agricultural tracts during droughts may let the land rest and may actually favor land restoration. By fostering migration and reducing rural populations, droughts may act as as a control mechanism that inhibits a further increase in the NEB rural population, and hence reduces environment stress.

Government Responses

Government responses in Northeast Brazil have been directed toward:

- assisting the stricken population during droughts;
- increasing infrastructure, especially for water resources;
- promoting economic growth.

In general, government programs fall into two categories, short-term, relief programs and more long term actions. The NEB relief programs constitute a rich and varied experience and, to a certain extent, a successful one. The main component of such programs has been job creation through public works of a Keynesian type.

Through this strategy, the government provides temporary work for those unemployed by the drought, assuring them a wage during the crisis. Though the wage is individually low (as is normal for the equilibrium wage in rural NEB), the volume of aggregated income from these wages is rather high. Compared to normal years, when a meaningful part of the subsistence production is directed to self-consumption and is not brought to market, in drought years the economy becomes more monetized. The general level of real income may be maintained, so that the fall in agricultural production is not transmitted to the economy as a whole. Consequently, the level of activity and level of employment in both the industrial and service sectors are not affected and may even grow. A study for Ceara has showed, for instance, a positive correlation between sales tax revenue (the main source of state income, a function of the general level of economic activity) and droughts (Arraes and Castelar, 1989).

During the extreme drought of 1983, there were more than 3 million workers employed in temporary government works.

Including their families, a total of approximately 10 million people may have benefitted directly from such action. In 1987, a year of partial drought, about 1 million workers were employed (Magalhaes *et al.*, 1989).

The type of work has varied with time. Traditionally, the work force has been employed in the construction of large and medium dams in the Semiarid, in so-called work fronts. These dams have enhanced drinking water supplies in both drought and nondrought years. More recently, fisheries and irrigation projects have being promoted. Other activities developed in work fronts include road construction, public buildings and community infrastructure. Many of these works become permanent infrastructure, and others are discontinued when the drought ends.

Another component of this strategy has been the distribution of drinking water. The water is captured mainly in dams and transported by trucks to local villages and disperse populations. In 1983, some 5,000 vehicles were used in this activity. A smaller number of vehicles were used in 1987 and 1990.

The concept of "permanent" or long term action has been included in regional and state economic development plans, where the objectives of promoting economic growth and reducing vulnerability to droughts have been frequently expressed. This strategy contemplates the construction of economic infrastructure and industrialization as the main components, so that the economy becomes less dependent on climatic variations. Irrigation and support for small farmers have also been present in such strategies. Less emphasis has been given to environmental protection and social development.

As a result, regional GDP has achieved a rather good performance by Brazilian standards. On the average, the NEB economy has been growing faster than the Brazilian economy during the last three decades. Economic infrastructure, such as transportation, communications, energy, and urban infrastructure, is reasonably developed. The new manufacturing sector is competitive, although its size is still small. Agricultural development and irrigation, however, still lag behind.

As a whole, the size of the NEB economy is too small in relation to its employment needs. The majority of the population (i.e., the rural population and the rural migrants that inhabit the urban peripheries) are excluded from stable sources of income. The NEB development strategies have not succeeded in reducing social vulnerability to droughts. As a consequence, each new drought strikes the poor population as heavily as ever, and requires ever in-creasing relief actions.

Lessons from Experience

Northeast Brazil is heavily affected by present climatic variations. Social, as well as economic and environmental, impacts of these climatic variations are intensified because of the region's underdeveloped condition. In the NEB, as in most developing regions, an extreme climatic event becomes a severe social problem, while in a developed region it has mainly economic consequences. Some conclusions and lessons can be drawn from the experience of Northeast Brazil and can be valuable for both policymakers and society in defining future policy answers to enhanced climatic variability due to anthropic climatic change.

Policy responses to climatic variations have been effective in alleviating the suffering of the most vulnerable population during droughts. But they have been ineffective in helping these populations to become permanently capable of adapting to droughts and other climatic adverse events. A definite policy answer to climatic variation should consider both short term relief activities and long term actions. Increasing self-capacity of each family or economic and social agent to cope with adverse climatic events is the appropriate policy objective. That means that the underdeveloped condition must be overcome for society to become resilient to climatic variations.

The experience of NEB constitutes a valuable input for the assessment of future impacts and possible responses to climatic change. Accordingly, future climatic change will be perceived as enhanced climatic variability and increased frequency of extreme events, an intensified picture of present variability.

Possible Scenarios of Climate Change

Regional predictions derived from global climate model contain a high level of uncertainty. No agreement exists on the direct effects of global warming in NEB with respect to rainfall and moisture, but it is accepted that there will be some temperature increase and a rise in sea level.

According to existing models, NEB may become more humid or drier, reflecting the afore-mentioned uncertainty. However, Kellogg (1988, p. 30), in his "attempt to present a best 'guess or scenario' of possible soil moisture patterns on a warmer earth," presents a map showing a wetter NEB.

On the other hand, the report of two workshops held in Villach and Bellagio in 1987 states that "future climatic changes could worsen the current critical problems of the semi-arid tropics" (Jaeger, 1988, p. 14). The IPCC Policymakers Summary Report (IPCC, 1990, pp. iii, 12, 13) indicates that "soil moisture will decrease. . . in areas that are already in marginal situation" and that "semiarid areas [are] the most sensitive areas." The IPCC report also emphasizes the lack of appropriate regional predictions. The general feeling of the IPCC report is that areas like NEB, with high present-day vulnerability, may be adversely affected. Unfortunately, the IPCC did not present a regional scenario for South America. The report of a recent workshop held in Boulder, Colorado, states

> there is no doubt that these marginal lands (i.e., semiarid lands)—covering fully one-third of the land and supporting one-fifth of the world's population—will be the hardest hit, socially and economically, by impending global warming. (UCAR, 1991, p. 1)

Under these circumstances, a study of the impacts of climatic change in regions like NEB must be based on hypothetical scenarios and use mainly qualitative, rather than quantitative, data. This problem has been faced by the IPCC work on impacts (IPCC, 1990, p. 9).

Since the workshop at Villach (Jaeger, 1988), the Semiarid regions have been assessed as among the most sensitive areas to climate change. This knowledge has recently been reassessed by scientists engaged in the IPCC studies, especially those studies dealing with the environmental, social and economic impacts of climate change.

For the purpose of this chapter, the author has considered a hypothetical scenario for NEB, with the following characteristics:

- average temperature increase of 0.6°-0.8 °C by the year 2025 (below the 1°C anticipated by the IPCC Group I "business as usual scenario" for the planet as a whole, considering that the warming will be less in lower latitudes);
- rainfall decline of about 10 percent to 20 percent, on the average, unevenly distributed (this is the most heroic assumption, since it is difficult to predict rainfall behavior);
- sea level rise of 20 cm by the year 2025 and 60 cm by the year 2100, as stated by the IPCC.

Climatic variability will increase. Droughts will be more frequent. Increased evaporation may cause heavier rains for short periods of time. Temperature peaks may be higher. Summer temperature in the Semiarid may become almost unbearable for human habitation.

For the analysis of climatic change impacts in the section below, I have taken as a reference mark the above "drier" scenario because of its compatibility with the IPCC report. The possibility of a "wetter" scenario would imply somewhat better predictions, but the likelihood of increased variability would bring severe impacts on the NEB population and economic activity.

To analyze likely impacts of climatic change, one has to rely on possible future socioeconomic scenarios of the economic and social reality of the region. These scenarios are presented in the sections that follow.

Assessment of Likely Impacts of Climate Change

Methodological Introduction

To assess the impacts of regional climatic change, information is required on:

- basic knowledge of the region;

- projections for key variables;
- future scenarios of climatic change.

Some regions are much more vulnerable to temperature warming. Other regions may not be vulnerable at all, while some others may even be net gainers. Vulnerability is dependent upon the characteristics of the region as, as shown in Table 15.2.

Having a basic knowledge of the region, it is now possible to project some key variables as a means to envisage future scenarios with respect to environmental, social and economic characteristics. The selection of key variables would include:

- population, employment, migrations, educational indicators;
- economic output and income;
- sectorial growth and productivity;
- energy production and consumption;
- trade relationships;
- environmental problems (land degradation, ecosystems disruption, desertification, industrial and urban pollution, etc.);
- land carrying capacity;
- social indicators (health, education, nutrition).

Projections of key variables are to be made under the assumption that present trends will prevail, taking into account present knowledge as to likely deviations from these trends. These variables can be projected through the use of statistical and econometric tools, as well as qualitative and quantitative data analysis and descriptive reasoning.

The projection should be made for a long period so as to be compatible with projections of climate change scenarios. It is proposed that projections be made for the year 2025 (compatible with the IPCC scenario), with intermediate points in years 2000, 2010 and 2020. Of course, the longer the period, the less confident the projection is. However, this is not a problem, if the interest of the assessor is predicated upon "what will happen if present trends persist." It is a projection, not a prediction. The construction of a future environmental-socioeconomic scenario based on such projections will be useful as a guide for society and policymakers in deciding what to foster and what to avoid in relation to the future regional development.

TABLE 15.2. *Characteristics that Determine the Vulnerability of a Region*

Category	Variable
Natural resources	Stability of existing ecosystems to exogenous variables such as human-induced climatic change and changing land-use patterns. Sources and uses of water resources, their spatial distribution and balance. Nature of soils and their actual and potential use. Characteristics of the vegetation, its use and management.
Social setup	Size and characteristics of the population and the labor force. Size and causes of migration. Education and health indicators. Poverty thresholds. Land-use patterns. Social organization.
Economic aggregates	Gross domestic product (GDP). Sectorial distribution of the GDP. Total and per capita income. Employment, unemployment and underemployment indicators. Technology, productivity and competition power.
Land-carrying capacity	Cross-sectional analysis of the interaction of the previous items, according to local characteristics.

Knowledge of the present situation and the projection of key variables provide the information needed for the construction of future scenarios of economic development and associated social and environmental dimensions. Consideration of the projected natural resource base, a function of its present situation and enhanced land-use practices, with a future increased population, is the main input for scenario building. The interaction of the basic elements of this scenario and the balance among social, economic and environmental variables will provide insights into future social and environmental imbalances that should be avoided.

In the case of NEB, the following future "business as usual" scenario for the year 2025 is envisaged, based on present trends and knowledge of possible deviations.

1. Total regional GDP will be multiplied 3 to 4 times, and per capita income will be 2.5 to 3 times higher than in 1990, reaching U.S.$3,200 annually. Income distribution will still be uneven, and poverty will persist.
2. Total regional resident population will be 74 million people, of which 80 percent will be living in cities. The rural population will number 15 million, with 50 inhabitants/sq km. Twenty million northeasterners will migrate to other regions between 1990 and 2025.
3. Illiteracy will still be a problem should present trends prevail. General levels of education, though somewhat improved, will still be low. Access to health and sanitation services and to shelter will still be limited for poor people. Since these social indicators are reflected in the quality of human resources (as to their productive capacity), labor productivity will still be below the country's average.
4. Sectorial distribution of GDP will change, with manufacturing increasing its participation from a present level of 26 percent to about 30 percent. Agriculture will decrease from 15 percent to less than 10 percent; services and the informal economy will be about 60 percent of the GDP;
5. Employment will also grow to a total labor force of about 25 million workers.

Of these, 40 percent to 50 percent will still be unemployed or underemployed workers, and hence poverty will still be a widespread problem.

6. Agriculture, despite a decrease in its participation in total GDP, will increase its production by 2 to 3 times. Food production (maize, rice, beans, manioc, meat) will grow, but there will still exist a regional deficit in food supply, as compared to consumption (see Magalhaes *et al.,* 1988, for projections of food deficit for the year 2000).
7. Without an increase in agricultural productivity, the land requirements for agricultural production will increase by 3 to 4 times. More marginal land will be used. There will be a pressure for more irrigation, but the potential for increased irrigation may be limited.
8. Electric energy requirements will grow from 27.3 thousand GWh to 88 thousand GWh. There will be a deficit in regional energy supply, and imports of energy will be needed (from the Amazon, mainly), unless new sources of regional energy production are introduced.
9. Land carrying capacity, which is already inadequate for the current population, will not increase as fast as population growth. Hence, overpopulation will be more widespread. Unemployment, rural-urban migrations, urban swelling and other social problems may increase.
10. Environmental problems will increase, with land degradation, desertification, rupture of local ecosystems, increased urban pollution, decreased agricultural productivity and increased production costs.
11. Coastal areas will become overpopulated, with the expansion of cities and of industrial and agricultural activities.

Estimation of Total Impacts

If current trends prevail, by the year 2025, the situation in Northeast Brazil is likely to be one of pronounced environmental and social stress, with increased vulnerability to exogenous causes of ecological imbalance, particularly climatic change. Though income will increase and relative poverty diminish, the absolute number of poor will increase. There is no indication that income distribution will improve.

As is currently the case, the highest vulnerabilities will be found in the poor rural populations and in the agricultural sector. Also, economic activity and populations of lowland coastal areas will be most vulnerable to seawater invasion. The big cities in the region—Recife, Fortaleza and Salvador—will be megalopolises of 5 million inhabitants or more, living at sea level.

The environment, as a whole, will be more fragile than today. Vegetation cover may almost disappear, if trends persist. Soil degradation will bring lower land productivity. Some parts of the Semiarid will become unsuitable for agricultural and cattle-raising activities. Water resources will be more limited than today because of increased demand and diminished supply. Hydroelectric production will also become more sensitive to droughts.

It is not possible to estimate social, economic and environmental impacts quantitatively, but in the absence of a corrective policy, it is possible to make qualitative assessments of such impacts. For this purpose, the "drier" climatic change scenario is the reference scenario.

Consideration of climatic change impacts is based on the analogy with present and past climatic variability as discussed in the section titled "Impacts of Climatic Variability and Policy Responses." Indeed, it is the increased variability in climate rather than climatic change itself that will be perceived by the population. Those with higher incomes will be only marginally affected by climatic change, although they could bear economic and financial losses, if part of their assets are inundated by seawater.

The main impact, as in present days, will be on the rural poor population. In general, the following consequences of climatic change in NEB are envisaged.

1. Dryland farming will be less productive and more risky, and will be more input-dependent. The majority of farmers will be affected, with the biggest impacts to small farmers.
2. The poor rural population—rural workers, small sharecroppers, and small farmers in the Semiarid—will be the most affected. With an increased risk of droughts, their living standards will become worse. Starvation and mal-nutrition will still be a problem. Child malnutrition will be reflected in poor adult intellectual and physical performance.
3. Agricultural activity will be heavily affected by climatic change, especially food production in the Semiarid. Increased risk will hamper farmers from investing in agricultural inputs. Productivity will continue to be low and production will be highly variable. Economic and financial losses will be more frequent, both at the level of the farmer and in the aggregate.
4. Irrigated agriculture will achieve higher productivity, but will suffer from water resource shortage and policy disconti-nuities. Water and soil restrictions currently limit irrigation to only 3 percent of the NEB area (approximately 5.0 million ha). Increased competition for water is likely to reduce this potential.
5. Manufacturing will be only marginally affected, except in the case of limitations in availability of energy. Industries that use agricultural inputs will be able to use imported inputs from other regions. Some agroindus-tries that are dependent upon irrigated production may be adversely affected in the case of extreme events, but may benefit from moderate events.
6. Water resources in NEB may be severely affected. More frequent extreme droughts and back-to-back droughts (i.e., two or more years of consecutive droughts) can cause the depletion of water reservoirs. In many semiarid areas, these reservoirs constitute the only source of water during droughts. This may be calami-tous for human water supplies in urban and rural areas, as well as to livestock and irrigated crops. Also, more frequent droughts in the regions of the São Francisco River may cause severe restriction in energy production and in irrigation, with decreases in industrial production (due to energy rationing) and in agricultural production (due to energy rationing and lack of water for irrigation).

7. The vegetation cover will suffer from water stress. Though the caatinga is adapted to the semiarid climate, an increased frequency and severity of droughts may cause imbalances that, together with intensive land use and occupation of more marginal lands, are environmentally disruptive.

8. Increased direct sunlight on the deforested land and, for short periods, intensive heavy storms, may increase soil erosion and cause desertification. Consequently, agricultural productivity will decline even more.

9. Climate change will increase environmental stress and reduce land carrying capacity as a result of human and economic activity.

10. As a result, poverty will persist in the rural NEB and migration will continue. With increased migration, poverty will continue to be widespread in the periphery of urban areas. Social problems associated with urban poverty, such as criminality and marginality, will persist and get worse. The existence of a large number of poor workers will hamper the growth of wages, limit the expansion of consumer markets, and perpetuate unequal and unfair income distribution. Inequities between rich and poor may be even more visible than today.

Conclusion

As a whole, the impacts of climate change in NEB, in conjunction with increased population and economic production, are likely to be highly negative. It can only be hoped that governments and society will be successful in adopting policies to reverse present trends. To prevent a future disaster, action must be taken now.

A Framework for Policy Response

Premise and Objectives

To increase the capacity of society and government to cope with, and overcome the social, economic and environmental adverse

consequences of climatic change, the following statements must be taken into account in this regard.

- Societal vulnerability is inversely proportional to the economic and educational status of the population. Poor and uneducated people are scarcely prepared to face climatic (and other) crises. For instance: Higher incomes allow for savings capacity, so that families will be able to overcome unemployment caused by droughts and have time for planning and adopting appropriate responses.

- Unproductive and environmentally disruptive technologies exacerbate agriculture vulnerability. Appropriate technologies, together with rational water use and watershed management, sound environmental practices and planned land use, may reduce agriculture vulnerability.

- Land use planning and regulation may lead to environmental conservation and to a reduction in the vulnerability of coastal zones to sea level rise.

The specific vulnerabilities of the NEB society to climatic change stem chiefly from its underdeveloped nature. Overcoming underdevelopment in an ecologically sustainable way will reduce or eliminate those vulnerabilities. Through sustainable development, it will be possible to

- eliminate poverty and improve income distribution;
- increase production and productivity, as well as income and employment opportunities;
- restore degraded ecosystems and preserve the productivity of natural resources;
- contribute to reducing the impact of global warming by avoiding or limiting emissions of greenhouse gases.

Toward a Sustainable Development Strategy

A Sustainable Development Strategy (SDS) requires a long term perspective. This depends on what present society thinks about the desired socioeconomic and environmental scenario for the next generations, say 30 to 40 years from now. The building of this scenario will be a basic step in defining an SDS. The scenario will provide the reference mark for the achievement of coherence between present policies and future goals.

From the NEB diagnosis outlined in previous sections, a possible desirable scenario for the region, for the year 2025, could include the following:

- the achievement of a more equitable society, free of extreme poverty;
- overall improvement in human resources, especially in terms of education (banish illiteracy), health and nutrition;
- universal satisfaction of other basic human needs, such as shelter, food, and self-realization;
- increased productivity and per capita income; more efficient use of economic and natural resources, especially of energy sources, land and water;
- progress toward an environmental equilibrium relating to the use of natural resources, economic activity and population growth.

The next step is to develop a strategy for adapting present societal and governmental actions and policies to conform to the desirable scenario. A planning process for sustainable development should be developed. The planning concept should not be seen as a centralized, rigid and authoritarian system, but rather as a decentralized, participatory, market oriented mechanism for achieving societal goals. It is a search for long term coherence between the present and the future and not a mandatory and inflexible decision and implementation process.

The development strategy that has been followed in NEB has encouraged economic growth with no regard to social and environmental considerations. Hence, social inequity and environmental degradation are the by-products.

Sustainable development, as proposed by the Brundtland report (World Commission for the Environment and Development-WCED, 1987), requires a radical change in the NEB development process and poses a challenge to policymakers and to society in general. In practice, it is very difficult to introduce social and environmental considerations into an open society, where the private rate of profit

has been the only guide to investment decisions. For this reason, an SDS for Northeast Brazil will benefit from similar programs in the country as a whole, as well as in the technology-exporting countries.

Another difficulty derives from the fact that no available methodologies exist for sustainable development planning and implementation. The theory of economic development has not yet incorporated environmental dimensions but has been mainly centered on the theory of capital accumulation and economic growth. More recently, the social dimension has been introduced, with studies on income distribution and target development strategies (see, for instance, Chenery *et al.*, 1974 and Streeten *et al.*, 1982). Introduction of the environmental dimension into economic development theory requires the difficult task of properly considering the analysis of environmental economic externalities and intergenerational equity as dimensions of the decision making process for the promotion of economic and social development. The general idea of SDS is accepted, but its implementation mechanism is not developed.

Up to now, environmental issues have been treated in two ways: first, as a separate sectorial strategy for preserving specific environments (ecological reserves, for instance, usually lacking enforcing mechanisms); and second, more recently, as impact studies legally required for specific projects, mainly infrastructure projects.

The SDS requires that:

- Each investment project or program should be decided on the basis of economic, social and ecological sustainability (not only economic) criteria.
- The environmental sustainability should be a dimension of each economic and social project. Specifically, each project should take into account emissions and absorption of greenhouse gases and other pollutants.
- Besides being a dimension of economic and social projects, environmental projects should be developed to protect specific ecosystems and endangered species.
- Cost-benefit and cost-effectiveness analysis of positive and negative environmental externalities should be

introduced as decision criteria.

- Use of appropriate instruments for creating awareness and inducing or enforcing population and economic agents to adopt the SDS should be fostered. These instruments include an appropriate planning process; environmental education; economic and financial incentives and disincentives; and regulation.

The SDS economic and social projects can be classified, in relation to their environmental characteristics, into neutral, improving or disruptive. Classification of the projects into each category can be done as a result of the environmental impact study (EIS) for each project. The EIS methodology still needs to be improved, especially with respect to enforcing mechanisms and to the introduction of limiting criteria on greenhouse gases and CFC emissions (which, in general, depend on federal and state regulation). In NEB the recent legal requirement for the introduction of EIS is still considered, in practice, as a bureaucratic requirement and not an integral part of the project. The 1988 Brazilian Constitution, however, provides the legal basis for the enforcement of regulations for environmental protection.

The SDS requires that economic, social and environmental issues be seen by decision makers and economic agents as interrelated matters and not as separate issues. Environmental criteria should be used in deciding and implementing any kind of project. The main instrument in this regard is the enforcement of EIS recommendations at the project level.

It must be recognized, however, that not only the actions of economic agents and the government affect the environment. Societal behavior in general, unplanned human settlements, population growth and many daily family decisions may affect the ecological equilibrium. Creating environmental awareness is crucial. This can be done through information dissemination and universal environmental education.

Once the requirements of the SDS are met, a strategy for overcoming present underdevelopment problems—of which increased resiliency to climatic change should result—could include the following:

- A strategy for increasing the region's economic base, as measured by the Gross Domestic Product (GDP), by a factor of ten between 1990 and 2025. This would mean an average growth rate of GDP of about 7 percent yearly, consistent with historical records. Per capita income will increase from its present value of U.S.$800.00 to about U.S.$5,000. Population will increase from its current level of 43 million (1990) to 65 million people. A special challenge will be the improvement of income distribution and elimination of underemployment. The dynamic factors associated with growth in GDP are to be found in industrialization, irrigated agriculture, agroindustries and tourism industries.
- A basic needs strategy targeted at the unemployed and underemployed in rural and urban areas so as to improve their living standards in the short term. The goals of such a strategy are summarized in Table 15.3, together with the predicted values under the "business-as-usual" scenario discussed in the section titled "Assessment of Likely Impacts of Climate Change." This strategy should be reinforced during drought periods, when the number of rural unemployed workers is drastically increased. Present experience in drought management provides useful lessons for the design and implementation of this strategy.
- An ambitious social policy where universal education is pursued as the highest societal and governmental priority. The children of the present generation should benefit from this action;
- A Coastal Zone land-use management strategy, incorporating environmental conservation and protection against, or adaptation to, possible seawater invasion. This strategy could be based on the detailed mapping of the area below 1 m above the highest sea tide level. An effort should be made to stimulate economic and human decentralization away from the coastal areas to higher and safer places.
- An ambitious environmental protection strategy, both as a dimension of the economic and social strategies and as a sectorial strategy with proper objectives and tools.

TABLE 15.3. *Northeast Brazil, Socioeconomic Scenarios*

	1990	2050	
		"Business-as-usual"	Desired
Population (millions)	42.8	74.0	60.0
Work force (millions)	15.2	25.0	21.0
Underemployed (millions)	10.0	12.0	1.5
Out-migration, balance (millions)	8.8	30.0	20.0
Illiteracy rate (percent)	47.0	30.0	0.0
Per capita income (US$)	800.0	3,200.0	5,000.0
GDP (US$ billions)	34.0	236.0	300.0
GDP Composition (percent)			
Agriculture	15.0	10.0	10.0
Manufacturing	26.0	30.0	30.0
Services	59.0	60.0	60.0
Electric energy use (1,000 GWh)	27.3	118.4	72.0
Per capita energy use (kWh)	636.1	1,600.0	1,200.0

The Planning and Implementation Process

The planning and implementation process of the SDS could be organized as sectorial but interdependent action plans according to the directions suggested below. Once the long-term objectives are given, actions should be organized into medium term action plans covering key activities of the SDS process. Two alternatives are considered:

- comprehensive action plans integrating economic, social and environmental programs and projects; or
- decentralized sectorial economic, social and environmental strategies consistent with a broad framework of sustainability.

The first alternative, though conceptually fine, requires centralized planning and a strong coordination effort. It faces inevitable resistance. The second alternative is appropriate for the scope of a democratic open society, but requires nonbureaucratic mechanisms for ensuring that the sectors follow the sustainability guidelines. Considering the existing NEB institutions and political organizations, the following suggestions are recommended:

- SDS framework: a regional legal and administrative framework where the philosophy of sustainable growth will be transformed into guidelines to be

followed by the states and federal agencies, as well as social and economic agents.

The main provision of the SDS framework should reflect, or be transformed into, federal or state laws, with certain goals that will be achieved through regulatory action, as well as through specific economic and financial incentives. The SDS framework for sustainable growth in Northeast Brazil should include regulations and enforcing mechanisms on:
- regional agroecological and economic zoning;
- urban and rural land use, pointing out coastal low zones;
- sectorial policy objectives and guidelines;
- environmental policy.

The lead for the elaboration of the NEB SDS framework constitutes a joint responsibility of the federal and state governments. Approval should be the responsibility of the National Congress and of the Houses of Representatives in the states. Mechanisms of societal and community participation should be widely utilized.
- Enact sectorial plans of action, both by the federal government and from the NEB states. The sectorial action plans will be the responsibility of the federal and/or state sectorial agencies and must conform

to the guidelines of the SDS framework. Once these guidelines are followed, there is no need for central coordination of the sectorial action plan. For the sake of their administrative feasibility, they must be decentralized and can even differ from one another in terms of specific duration and implementation mechanisms and procedures.

Following the general guidelines, each sectorial plan of action should include, in addition to specific goals, environmental objectives. Both present and future relationships between the sector and the environment must be explicit and specific goals must be established for:

- how the sector is presently affected by environmental variables, such as climatic variability, and how it might be affected in the future by climatic change;
- how the environment and climate (in particular) are or may be affected by the sector activities;
- what specific strategies must be followed to increase the contribution of the sector for the betterment of the environment and to nullify its contribution to climate change.

• Follow-up and evaluation of programs, projects and activities within the sectorial action plans, as to:

- adherence to the sustainable development regulations and to the SDS framework directions;
- effectiveness and efficiency as to the achievement of objectives and goals.

Follow-up activities can be done by the executing agencies, and related information should be made available to the evaluation agencies. Evaluation could be centrally coordinated, at the federal level by the regional development agency, and at the state level by specific state agencies. This work should be done by evaluation teams linked to the regional universities and to regional and state research centers, under contract with the federal and state governments. The result of the evaluation assessments should be made available to governments, political representatives and society in general, and it should constitute a feedback mechanism for

improving the performance of the sectorial agencies.

The Political and Cultural Dimensions

A technically sound SDS strategy is a most important step, particularly if the planning and implementation process is legitimized by community participation and by required negotiations between the affected social and economic agents. This is not enough, however.

The SDS must take into account the cultural and political factors that affect the regional society. The development process is dependent not only on investment decisions but also on the cultural and political characteristics of the population. Society's values and attitudes determine the willingness to progress, as well as the possibility of postponing present consumption in favor of investment decisions. In the case of NEB, where most of the people live below poverty thresholds, the cultural setup favors a short term horizon with daily survival as the main objective.

On the political side, the regional social inequalities influence, and are a result of, the regional political power. Politicians represent the region's elite, whose interests differ from those of the population as a whole. The elite are able to profit from the underdeveloped situation and even from severe climatic variability, like drought. Hence, the politicians will not normally favor a development strategy that may, in the short run, alter the regional distribution of power. However, this situation is changing as a result of urbanization, industrialization and development of communications. This process of change can be hampered by the presence of the traditional elite, illiteracy and extreme poverty, constituting a situation of circular causality: sustainable development is necessary to overcome poverty, illiteracy and environmental degradation; but poverty, illiteracy and environmental degradation hamper development possibilities.

Disrupting this vicious circle requires external interference in order to reinforce regional and local modernization movements, make education feasible, promote consciousness building and disseminate sustainable development experiences.

The International Perspective

International technical and financial cooperation, as an external factor, will be of utmost importance for overcoming local resistance and promoting sustainable growth. Besides its role in financing investment projects, another important role may be more qualitative than quantitative, through its contribution to increasing consciousness, promoting modernization and market competition, introducing sustainable technologies and training human resources. The introduction of modern elements into local society will be an important outcome.

Another outcome will be the opening of foreign markets to regional products. While local wages are low—owing to the large number of unemployed and underemployed workers—increased regional production will require extraregional markets. In its turn, increased production means increased labor demand and finally a pressure for increased wages; hence, internal regional market and trade will grow. In the long run, everyone wins. In the short run, there will be winners and losers.

Conclusions

The implications of, and responses to, climatic change have become an extremely important issue for humankind. This chapter has examined this issue from the viewpoint of a developing region, Northeast Brazil (NEB). Though a region within a country, NEB is a very large area, comparable to some of the largest countries in the world in terms of population and size. Its underdeveloped nature is similar to that of many underdeveloped nations, with low income levels, adverse social indicators and poor quality of natural resources. The semiaridity of 60 percent of its area is comparable to similar conditions in several African, Asian and North American developing regions.

As in many other developing countries, NEB faces high climatic variability, with droughts being a permanent concern for government and society. Several types of policies, including relief policies and permanent policies, have been practiced for many years and can now be evaluated. Many lessons can be drawn from that experience and

may, by analogy, provide a valuable database for decision makers when deciding how to react to future climatic change.

Present impacts of climatic anomalies in Northeast Brazil are very severe. The underdeveloped condition of the region makes it much more vulnerable to droughts and other climatic events. The social effects of droughts are intense and, for the majority of the population—some 25 million rural and urban poor—may affect their ability to survive. The impacts of droughts are not the same for everyone. The poor rural population is much more vulnerable than the urban population. By the same token, agriculture is more vulnerable than manufacturing. Indeed, some social groups may even benefit from droughts.

This chapter has tried to develop a possible climatic change scenario for NEB. Unfortunately, regional climatic change predictions are still not reliable. For NEB the information is even more scarce. Examination of the climatic change literature in search of impacts on NEB does not permit one to advance beyond the establishment of some hints on possible future scenarios, and these scenarios may contradict each other, depending on the sources of information one is using. For the purpose of this chapter the author has relied upon the reasoning developed by Group II of the Intergovernmental Panel on Climatic Change (IPCC, 1990), where it is suggested that the effects of climatic change will be more severe in semiarid areas. If this is true, NEB will get drier and extreme droughts will become more frequent.

Assessment of past and present policy responses to climatic variability in Northeast Brazil leads one to the following main conclusions:

1. These policies have been effective as a means for reducing the suffering of the poor population during extreme climatic events. In this regard, the practicing of a Keynesian (temporary) type of employment policy is an effective tool that is aimed at maintaining an income stream for the large numbers unemployed by the droughts and keeping the regional economy alive. Associated with this policy, the building of a water resources infrastructure and other public works (by

the so employed labor force) has played a major role in improving the living conditions in rural Northeast Brazil.

2. Such policies have not been effective, however, in permanently increasing the region's resiliency to droughts or any other type of adverse climatic event. Indeed, each new drought finds the region even more vulnerable because of population increase and added environmental stress. The permanence of the underdeveloped condition is the main explanation for the extreme vulnerability of NEB to climatic variability.

How can these conclusions be useful for the definition of policy responses to climatic change in NEB? If one considers that climatic change will be manifested mainly as an altered pattern of future climatic variability—besides the invasion of seawater in low coastal zones—one can conclude that the type of relief action that has been practiced up to now is still valid. Social and economic consequences of future climatic variability can be faced through employment policies and public works construction. The need for such policies will be more frequent, however, and in some areas the relief action may become a permanent, rather than temporary, requirement.

In a developing region like NEB, there is only one type of answer that can effectively address climatic change. It is the promotion of sustainable development, as proposed by the Brundtland Report on *Our Common Future* (WCED, 1987), in terms of poverty eradication, increased per capita economic output and environmental conservation. Successful implementation of a sustainable development strategy will provide to the developing NEB population the necessary tools to face future climatic crises. It is only when each social unity—each family or each individual—becomes able to respond to the crises that menace their way of life, like the interruption of the family income stream, that society will be prepared to face the effects of climatic change. The difference between developed and developing regions, as to climatic change impacts, relies heavily on this point. Whereas in developed regions a severe drought may impose heavy economic losses, the society is equipped to mitigate the impacts through individual mechanisms, such

as personal savings and high creativity, and social mechanisms, such as unemployment insurance. In developing regions, however, these mechanisms are not available, and the capacity of society to face the drought is nonexistent. Not only economic losses are incurred, but social losses are imposed on the majority of the society.

A sustainable development strategy is the appropriate response to climatic change in developing regions. For some time, these regions must be allowed to use more energy, but energy efficiency must be improved from the beginning. The Brundtland Report (WCED, 1987) is clear about this issue. It lays the political and ethical, as well as technical, foundations for sustainable development. It does not, however, delineate the mechanism for the practical implementation of a sustainable development strategy. Incorporation of the environmental dimension into the development theory is also still to be done. This is an important step, required for the dissemination of the sustainable development strategy into the practice of university teaching and government and societal planning.

This chapter has tried to discuss a proposal for the sustainable development of NEB. The strategy must rely on the specific conditions of the natural resources, the economy, the condition of the environment, and on the socioeconomic and cultural characteristics of the population. The need for improving the quality of the human resources is pointed out. There will be no sustainable development unless the entire population becomes educated and well nourished. Introduction of the environmental dimension requires a radical change in mentality, which will only be achieved through widespread environmental information dissemination and education. It has also been emphasized that, although the main effort for development must be domestic, international and extraregional cooperation will be a necessary condition for overcoming the social, economic and cultural restrictions that hamper the sustainable development process.

Although this chapter has been based on the specific situation in Northeast Brazil, its main points can be applied to many of the developing regions of the world. Some of the conclusions that can be more widely applied are the following:

- Adverse impacts of climatic warming will be intensified in developing regions or countries, because of
  - higher social, economic and environmental vulnerabilities; and
  - limited adaptation capability;
- A definite response to climatic warming in developing regions will only be possible through the implementation of sustainable development strategies that simultaneously consider economic growth, social justice—especially poverty eradication—and environmental conservation.
- Promotion of sustainable development strategies requires
  - development and dissemination of methodologies for sustainable development planning, as well as technologies for sustainable development implementation;
  - political decisions for implementing the sustainable development strategies that reflect increased awareness by society and policymakers;
  - international technical and financial cooperation.

Finally, as recently stated by scientists from both developed and developing regions (Glantz, 1990), "global concern about climate change should also be translated into global responsibility for the sustainable development of the developing countries through international financial, technical, and scientific cooperation."

References

Arraes, R.A., and I. Castelar, 1989. The effect of drought on public finances in the state of Ceara. In: *Socioeconomic Impacts of Climatic Variations and Policy Responses in Brazil*. A.R. Magalhaes and E. Bezerra Neto (Eds.). UNEP/SEPLAN-CE, Fortaleza, Brazil (unpublished).

Barreira, C., 1989. Drought: Power and survival. In: A.R. Magalhaes and E. Bezerra Neto (Ed.), *op. cit.*

Chenery, H., M.S. Ahluwalia, C.L.G. Bell, et al., 1974. *Redistribution with Growth*. Published for the World Bank and the Institute of Development Studies, University of Sussex. Oxford University Press, London.

Furtado, C., 1964. *The Economic Growth of Brazil*. University California Press, Berkeley.

Glantz, M.H. (conference organizer). Workshop on "Assessing winners and losers in the context of global warming," Saint Julians, Malta, June 18-21, 1990. ESIG/NCAR, Boulder , Colo. (in press).

IBGE-Instituto Brasileiro de Geografia e Estatistica, 1988. Anuario Estatistico, 1987. Rio de Janeiro.

IPCC-Intergovernmental Panel on Climate Change, 1990. *Overview and Conclusions. Climate Change: A Key Global Issue*. WMO/UNEP, Geneva (draft report).

Jaeger, J., 1988. Developing policies for responding to climatic change. A summary of the discussions and recommendations of the workshops held in Villach and Bellagio, under the auspices of the Beijer Institute, Stockholm, WMO/UNEP/WCIP (WMO-TD 225).

Kellogg, W.W., 1988. Human impact on climate: The evolution of an awareness. In: *Societal Response to Regional Climatic Change*. M.H. Glantz (Ed.). Forecasting by Analogy, ESIG/NCAR, Westview Press, Boulder, Colo. pp. 9-39.

Magalhaes, A.R. and O.E. Reboucas, 1988. Conclusions. In: A.R. Magalhaes *et al.* (1988), op. cit.

Magalhaes, A.R., H.C. Filho, F.L. Garagory, *et al.*, 1988. The effects of climatic variations on agriculture in Northeast Brazil. In: *The Impacts of Climatic Variations on Agriculture. Vol. 2: Assessments in Semiarid Regions*. M.L. Parry, T.R. Carter, and N.T. Konjn (Eds.). UNEP/IIASA, Kluwer Academic Publishers, Dordrecht, The Netherlands, pp. 273-380.

———, and E. Bezerra Neto (Eds.), 1989. *Socioeconomic Impacts of Climatic Variations and Policy Responses in Brazil*. UNEP/SEPLAN-CE, Fortaleza, Brazil (unpublished).

———, J.J.S. Lemos, J.A. Pereira, *et al.* 1988. A questao da producao e do abastecimento alimentar no Brasil.

Diagnostico regional- Regiao nordeste. In: *A Questao da Producao e do Abastecimento Alimentar no Brasil: Um Diagnostico Macro Com Cortes Regionais*. M.N. Aguiar (Ed.). SEPLAN/IPEA/PNUD/ABC Brasila, pp. 137-233.

———, J.R.A. Vale, A. B. Peixoto, *et al.*, 1989.   Governmental strategies in response to climatic variations: Droughts in Northeast Brazil. In: A.R. Magalhaes and E. Bezerra Neto, *op.cit.*, pp. 6.01-6.36.

Streeten, P., with S.J. Burki, M. Ul Haq, *et al.*, 1982. *First Things First. Meeting Basic Human Needs in Developing Countries*. World Bank, Oxford University Press, London.

SUDENE-Superintendencia do Desenvolvimento do Nordeste (1985). Aspectos do Quadro Social do Nordeste. Recife, Brazil.

UCAR Office for Interdisciplinary Earth Studies, 1991. *Arid Ecosystems Interactions: Recommendations for Dry-lands Research in the Global Change Program*. Report OIES-6, June 1991 (Report of a workshop), Boulder, Colo.

World Commission for the Environment and Development-WCED, 1987. *Our Common Future*, Oxford University Press (Brundtland Report), Oxford and New York.

16. REGIONS OF RESILIENCE: AN ESSAY ON GLOBAL WARMING

Ian Burton

Response to Global Warming

The Emergence of an Issue

As the islands of scientific knowledge expand in the surrounding seas of ignorance, the greater becomes the boundary with the unknown. It is thus a perennial feature of the scientific enterprise that the more that is known the greater the consciousness of ignorance. Standing now on the seashore frontier of the infant island discipline of climatology, there is a new realization of how little is known about the forces that shape Earth's climate, how little is known about how climate in turn shapes the processes of the biosphere on Earth's surface, and how even more unreliable is the meager knowledge of the interactions among climate, biosphere and human activities.

Previously such ignorance has mattered little. Earth's climate, and of its climatic regions, had perforce to be taken as a given. True, it has been a long-standing human ambition to control the weather. In many human societies there are well-rehearsed rituals designed to influence the weather. The rain dances among Indian tribes of the American Southwest are legendary. Ancient Greek sailors made sacrifices in an attempt to influence the gods to produce a favorable wind. These and other exercises achieved little except perhaps for those who had sufficient faith in them. Climate has been a fact of life, no more subject to human choice or control than the gender of an expected child, or the results of next week's lottery.

Now suddenly it is realized that human activity is set to have a major, albeit unintended, impact upon the atmosphere, threatening to alter irrevocably the sequence of daily weather events that, taken together, add up to the statistical abstraction we call "climate." Now suddenly humankind is acutely aware of realms of ignorance that are of great consequence.

There has been speculation for some time that human activities, especially in releasing growing quantities of carbon dioxide into the environment, could heat the atmosphere and so contribute to climate change. Arrhenius drew attention to this as long ago as 1896, but for decades it remained little more than a scientific curiosity (Arrhenius, 1896).

In 1958 the Mauna Loa Observatory in Hawaii began its series of CO_2 measurements. As the record grew year by year and the rising curve began to take shape (Figure 16.1), the phenomenon attracted the attention of growing number of first atmospheric, and then other, scientists. A major advance took place in 1987 when the Vostok 5 ice core data from Antarctica were published in *Nature* (Barnola *et al.*, 1987). The key graph (Figure 16.2) shows that over the past 160,000 years temperature changes in Antarctica have been closely correlated with atmospheric CO_2 concentrations. These have generally declined from a high of some 280 parts per million (ppm). 130,000 B.P. to 200 ppm at the end of the last glaciation, about 20,000 B.P. By 1800, atmospheric CO_2 concentrations had risen back to 280 ppm, a 160,000-year high. Some 25 to 30 years ago there were suggestions that the warmer interglacial period, which we now enjoy, might be coming to an end; that Earth might be about to experience a long slide of temperature levels into another glaciation.

The Mauna Loa and the Vostok data have altered all that. The change since 1800 can only be shown on a graph with the scale of Figure 16.2 as a vertical line. Carbon dioxide concentrations now stand at 350 ppm and are set to grow to 560 ppm (or double the preindustrial levels of 1800) by the middle of the next century or sooner.

257

FIGURE 16.1. Carbon dioxide levels at Mauna Loa and fossil fuel CO_2 emissions. The solid line depicts monthly concentrations of atmospheric CO_2 at Mauna Loa Observatory, Hawaii. The yearly oscillation is explained mainly by the annual cycle of photosynthesis and respiration of plants in the northern hemisphere. The steadily increasing concentration of CO_2 at Mauna Loa since the 1950s is caused primarily by the CO_2 inputs from fossil fuel combustion (dashed line). Note that CO_2 concentrations have continued to increase since 1979, despite relatively constant emissions; this is because emissions have remained substantially larger than net removal, which is primarily by ocean uptake. (Source: U.S. Environmental Protection Agency, 1989a, chapter 1, p. 10.)

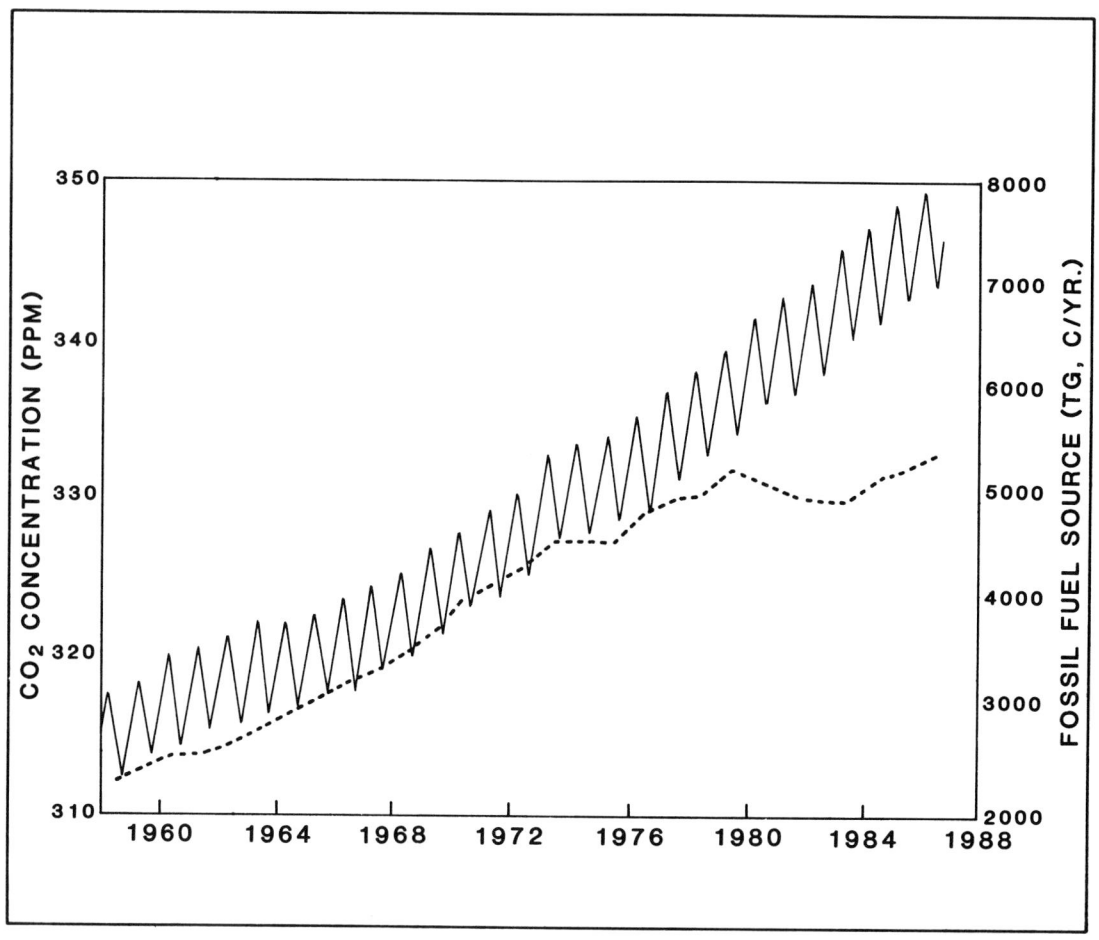

Over the past 160,000 years neither the sources nor the sinks of CO_2, nor cloud cover, nor ocean influences, nor the behavior of polar ice sheets, have intervened to separate the trends in atmospheric CO_2 from the trends in temperature. What perhaps is a little surprising, therefore, is that global mean surface air temperature has only increased by an estimated 0.3°C to 0.6°C over the past 100 years. As the cumulative effect of dramatically increased CO_2 concentrations becomes stronger, not only is the atmosphere poised to get warmer, but it may suddenly do so at a much faster rate.

Clearly the threat is there, and the risk of a major disruptive impact upon human society is high. Given a long enough time perspective it approaches certainty. The scientific community is, therefore, right in drawing attention to the danger, and in suggesting that some response is needed now. To delay is to risk catastrophic consequences—the complete disappearance of some island nations like the Maldives; the

FIGURE 16.2. Carbon dioxide levels and temperature over the last 160,000 years from Vostok 5 ice core in Antarctica. The temperature scale is for Antarctica; the corresponding amplitude of global temperature swings is thought to be about 5°C (Source: Barnola *et al.* 1987).

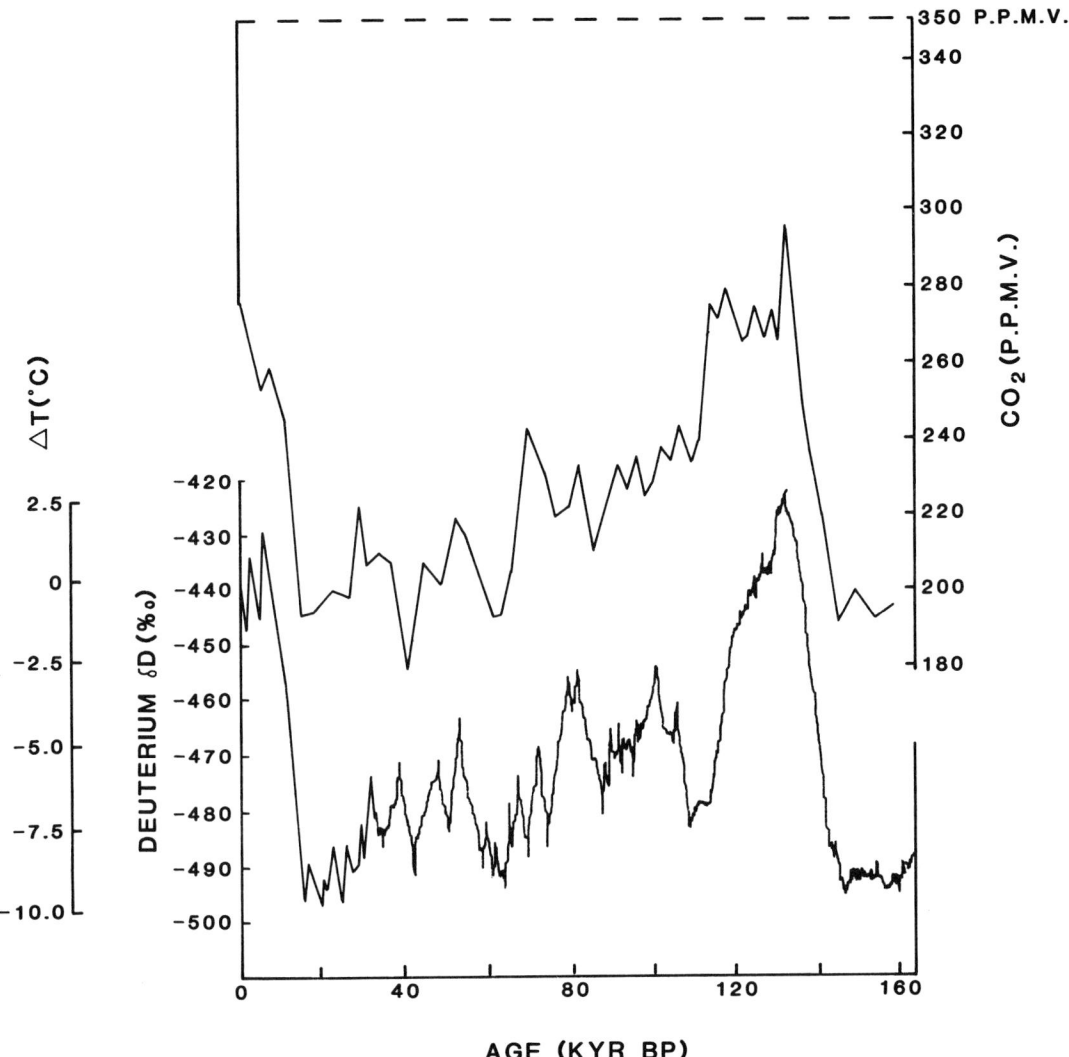

dislocation of millions of peasant farmers on the low-lying deltas of the Nile, Ganges-Brahmaputra, Mekong, Yangtze and others; the radical disruption of world agricultural production and trade; the loss of enormous areas of forest and forest species; the spread of diseases and pests into new unprepared regions; changes in hydrological regimes and patterns of water distribution affecting electric power production, water supply and commercial fisheries; changes in ocean currents; in short, a total transformation of

the pattern of climate and biosphere resources. In addition to these direct and probably incremental changes, the likelihood exists of greater climatic variability and more frequent extreme events such as floods, droughts, and tropical cyclones.

Herein lies the conundrum of climate change. While the consequences are potentially catastrophic, the magnitude of climate change, the rate of change, and its differential character and impacts in the various regions of the globe are poorly understood. There is pervasive uncertainty. What possible actions can be taken under these circumstances? This is the public policy issue that has now emerged full-blown onto the agenda at all levels of government. How do we now respond to the risks given the inevitable and continuing uncertainty?

To address the issue, and to draw it to public attention, a major international conference was held in Toronto in June 1988, sponsored by the United Nations Environment Programme, the World Meteorological Organization, and the Government of Canada. The conference statement adopted by the participants dramatically announced that, "Humanity is conducting an unintended, uncontrolled, globally pervasive experiment whose ultimate consequences could be second only to global nuclear war" (World Meteorological Organization, 1988, p. 292). In its most frequently quoted call for action, the conference recommended that governments and industry:

> Reduce CO_2 emissions by approximately 20 percent of 1988 levels by the year 2005 as an initial global goal. (WMO, 1988, p. 296).

This clarion call has been widely hailed as far-sighted and as a major achievement of the conference. It has also been severely criticized as being unsoundly based. According to some of the critics, steps to reduce significantly the emission of greenhouse gases promise to be extremely expensive for those who adopt them, and to create tensions and conflicts with others who can be relied upon to resist such action. Until both the future magnitude and the rate of climate change are better understood and predicted in sufficient spatial detail, it is virtually impossible to make credible estimates of the costs of inaction, although some attempts have been made (e.g., Nordhaus, 1990). It is also clear that the costs of inaction are necessarily increasing with every day of delay. This is because the magnitude of the resulting climate change will be greater, and the resulting impacts more severe. Moreover, the costs of action to reduce greenhouse gas emissions are steadily increasing as greater use is made of fossil fuels. The longer the world community rides on the tiger of fossil fuels the more expensive it becomes to dismount.

In the face of the uncertainty, and as recognition of the need for some "response now," the international community of experts, policy analysts, diplomats and political leaders has converged on the notions of "precautionary response" and of a "no-regrets" policy. This means that action should now begin that can be justified, and should avoid excessive response that might in the future be seen to have been more costly than necessary.

The classic example of precautionary response is a return to the promotion of energy conservation and energy efficiency. There are many ways in which the energy intensity of the industrial economies can be reduced at little or no net cost (Goldemberg *et al.*, 1987). The same appears to be true in developing countries. The effect of such action would be to reduce carbon dioxide emissions from the burning of fossil fuels compared with what they would otherwise be. Such "precautionary responses" when they can be clearly identified, represent "win-win" situations in which the threat of climate change is delayed or reduced by some amount, and is achieved at practically zero cost. The benefits of enhanced energy efficiency are by now so widely accepted that it should be adapted anyway without the stimulus of global warming.

What then are likely to be the other elements of so-called precautionary response, and how can they best be identified and adopted? The major thesis of this chapter is that many appropriate precautionary responses are different in different places, and they can best be developed at the regional level. It is at this scale that responses can best be evaluated against the uncertainties of future economic conditions as well as of climate change. Thus, the probability of avoiding regret is greater at the regional level, and the danger of making costly wrong responses, or of too much or too little response, too late or too soon, can be minimized.

Hence, a policy orientation is advocated. It is based on the idea that regions should seek to understand, for themselves, the threats and opportunities inherent in climate change. In so doing, regions should take account of both climatic and socioeconomic uncertainty. Regions should seek to enhance their own adaptive capacity to deal with climate changes whatever they may prove to be. The phrase "resilient regional response" is a shorthand expression to indicate this approach. It means the better positioning, and the shaping of the regional economy and society in order to minimize the adverse effects of climate change, and to take advantage of such opportunities that may arise. It means the creation and strengthening of capacity for adaptive and flexible response. It means identification and reduction of vulnerability.

The term "regions" in this context refers to a scale of analysis and decisionmaking that lies between the continental and the local. Thus, in North America, one might identify such regions as the Great Lakes Basin, New England, the Southwest, the Great Plains, and so forth. In South America, one might identify the Andean region, the Amazon Basin, the southern cone and the like. It is clear from these examples that regions may be subdivisions of nation-states, where a nation has large territorial extent, or may include all or parts of several nation-states especially where such nations are more limited in size. At the same time, a region in this context can also be an area with common geobioclimatic characteristics, as well as an economic and social unit.

No single level of government or other decision-making structure necessarily corresponds to the geographical boundaries of such regions. The term "response" therefore refers to actions that could be taken within the region by governments (state, provincial, local) business, industry, and other private- and independent-sector organizations, as well as by informal communities, families, and individuals. Decisions within regions also involve cooperation with other jurisdictions at national and international levels. For present purposes, therefore, no benefit is to be gained from being pedantic about the precise definition of a hypothetical region. It is recognized, however, that some regional-level decisions may require very great precision with respect to who is and is not included within a particular program or response strategy. This is especially the case for the purposes of assessing benefits and costs, or of adopting regulations and other market interventions.

Before proceeding to an elaboration of the strategy of regional resilient response it will be helpful to revisit some of the more fundamental distinctions than can be made about the nature of the relationships between climate and human society. These distinc-tions concern the nature of "climate impacts," limitation and adaptive responses, and the harmonization of responses at different levels of decision making. As a further preliminary consideration it is also desirable to cite the most authoritative judgment concerning what is really known about climate change, and to describe the necessity for response as a matter of urgency.

The Stimulus

In any discussion of response to climate change it is first necessary to know about the stimulus. In this case, the stimulus has two properties that, taken together, are difficult to handle. First, climate change incontrovertibly exists. Second, practically everything else about it is uncertain. Although there remain some skeptics about the fact of global warming, these are now to be found almost entirely outside the scientific community. Among atmospheric scientists there is no fundamental disagreement. For example, the final report of Working Group 1 of the Intergovernmental Panel on Climate Change (IPCC) recently concluded:

> WE ARE CERTAIN OF THE FOLLOWING:
> - there is a natural greenhouse effect which already keeps the Earth warmer than it would otherwise be.
> - emissions resulting from human activities are substantially increasing the atmospheric concentrations of the greenhouse gases: carbon dioxide, methane, chlorofluorocarbons (CFCs) and nitrous oxide. These increases will enhance the greenhouse effect, resulting on average in an additional warming of the Earth's surface. The main greenhouse gas, water vapor, will increase in response to global warming and further enhance it. (IPCC Working Group I, 1990, Executive Summary, p. 1)

Thus, it is clear that Earth is set to get warmer as a result of human activities, and that the warming process itself will give rise to other changes that will further enhance it.

Beyond this, the rate of change, its magnitude and the regional variations of climatic parameters such as temperature and precipitation can only be predicted on the basis of current models. These are generally acknowledged to be highly approximate and in need of substantial improvement before much reliance can be placed upon them. The models indicate a rate of increase of global mean temperature during the next century of about 0.3°C per decade with an uncertainty range of 0.2°C to 0.5°C. A further element of uncertainty is introduced by the growing realization that the relationships are not necessarily linear. While Earth is likely to get warmer at an average rate of 0.3°C per decade, the change could occur suddenly. The atmosphere is notoriously unstable. It never attains a state of equilibrium, and, as has now been shown in many areas of science, systems in a far-from-equilibrium state are prone to sudden restructuring (Prigogine and Stengers, 1984; Jantsch, 1980).

On top of the undeniable reality of global warming, therefore, must be set the possibility that the estimates are conservative and that the changes may occur suddenly at any time. On the other hand, cautious atmospheric scientists are advising that, on the basis of actual weather observations, it may not be possible definitively to detect global warming for a decade or more. Climate variability is such that detecting the signal of warming from the background noise of fluctuations will take more time and more research.

Decision-makers in search of firm ground upon which to make choices are frequently insistent that they be given unequivocal answers. It is frustrating for decision-makers, and for the general public, that scientists sometimes stress the need for long-term research, pointing to the possibility of sudden dramatic change. Both positions are valid. They are not incompatible, nor is it yet possible to attach probabilities to specific scenarios of climate change. Such a stimulus makes the choice of action very difficult. Fresh and innovative approaches to decision making are required.

The Necessity of Response

It is tempting to think that no really substantive response is necessary at this stage. Let the research community get on with the task of reducing the uncertainties. There will be time enough to take action once global warming has been clearly demonstrated and more precisely quantified.

Such a wait-and-see attitude could have disastrous consequences. On the other hand it might not. A judgment call is needed on the balance of risks and benefits from action and inaction. There is little firm evidence on which to base such judgment. On the other hand, many overt actions can indeed be taken now that are virtually cost-free or that fall so solidly into the policy direction of "sustainable development" that the only difficulty is to explain why they have not as yet been adopted. The costs of inaction, while of uncertain magnitude, certainly exist and could be enormous.

Under these circumstances, not to take some action would surely be a dereliction of responsibility, for government, business and private citizen alike. But what actions should be taken?

Let us suppose that the world community were to agree that the only equitable and acceptable strategy would be to stabilize atmospheric concentrations of long-lived greenhouse gases at 1990 levels. According to the IPCC, this would require immediately reducting by 60 percent emissions from human activities. The possibility of reducing CO_2 emissions immediately by 60 percent is clearly far beyond the bounds of practicality. The most that the Toronto Conference dared to suggest was a 20 percent reduction over 1988 levels by the year 2005. Yet if some action is not taken the issue will grow steadily more difficult to address. Some action is clearly necessary as an act of prudence.

Policy Orientation

This brings us back to our policy orientation. The proposal is that regions act to protect themselves from the impacts of global warming by adopting a mixed strategy of limiting greenhouse gas emissions and adapting to climate change. To design and select an appropriate strategy will depend upon the characteristics and qualities of each region. For this, much research and reflection is necessary. Precise policy prescriptions require detailed research, not only on climate and its impacts, but also upon the changing economic and social prospects of particular regions.

The Nature of Impacts

The power of modern science since Bacon and Descartes has hinged upon an understanding of cause and effect. In this classical paradigm, scientific advance depends largely upon the isolation of cause/effect relationships and their verification in exact and replicable experiments. It is an axiomatic reaction of scientists therefore when confronted with the notion of responding to climate change to begin to construct a chain of causal relationships. Emissions of carbon dioxide into the atmosphere result from the human use of fossil fuels. The carbon dioxide "traps" long-wave radiation and thus contributes to an enhancement of the greenhouse effect. A warmer atmosphere will in turn have effects on the biosphere and hence on human economy and society. It is a small step from this mode of reasoning to think in terms of climate impacts, and to design research accordingly. Thus, climate change is seen as causing first order impacts on the biosphere, and this results in turn in second-, third- and nth-order impacts cascading through social and economic activities. This reasoning has given rise to a generation of impact studies. Figure 16.3 (from Cohen, 1986) well illustrates the sort of research framework that is widely followed. The methodology available for such research has been fully described (Kates *et al.*, 1985) and the recent IPCC report (IPCC Working Group II, 1990) summarizes and synthesizes a considerable volume of research.

As has been pointed out by many critics of this research (see, for example, Timmerman, 1989), it has necessarily been limited in its assessment of potential climatic impacts. Virtually all studies have been based upon such linear impact models. Consequently its results have limited value. The authors of the IPCC report concede that

> This report does not attempt to anticipate any adaptation, technological innovation or any other measures to diminish the adverse effects of climate change that will take place in the same time frame. (IPCC Working Group II, Policymakers Summary, p. 2)

In the area of agriculture, the economic sector most directly dependant upon climate,

the report concludes that

> Sufficient evidence is now available from a variety of different studies to indicate that changes of climate would have an important effect on agriculture and livestock. Studies have not yet conclusively determined whether, on average, global agricultural potential will increase or decrease. . . There may be severe effects in some regions. . . (IPCC Working Group II, Policymakers Summary, p. 2)

The report recognizes that other variables beyond climate are subject to change, but makes no effort to take these seriously into account. It is recognized that "the predicted population explosion will produce severe impacts on land use and on the demands for energy, fresh water, food and housing, which will vary from region to region according to national incomes and rates of development" (IPCC Working Group II, p. 1, Policymakers Summary). But the human capacity for creative and adaptive response is given short shrift.

The overall impression is one of emphasis on the potential negative impacts of climate change, the high uncertainty associated with the impacts, and the almost complete absence of attempts to deal in any satisfactory way with the process of adaptation to climate change so as to mitigate its effects.

It is not surprising, therefore, that among its suggestions for future action the report calls for more research and, significantly, "development of methodology to assess sensitivity of environments and socioeconomic systems to climate change" (IPCC Summary Report, Working Group II, p. 45).

Limitation and Adaptation

There are two broad kinds of response to climate change. We can seek to limit the change itself by reducing the rate of increase of greenhouse gas concentrations in the atmosphere, or we can seek to adapt human society, economy, and behavior to a warmer Earth. Since some global warming is already inevitable it is clear that both adaptation and limitation are needed. Only a mixed strategy makes sense.

Here the complexities become legion. The IPCC reports are heavily influenced by

FIGURE 16.3.   An example of the research framework widely followed.   It is based on a view of climate change causing first order impacts on the biosphere, resulting in turn in second, third, and nth order impacts cascading through social and economic activities. (Source: Cohen, 1986.)

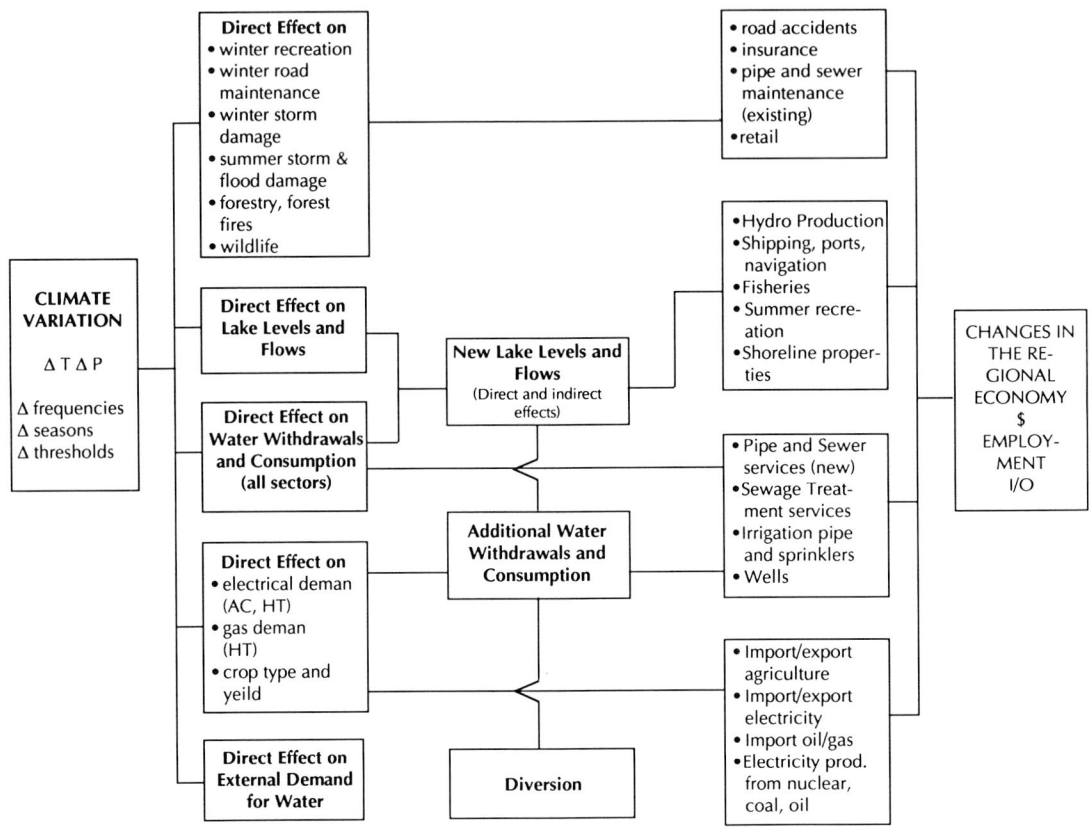

concern for the negative impacts of climate change. Hence there is an emphasis on limitation. It is recognized, however, this will be difficult to achieve in the short run.   The world economy is heavily dependent upon fossil fuels.   Many developing countries are counting upon increasing their use of energy, and they are looking to fossil fuels as the major source.   Alternative sources of energy are more costly; are associated with risks that people are reluctant to accept (nuclear energy); or are not yet well enough advanced in the technology of production to be major sources in the near future (wind, solar, tidal, geothermal). Hence, continuing emissions of CO_2 are inevitable. Under the present economic and technical conditions, some reductions

in the rate of growth of carbon dioxide concentrations in the atmosphere can no doubt be achieved.   As previously cited, in developing and especially in some developed countries, much can and should be achieved by improvements in energy conservation and energy efficiency (Goldemberg et al., 1987).

Levels of Decision Making

An approximate correspondence exists between kinds of decisions about global warming and the level at which they are likely to be made.   Broadly speaking, limitation strategies require decisions at national and international levels.   This is because environment and energy policy decisions in most countries rest at the national or central level of government. There is also a commonly held perception that reducing the emissions of greenhouse gases will be expensive, and that there is little incentive to adopt such action to protect a common property resource like the global atmosphere unless all nations agree to act in concert.

Indeed, steps are now being taken at the international level to draft a Climate Convention. It is hoped that such a convention can be ready in time for signature at the United Nations Conference on Environment and Development to be held in Brazil in 1992. It is also proposed that specific protocols will be prepared to deal, for example, with particular greenhouse gases.

Nevertheless, some actions to limit the emission of greenhouse gases can be taken at regional and local levels, and some jurisdictions are already beginning to do so. There is in fact no necessity for regions to wait for national policy to take shape or for international agreements to limit emissions, although in most cases some detailed study will be required at the regional level to identify a range of policy options and to select from among them the most appropriate choices.

Adaptation strategies fall much more heavily within the ambit of regional-level decision making. Protection of vulnerable areas, design of public works, land-use planning, building codes and standards, and related measures are usually considered to be matters for regional and local decision making. Clearly some policy options are adaptive and can be made at the national level. For example, decisions with respect to agriculture, especially price supports, often require national government authority and are also subject to international negotiations under the General Agreement on Tariffs and Trade (GATT).

Nevertheless, from a regional perspective there is much to be said for concentrating on the development of adaptation strategies, for the region itself, and seeking externally to influence national decision making toward appropriate policy for limitation. There can be no doubt that limitation strategies at the regional and local level would benefit from national guidance and possibly direct financial assistance.

It is important, however, to recognize the capacity and indeed the propensity for spontaneous, self-organizing behavior. Not everything has to be mandated by government. After reviewing the research on climate impacts in the Great Lakes Basin, Hare and Cohen note that

> In practice, a market economy of the kind both countries practice (U.S. and Canada) adapts continually to perceived changes; it does not wait for official action or sanctions before taking precautionary measures. Once it becomes the received wisdom that the greenhouse effect is really in progress, the tasks of governments will be to keep up with the adaptive measures that will be taken spontaneously. Individual institutions and consumers will then make the pace—as the individual voter will do on election days (Hare and Cohen, 1989, p. 58).

The role of government is properly circumscribed by this observation. But some broad social guidance is nevertheless necessary. The role of governments at the regional level legitimately includes the development of options for adaptive response, and this in part requires the removal of obstacles or disincentives that now exist. Governments can also legitimately influence the direction of adaptive response by encouraging those that have greatest social utility in a particular regional context.

The Sinews of Resilience

The Concepts of Resilience and Vulnerability

To be resilient means to have the capacity to bounce back, to have the power of recovery. Dictionary definitions frequently also refer to a "return to an original position." This is not what is meant by resilience in this context. Resilience here refers to the ability of a region to absorb negative impacts or changes and to recover from them in ways that enhance the health and well-being of the economy and the society, but not necessarily by returning to the original condition. Indeed, since human society is in a continual state of evolution, attempts to return to the status quo can be counterproductive.

It is a common experience after a disaster

that communities seek to return as fast as possible to pre-disaster conditions. This can be seen in the process of reconstruction after an earthquake, a major flood, or a war. The reason for this is often that it is seen as the best means to restore confidence and to heal the wounds, both physical and mental, that have been suffered.

Such actions can be counterproductive because they have a tendency to re-create the conditions that gave rise to the disaster in the first place. Thus, an opportunity to rebuild and to create anew a community that is less vulnerable may be lost.

Vulnerability is the obverse of resilience. It means susceptibility to damage. Both vulnerability and resilience are used here to denote properties of systems. In this case the "system" is the aggregate of human population, physical property, social and economic activity, and the intangible qualities of culture that collectively constitute the society of a region. Each of these dimensions has its own capacity for resilience and its own vulnerability.

The strategic and policy orientation proposed is that regions should seek to enhance their resilience and simultaneously reduce their vulnerability in the face of climate change. In theory, the way to do this might be to create measures of aggregate resilience and vulnerability and to seek to identify choices that would result in increases in the quantitative value of resilience, and reduction in the quantitative value of vulnerability. Possible practical approaches to this will be described shortly in relation to the specific properties of social systems. As will be seen, these cannot readily be accumulated into total measures.

Sustainable Development

The concept of resilience is therefore somewhat similar in function to the currently fashionable concept of "sustainable development" popularized by the Brundtland Report (World Commission on Environment and Development 1987). It represents an idea that can be loosely defined but which it is not possible to quantify in precise terms and is, therefore, in an analytical sense nonoperational. The fact that this is so does not invalidate the concept. On the contrary, a certain fuzziness

is to be preferred since it then avoids the misleading sense of precision that is sometimes created by quantitative analysis in the social sciences. Sustainable development has been defined by the Brundtland Commission as development that "meets the needs of the present generations without compromising the ability of future generations to meet their own needs."

While the lack of a quantitative definition of sustainable development or resilience makes it difficult to know how much resilience has increased or decreased, the "fuzzy" quality of the concept helps to ensure that choices will be seen as partly political and not solely technical. How a regional level of resilience is identified and maintained is essentially a matter of political judgment. It is a matter of determining what society really values and what sorts and magnitudes of risks people are prepared to tolerate.

Indeed, in an important sense, resilience and sustainable development can be regarded as synonymous. The idea of sustainable development brings together the three concepts of environment, futurity and equity. They are integrated in the general proposition that future generations should be compensated for reduction in the endowments of natural resources brought about by the actions (e.g., consumption patterns) of present generations. The underlying logic of this proposition is simple. "If one generation leaves the next generation with less wealth then it has made the future worse off. But sustainable development is about making people better off" (Pearce *et al.*, p. 3).

The wealth that is left by one generation to the next consists of two components or two types of capital. One is natural capital in the shape of natural resources. The other is the social capital in the shape of wealth, health, infrastructure, know-how, social cohesion and so forth. It is generally assumed that sustainable development is being practiced when the drawdown of natural capital (e.g., depleting coal or oil) is compensated for by the creation of an equivalent amount or more of social capital.

In theory, sustainability might be measured by adding the stock of natural and social capital and ensuring that the total increases over time. In practice, this measurement is impossible, not only because this would require the addition of non-commensurables but also because social capital cannot substitute perfectly for natural capital. Once destroyed, the ozone layer cannot be replaced within the range of human financial resources, technological know-how and time scales. Similarly, old growth tropical and temperate rain forests cannot be replaced by manmade capital even though there might be substitutes for the products of those forests. There is no substitute for the forest ecosystems themselves. It should also be recognized that the growth of manmade capital in the form of knowledge, scientific understanding and technology can, in fact, increase or create natural capital. Oil buried beneath the seabed is not part of the natural capital stock until its existence is known and until such time as the technology exists to discover it, to bring it ashore, and put it to use. Similarly, the energy contained in an atom of uranium is not an effective natural resource until the scientific understanding of atoms is developed and the technological capacity to split the atom and harness the released energy is available. Future advances in technology can be confidently expected to create new resources out of "neutral staff." Natural capital can also be increased. In effect, climate change can contribute to the enhancement of natural capital as well as its depletion.

The concept of resilience of systems can therefore be used as a surrogate for the combination of natural and manmade capital. In other words, the total wealth of the planet, or of a region within it, can be viewed in terms of the capacity of the socioeconomic system to bounce back and recover from adverse change. Whereas the concept of sustainable development directs attention to the preservation of natural capital, the concept of resilience focuses on the capacity to use human capital in fashioning adaptive response to change.

A feature of the concept of resilience in addressing the issue of climate change is that it does not treat climate as a natural resource which is being depleted. Rather, it addresses the capacity of the human system to interact with climate in a way that enhances the aggregate value of natural and social capital. Thus, while not being subject to precise and arbitrary quantification, the objectives of increasing resilience and reducing vulnerability converge with the objectives of sustainable development.

Resilience Versus Resistance

To clarify further the concept of resilience, it is useful to make a distinction between *resilience* and *resistance* (Burton, 1983). Resistant systems rely on strength and control. Resilient systems rely on flexibility and adaptability. The distinction is easy to visualize in the case of materials. A car windshield is resistant to shock but it is not resilient. A flying stone will usually bounce off and no harm will be done. The glass has been made especially strong to resist. But sometimes because of the speed, the angle of impact, the size of the stone or a flaw in the glass, the windshield will shatter. If the windshield were made of transparent rubber, it would behave more like a trampoline. When the stone struck the rubber it would absorb the shock and then bounce back—resilient and not resistant.

Flexibility and adaptability are properties of social systems that are appropriate for responding to climate change. A resistance-inspired strategy is not appropriate for managing the interaction between climate and society. It is not practicable to speak of control or stabilization when it comes to atmospheric processes that are by their nature restless, and ever changing. Even the goal of stabilizing atmospheric concentrations of CO_2 seems far fetched when examined in the light of Figures 16.1 and 16.2. It seems better to recognize that a large part of the equation will necessarily involve flexible and adaptive strategies.

Much more is involved in response to climate change than the properties of materials. We need to consider the properties of systems of agricultural production, of forest, water and soil management, of electrical power generation and so on. In seeking to enhance the property of resilience in such systems, the nature of the stimulus or perturbation has to be taken into account.

Much can be learned in this context from response to climate variability as distinct from long-term change. For example, traditional farmers in some tropical regions plant a wide variety of crops all mixed together. The practice of bush-following and intercropping looks untidy and haphazard to the untrained eye, but when a drought hits or insects or plant diseases invade the plot, some crops may be lost or damaged but others survive. The variety provides a built-in safeguard. Resilient but not resistant. By contrast, large areas of a single crop, as practiced in plantation agriculture, can be commercially more attractive but are subject to the risk of complete devastation by drought, pests or disease. Response to such events usually results in attempts to add more control by irrigation, or addition of chemicals for pest and disease control. The choice of strategy depends upon the character of the system and the nature of the apprehended perturbation. There is often a trade-off. Greater resilience entails the risk of more frequent small scale losses. Greater resistance reduces the risk of small scale losses at the expense of low probability catastrophic events.

The choice between these two extremes is not always self-evident, and it is a useful function of research to make it apparent. The choice of risks can then become a matter of public debate and decision, based on the values and preferences of the entire society. In the case of climate change, the picture is so unclear and the uncertainty so great, that the risks associated with limitation strategies and those associated with adaptive strategies cannot be clearly and unambiguously defined and are certainly still some distance from precise estimation.

Under the circumstances, it seems prudent to search for responses to climate change that have the advantages of resilience. It is, after all, more difficult to design resistance into social systems when it is not clear what is to be resisted.

The Properties of Social Systems

In conventional impact studies it has become the norm to adopt a sectorial approach. Thus studies are made of climatic variables and agriculture, forests, home heating and cooling, hydroelectric power generation and so forth. The effect of adopting a "resilience approach" is to direct attention to the properties of social systems, especially in this case regional socioeconomic systems, and the ways in which they may be vulnerable or resilient.

It is possible in theory to identify and classify a very large number of systemic properties. For present purposes five categories are proposed: (1) economic, (2) biological, (3) structural, (4) functional and (5) psychological.

Economic. The single most significant property of a social system for determining its resilience is its accumulated wealth and economic well-being. Richer means safer. Better-off means more adaptable. It follows that in any region the poorer segments of society and the economic activities that are weakest are likely to be most vulnerable. The economic resilience of a region can be approximately measured by its gross regional product, less outstanding debts. As the wealth of the region grows so does its capacity to respond resiliently to perturbations and to protect the vulnerable. Distribution of wealth is also a factor. A distribution that concentrates a lot of wealth in a few hands, and leaves many people poor, encourages greater vulnerability.

On a global scale it is now widely accepted that climate change is likely to have greater impacts on the poorer countries and regions. Hence, economic development, here, as in so many other cases, is the single best prescription for response to climate change.

Biological. The biological properties of a social system are the human population, its demographic and health characteristics. There are many possible indicators. Rapid population growth, for example, strains the capacity of the system. A population with younger demographic structure can be expected to respond more energetically than an older one, and so forth. Clearly, the prevalence of disease is also an important factor in vulnerability. A healthy population is less vulnerable and vice versa.

Steps to increase the biological health of a social system are as important as the attainment of economic health. Indeed, the two are closely correlated. It is difficult to have a healthy society that is not also wealthy. Although wealth does not guarantee health, it helps a lot. In considering biological health, and climate change, specific attention is needed to the interaction of climate parameters and disease. A warmer planet is likely to mean changes in the distribution of disease vectors.

Structural. The infrastructure of a social system refers to physical structure—buildings, roads, transport and communications facilities, factories, schools, hospitals, dams, irrigation systems and the like. The stronger and more highly developed this infrastructure, the better able is the system to withstand shocks and stresses. There is no readily available measure for total infrastructure, but it can be catalogued, and, if necessary, valued. Its mere existence is important, but it needs to be qualified by its state of repair (depreciation) and environmental and social compatibility. From the perspective of climate change, specific attention is necessary to the adaptability of infrastructure design.

Functional. Another property of a social system is its capacity to function or to work well. This refers to its organization, its administrative arrangements and other elements, the absence of which impedes effective response to shock or stress. The functional strength of a social system is hard to measure, but various indicators can be developed.

For the most part, it is not important to try to identify functional properties that are specific to climate change. One area that is an exception to this is disaster preparedness or emergency response. Since climate change is likely to be associated with a greater frequency of extreme events, the functional response capacity of society to natural disasters is likely to become more important. In general, a well-functioning society is better able to adapt to climate change along with a host of other perturbations.

Psychological. The state of mind of a population significantly affects its ability to respond. Where people have confidence in the society to which they belong, where they share common values and expectations, there is a greater capacity to withstand shock and stress. When such features are weak or absent, societies are prone to internal conflict, and cohesive response to crisis can be impaired.

The threat of climate changes can in fact have an adverse psychological effect on a society. The effect stems not so much from the actual physical or economic impacts of climate change, but from the uncertainty of the threat and the uncertainty about response. Psychological research has demonstrated a phenomenon of denial in the face of great and unpleasant threats. There is therefore a danger that overdramatization of the consequences of global warming could lead to an ostrich-like denial of the problem. Such a reaction would be incapacitating and distinctly unresilient.

The Design of Social Systems

In the works of Plato, Thomas More, Thoreau and Karl Marx, among many others, can be found evidence of the strong utopian tradition in Western social thought. The policy issues posed by the threat of climate change challenge the whole of humanity to think once again about future directions and objectives. In its own way, the Brundtland Commission's report *Our Common Future* (World Commission on Environment and Development, 1987) is in this tradition. It does not however, lay down a precise blueprint for future society. Rather, it is content to propose a process for arriving at a "sustainable future."

In concurring with the Brundtland Commission, the process advocated here is a design process. It suggests that governments, private-sector groups and others should set about the design of social systems that are resilient. Resilience is not something that can be designed once and for all, put in place, and forgotten about. Rather, the design of resilience is a learning process and so also is the reduction of vulnerability. It requires more or less continual re-examination. As the climate changes is unanticipated ways, as the whole global systems of environment and economy interact with unpredicted and unpredictable consequences, there is a continuing need to adopt responses that are flexible and progressive.

Identification of such responses requires research. A few responses are now more or less self-evident. For the most part, however, detailed investigations are needed. In general, it will be found that steps to enhance resilience in relation to climate change are concordant with the broader view of resilience of social systems described in this chapter. They are also consonant with the objectives of sustainable development. The frontline scouts in the search for resilient regional response to climate change have many allies.

A good place to start the search is in the area of climate sensitive natural resources. It is evident that management of natural resources still lags in many regions, far behind the requirements for sustainable development. Improved management of forests, soils and water can also make the economic activities dependent upon such resources less vulnerable to climate change. For example, if water is supplied to the farm or the factory at subsidized prices or at virtually zero cost, then it will not be surprising to see wasteful practices persist. Where the effects of climate change are in the direction of greater aridity, the potential impacts of climate change can be diminished now by moving to more efficient and less wasteful uses of water. The evidence suggests that in many areas economic gains to the community could be achieved by the immediate adoption of improved practices in natural resource management.

The concepts of resilience and vulnerability, as described here, provide a framework in which to assess policy directions in natural resource management and other fields. While it is helpful to think in terms of a "resilient economy" or a "biologically less vulnerable population," it is not possible to design such features into a society in the aggregate. They have to be built in piece by piece. It is here that much detailed research is needed. It is here also that individual behavior and community action come into their own.

The person who chooses to ride a bicycle rather than drive a car; the household that recycles newspapers and other domestic refuse; the community that subsidizes public transport and puts a tax on downtown parking—are all taking actions that promote resilience in the face of climate change as well as other environmental threats. In this sense, response to climate change is not something apart. Rather, it is an integral element in a broad social strategy of harmonizing human activities with the constraints of the environment.

Resilient regional response emphasizes the opportunities for creative as well as defensive action. It can also provide the framework for a mode of analysis that can serve as an aid to decision-makers. In complex societies policy choices cannot be made on the grounds of ideology alone; they need also to be defended and assessed in practical and pragmatic terms. The public rightly demands reassurance that public funds are not frivolously spent in pursuit of some poorly-grounded notion of the public good, nor on the basis of some narrow sectorial advantage. A comprehensive assessment is required. Ultimately, it needs to be comprehensive at a global scale, and this is where the current studies and negotiations stimulated by the Intergovernmental Panel on Climate Change (IPCC) are designed to lead.

In the present context, there is a need for a comprehensive assessment at the regional level to guide both public and private choices.

Toward a Comprehensive Mode of Analysis

The Need for an Integrated Overview

The range of policy options for responding to climate change is enormous and includes many possible options for limiting greenhouse gases; it also includes all the steps that can be imagined to strengthen resilience and reduce vulnerability in the economic, biological, structural, functional and psychological properties of social systems. Practically nothing is excluded. The report of the IPCC, for example, states that

> Because a large projected increase in world population will be a major factor in causing the projected increase in global greenhouse gases, it is essential that global climate change strategies take into account the need to deal with the issue of the rate of growth of the world population. (IPCC Overview Report, 1990, pp. 12-13)

Once such an argument is admitted, there is no basis for excluding macroeconomic issues, trade issues, technology transfer issues, issues of geographical and intergenerational equity, and so forth. And in theory, at least, all these questions need to be asked simultaneously, in an integrated and comprehensive fashion. This leads on to such questions as: "Where is the world going?" "What is the future of humanity?" "What is progress?" "Are there social limits to growth?" "Is sustainable development with equity possible?"

The answers developed to such questions have never served as a pragmatic guide to action, yet it is clear that without some sort of integrated view, there is no guidance in choosing a judicious blend of policies. Strategies thus tend to become mere aggregations of separate actions developed in response to local or sectional interests. The development of a multitude of separate actions is highly desirable, but to guide the selection of priorities more comprehensive modes of analysis are required. What means are available for integrating the many different factors that will bear skeptical scrutiny under conditions of great uncertainty?

Economic and Other Analysis

Money provides the only commonly recognized medium in which the advantages or disadvantages of various policy options can be weighed. Hence the development of techniques such as cost-benefit analysis designed to reduce future streams of benefits and costs to some common terms whereby alternative strategies can be evaluated.

Cost-benefit analysis is an economic technique designed for and most appropriately used at the project level. At this level its shortcomings are well known and allowances can sometimes be made for its distorting effect on decisions, including the effect of discount rates in reducing time horizons and failure to incorporate all environmental costs. The larger the scale and scope of the policy under review, the less attractive cost-benefit analysis becomes. At the global scale of climate change, the technique is of little practical value. There is no way that total world costs and benefits of climate change can be assessed and meaningfully compared with the costs and benefits of action and inaction. Indeed, there is no possibility that a meaningful analysis of this kind can be made at the regional level. The present uncertainties about climate change and social response are simply too great. There is no reason, however, why specific cost-benefit analysis studies cannot be used to identify particular sensitivities or areas of vulnerability, as well as opportunity. Cost-benefit and other quantitative analysis need to be used with circumspection and integrated into a broader decision framework. This is called SWOT analysis.

SWOT Analysis

It is proposed that to deal with an issue of the magnitude and uncertainty of global warming, a new method of analysis be developed for application at the regional scale. The required method should be broad and eclectic enough to be able to incorporate specific

quantitative analyses of the kind that are now carried out in climate impact studies, as well as cost-benefit analysis. Moreover, the required method should thus be additional to current practice and not a replacement for it. The required method should be simple, readily comprehensive to decision-makers and the general public. It should also serve as a modest aid in formulating judgments rather than a precise formula that allows no latitude for interpretation. Indeed, a guide to systematic formulation of judgments in complex systems is both the least and the best that can be expected.

Adopt for a moment the Olympian view of a person or agency at the pinnacle of power and responsibility for the well-being of a region. As the regional domain is surveyed in all its variety and complexity, all of the evidence that has to be taken into consideration can be assembled in an orderly fashion in response to four questions:

- What are the threats to the future well-being of the region?
- What opportunities exist or can be created for the region to improve the lives of its people?
- What strengths are found within the region to enable it to take advantage of opportunities and minimize the threats?
- What weaknesses are found within the region that may expose it to threats and reduce the capacity to seize opportunities?

In some corporate boardrooms these four simple questions are summarized in the acronym SWOT (strengths, weaknesses, opportunities, threats). There is no necessary logic to the order of these words or the questions they represent. They are, however, highly suggestive of a way in which evidence can be systematically marshalled for the purpose of making judgments in complex situations. They are also a stimulus to the formulation of more specific questions.

How might this apply to global warming at the regional scale? Clearly the research results of studies of changing climate parameters can be entered under both the "threats" and the "opportunities" column. Most regions will experience a mixture of the two. If the climate of a region becomes warmer, this may have an adverse effect on those economic activities that require somewhat cooler conditions. On the other hand, the greater warmth will also certainly open the door to activities that were previously excluded.

The ability to protect the region from the adverse effects and to take advantage of the open door will depend upon its relevant capacities (strengths) and incapacities (weaknesses). These need to be systematically described and understood before a sense of possible future directions can be developed. It will normally be simpler to organize such an inquiry along sectorial lines.

There are as yet no guidelines for SWOT analysis, nor any handbooks of methods and procedures. Many of the ingredients are already available, however. These include methods of regional economic analysis and a whole array of other social science techniques useful in studying regions. Such studies might well be organized in categories that are not only sectorial, but that are addressed to the properties of social systems described earlier.

A next step in the development of this methodology is to learn by doing. In fact, several regional-level studies that will prove grist to this mill of methodological development are now proposed and others are being developed. One such study is proposed for the Great Lakes Basin.

The Great Lakes Basin

A report on the potential effects of global climate change in the United States includes the results of studies of the American portion of the Great Lakes Basin. (U.S. Environmental Protection Agency 1989b). The report finds that:

> Global climate change could affect the Great Lakes by lowering lake levels, reducing ice cover, and degrading water quality in rivers and shallow areas of the Lakes. It could also expand agriculture in the Northern States, change forest composition, decrease regional forest productivity in some areas, increase open water fish productivity, and alter energy demand and supply.

The report provides substantial detail on each of these changes, mostly in physical terms. There are a few attempts to provide estimates of costs in dollar terms. For example, it is estimated that under some assumptions, the costs of additional electric power generation (mostly for summer cooling) would increase by $300 million by 2010 and by $25 billion to $35 billion by 2055. These figures can be compared with the costs of additional capacity that is needed to meet GNP and population growth without climate change ($488 billion to $715 billion).

The value of such figures is that they show that the additional climate-induced electricity generation costs would be in the order of 5 percent over the next 65 years. Such an estimate does not suggest the need for urgent action now in response to climate change. On the other hand, it does indicate the potentially very large demands for additional electricity based on expected population and economic growth.

This increase might be met by fossil fuels plants, thus adding more pollutants to the air and in particular increasing the emissions of CO_2. Since lake levels are predicted to fall, there will be less hydropower production, thus further increasing the demand for energy from other sources.

What could be the character of resilient response in such a case? It points to the need to search for ways to reduce future electricity demand and to find alternative sources. Small scale sources of energy for industrial, and household use might be more vigorously explored, such as wind and solar power and generation from small dams. There is an argument to be made for a more decentralized system of electric power production. Such a strategy might promote resilience at the same time as reducing the vulnerability to climate change. Here is an example of where policy shifts in the direction of demand management and alternative supply probably meet simultaneously the requirements for adaptation to climate change, sustainable development, economic efficiency, and limitation of emissions of greenhouse gases.

Much the same applies to the management of lake levels, navigation on the Great Lakes, water supply, agricultural policy, forestry, fishing and tourism. An objective of research on such aspects of the regional economy is, therefore, to identify such policy options and make them available to decision-makers, both governmental and nongovernmental.

While policy choices and actions can be developed on a piecemeal basis, it will be better if they can be eventually assembled within the framework of a SWOT analysis. This would give both decision-makers and the public an opportunity to develop a more comprehensive overview and understanding of the sinews of resilient regional response and to choose their own priorities among them.

The Ethics of Resilient Response

This chapter has been wholly devoted to the development of a diagnosis of the global warming problem from a policy perspective, including the presentation of a framework for response and the initial development of a method of analysis. If the assumptions and arguments developed here are valid, and if they are imaginatively and industriously applied in any given region, then that region should gain some advantage. The prospect is thus raised of regions that over time become more resilient and less vulnerable to global warming. To the extent that policies for resilient regional response are successful, it may be argued that the most favored regions will gain, and those that are initially disadvantaged may become more so.

Are these then ethical concepts that have been formulated in this discussion? Might not regional resilient response in some places increase global inequities still further?

There are two defenses against this criticism. First, nothing in the content of seeking to develop an effective regional response to global warming prevents or diminishes the possibilities for others to do likewise. Nor does it inhibit the drive toward international collaboration and agreement. On the contrary, it should facilitate it.

Second, the stronger the region—any region—the more able it is to help others. A world composed of resilient regions each doing their utmost to use the mix of limitation and adaptive strategies best suited to their own circumstances would also be a resilient world.

There is, in effect, a new avenue for co-operation here. It could begin at the research level, where regions work with each other in collaborative and comparative studies to develop, and apply, the sorts of frameworks and methodologies proposed here. It has been a common practice since 1945 for cities to be "twinned" with counterparts elsewhere in the world. Perhaps this would also work for regions, especially regions that have complementary economies, and complementary needs in the development of resilience and the reduction of vulnerability. Technology transfer does not necessarily have to be orchestrated only on a global basis. Equity is more readily achieved where understanding is present.

In the last analysis, how humanity comes to terms with global climate change, as well as other global environmental and economic changes, will depend upon the development of common understandings, a willingness to share a common environmental heritage in an equitable fashion, and a political will to take the necessary action. The most important regions of resilience are in the individual and collective minds of humanity. These are the regions of intellect, imagination and, yes, I will say, love.

References

Arrhenius S., 1896. The influence of the carbonic acid in the air upon the temperature of the ground. *Philosophical Magazine*, **41**:237-76.

Barnola, J.M., D. Raynaud, Y.S. Korotkevitch, *et al.*, 1987. Vostok ice core: A 160,000-year record of atmospheirc CO_2. *Nature*, **329**:408-14.

Burton, I., 1983. The vulnerability of cities. In: *Approaches to the Study of the Environmental Implications of Contemporary Urbanization.* UNESCO MAB Technical Notes 14, Paris, pp. 111-17.

Cohen, S.J., 1986. Impacts of CO_2 - induced climate change on water resources in the Great Lakes Basin. *Climate Change.* **8**:135-53.

Goldemberg, J., T.B. Johansson, A.K.N. Reddy, *et al.*, 1987. *Energy for a Sustainable World.* Wiley Eastern Ltd., New Delhi, 1988; and World Resources Institute, Washington D.C., 1987.

Hare, F.K., and S.J. Cohen, 1989. Climate sensitivity of the Great Lakes system. In: *Report of the First U.S. - Canada Symposium on Impacts of Climate Change on the Great Lakes Basin.* Canadian Climate Centre, Downsview,

Ontario. U.S. National Climate Program Office, Rockville, Md, pp. 49-60.

Intergovernmental Panel on Climate Change (IPCC), 1990. *Reports of Working Groups I, II, and III, and Policymakers Summary.* Geneva, Switzerland.

Jantsch, E., 1980. *The Self-Organizing Universe.* Pergamon International Library of Science, Pergamon Press, Oxford, U.K.

Kates, R.W., J.H. Ausubel, and M. Berberian (Eds.), 1985. *Climate Impact Assessment, SCOPE 27.* John Wiley, Chichester, U.K.

Nordhaus, W.D., 1990. *To Slow or Not to Slow: The Economics of the Greenhouse Effect.* Unpublished Paper, Yale University, New Haven, Conn.

Pearce, D., A. Markandya, and E.B. Barbier, 1989. *Blueprint for a Green Economy.* Earthscan, London.

Prigogine, I., and I. Stengers, 1984. *Order Out of Chaos.* Bantam Books, Toronto.

Timmerman, P., 1989. Everything else will not remain equal: The challenge of social research in the face of global climate warming. In: *Report of the First U.S. - Canada Symposium on Impacts of Climate Change on the Great Lakes Basin.* Canadian Climate Centre, Downsview, Ontario. U.S. National Climate Program Office, Rockville, Md, pp. 61-77.

U.S. Environmental Protection Agency, 1989a. Policy Options for Stabilizing Global Climate, Report to Congress. USEPA, Washington D.C., p. I-10.

U.S. Environmental Protection Agency, 1989b. The Potential Effects of Global Climate Change on the United States, J.B. Smith and D. Tirpak (Eds.). Report to Congress. USEPA, Washington D.C., pp. 287-321.

World Commission on Environment and Development, 1987. *Our Common Future.* Oxford University Press, London and New York.

World Meteorlogical Organization, 1988. *The Changing Atmosphere: Implications for Global Security.* Conference Proceedings WMO-OMM No. 710, Geneva 1988.

17. REGIONAL BOUNDARIES AND GLOBAL CLIMATE CHANGE: NORTHEAST NORTH AMERICA

William R. Moomaw

> It is easier to think globally than it is
> to act locally.
> Madeline Kunin, former Governor of
> Vermont

When Governor Kunin made her comment on March 7, 1990, she was hosting a meeting of the Conference of the New England Governors and the Eastern Canadian Premiers on Sustainable Development for the Northeast Region. Just the day before, one-quarter of Vermont's towns had rejected her hard won comprehensive land use plan. She was expressing some of the frustrations of a regional leader who has been a prominent spokesperson on global climate change, stratospheric ozone depletion, energy policy, transportation and land-use planning. As one of the few senior, elected officials in the United States to have read *Our Common Future* (WCED, 1987), and, because of Vermont's strong home-rule provisions, she understands better than most that achieving sustainable development or addressing global change ultimately requires local action. This chapter examines global climate change from the perspective of the northeastern part of North America to determine whether a regional response to a planetary problem might help resolve the dilemma so clearly drawn by Governor Kunin.

Concern over human induced climate change arises because our technological and economic activities are rapidly increasing heat trapping greenhouse gases in the atmosphere. Some of these gases, like carbon dioxide and methane, are natural atmospheric components that, along with water vapor, trap sufficient radiant heat to make the planet habitable. Growth in other natural greenhouse gases such as hydrocarbons and lower atmospheric ozone are also related to increases in conventional air pollution, while nitrous oxide appears to be increasing more from the use of nitrogen fertilizer than fossil fuel combustion. The Intergovernmental Panel on Climate Change calculates that 57 percent of increased global warming comes from the energy sector, and 10 percent to 30 percent from land use changes including deforestation. Industrial gases that did not previously exist, such as the CFCs, halons and other chemicals, now account for about one-fifth of annual additions to heat trapping. Stabilizing the composition of the atmosphere at current levels (already well above historic concentrations) is estimated to require a 60 percent reduction in carbon dioxide, a 15 percent to 20 percent reduction in methane, and a near total elimination of synthetic chlorofluorocarbons (CFCs) (IPCC, 1990).

Today, the principal focus of global climate change policy is in the international realm addressing questions of treaties, resource transfers and compliance issues. This global perspective is exceedingly important, but it ignores the pressing issue of the appropriate organizational level for implementing whatever international agreements are signed. While a comprehensive, worldwide climate change agreement may be achieved politically, a hierarchy of responses seems essential if any effective action is to follow. This analysis focuses on the role that regional schemes might play in implementing climate agreements.

Individual nation-states are the world's operational political entities, but their boundaries often bear little relationship to relevant ecological regions. Furthermore, not all climate change responses are best made at the same political level of organization or even within the same spatial boundaries. In other words, the boundaries of decision making and implementation of global climate change responses may vary with the particular problem and strategy. The challenge is to determine the ecological, political and economic factors that define the appropriate scale for implementing the most effective response.

In the analysis that follows, it is assumed that, unless actions are taken, the estimates given by the Intergovernmental Panel on Climate Change represent a reasonable scientific consensus for likely future warming, sea level rise (SLR) and other consequences of climate change. Despite uncertainties, it therefore seems prudent to develop a plan that anticipates an average global temperature rise of about 3°C (perhaps 6°C in this region) and an increase in sea level of one-third to one-half meter by the middle of the next century (IPCC 1990).

Defining the Region

The region is first defined geographically as Northeast North America, an area that overlaps the political boundaries of Canada and the United States and several important natural biomes. This analysis will focus principally upon the political subdivisions of the six New England states, Vermont, New Hampshire, Maine, Massachusetts, Rhode Island and Connecticut, and the five eastern Canadian provinces, Quebec, New Brunswick, Nova Scotia, Prince Edward Island and Newfoundland (see Figure 17.1). Although this choice is convenient and conforms to an existing regional political organization, these provinces and states also belong to numerous additional groupings for a variety of political, administrative and economic development purposes. Even within the context of global climate change, it will sometimes be necessary to alter the regional boundaries of concern to obtain an effective solution; for example, by including the large and economically powerful neighbors, New York and Ontario.

Examples of the flexible political definition of the northeast region are plentiful. The New England governors and eastern Canadian premiers have formed a long standing association that addresses a range of economic and environmental issues of mutual concern. The organization developed, and the individual states and provinces enacted, a common regional response to acid rain well in advance of national legislation. Several Canadian provinces are negotiating

independently with New England and New York to provide coal, nuclear and hydro generated electricity.

Within the United States, the failure to amend the Clean Air Act for more than a decade prompted the six New England governors, along with their counterparts in New York and New Jersey, to develop a common set of standards on gasoline volatility to reduce the formation of smog that often blankets the Northeast region in summer. In 1991 this same group of states concluded that the region would need to adopt the more stringent California emission standards for automobiles in order to meet air-quality standards. In another example, 13 northeastern states and the District of Columbia have been developing a regional capacity assurance plan for hazardous waste, although New York has recently dropped out of the consortium.

New York State and the city of Toronto have set goals for a 20 percent reduction in carbon dioxide, and Vermont has set carbon dioxide and CFC reduction goals. In 1990 Connecticut became the first state to enact global warming prevention legislation. The 10-state Eastern Regional Group of the Council of State Governments and the New England Governors have prepared reports on climate change and have drafted recommendations for legislation.

The Northeast region is ecologically homogeneous enough that it faces a common set of problems from global climate change, yet it is sufficiently varied economically that various response strategies are possible. Several parts of the region are complementary to one another in terms of resource supply and shortages. Politically, there is a history of cooperation among the political subdivisions on transboundary problems. The cooperative, regional approach to addressing climate change described here may provide a useful model for other regions that are evolving new forms of affiliation such as the European Community, Eastern Europe and the former Soviet Union.

Regional Consequences of Global Climate Change

The Northeast region faces a serious set of consequences from global climate change

FIGURE 17.1. Major political boundaries and cities of Northeast North America.

because of the pattern of settlement and the particular ecosystems found there. The population is concentrated in high density coastal cities and suburbs that run nearly continuously from New York City through Providence to Boston and Portland. A second parallel band of high density population exists in southern Canada along the St. Lawrence River Valley and the Great Lakes from Ontario in the west to Quebec City to the east. Additional cities lie along the Hudson and Connecticut river valleys. The Atlantic provinces and Maine are characterized by small coastal cities and towns and a sparsely populated interior that consists of small villages, forests and farms. The vast

northern expanses of Quebec and Newfoundland consist of boreal forest, lakes and, in the most northerly parts, tundra and permafrost. These regions are inhabited by relatively small numbers of indigenous people and others who rely heavily upon the natural resources of the region. Figure 17.1 shows human habitation patterns by major cities.

Sea-Level Rise

All but one of the 11 political subdivisions (Vermont) are coastal and hence SLR poses major problems as it affects coastal land use patterns, wetlands, estuaries and fisheries.

Even some interior regions would experience the consequences of SLR because of the tidal, gradually sloping river valleys. Coastal areas are also the most vulnerable to any intensification or increased frequency of tropical storms. The combination of storms and higher sea levels could cause extensive damage to the large number of coastal cities, highways, sewage treatment plants, nuclear and fossil fuel fired power plants, port facilities and other urban infrastructure.

The most vulnerable areas are the coastal wetlands, rich estuaries and soft sand islands. It is estimated that Cape Cod is currently losing 50 acres of land to the sea each year. In addition, about half of the original coastal wetlands of New England has already been drained, and much of the remainder is boxed in by urban development. This makes it particularly difficult for new wetlands to form further inland as sea level rises. The USEPA has estimated that two-thirds of the remaining wetlands could be lost should sea levels rise by 1 meter (Smith and Tirpak, 1989, pp. 132-133). This would have a devastating impact on the important Northeast fisheries and other wildlife that depend upon these coastal areas as habitat and nursery.

Much less of the Canadian coast has been developed, but significant losses are expected to occur for relatively flat coastal areas such as Prince Edward Island and other Atlantic provinces. Because much of the coast of Canada and Maine is less heavily developed and is generally rocky and steeper, problems are likely to occur more gradually than elsewhere in the region. The implications of SLR for the massive tidal swings of the Bay of Fundy are another unexplored problem area.

Forests and Tundra

While the world's attention is centered upon tropical forests, the forestlands of Canada cover the third largest area of any nation, and its standing plants and soils comprise one of the principal terrestrial carbon reservoirs in the world. The United States is currently examining the future of its much smaller but still significant forests in Maine, New Hampshire, Vermont and New York through the Northern Forest Lands Study being conducted by the U.S. Forest Service.

Although not included within the original scope of study, the implications of climate change and acid rain are considered in the final report (U.S. Forest Service, 1990, p. 21) as the result of public and expert comment.

A major adverse environmental consequence likely to accompany global warming is the rapid decline of the region's two principal forest types, mixed hardwoods in the south and boreal evergreens in the north. Both warmer temperatures and altered rainfall will contribute to the demise. The potential loss is important in its implications for biodiversity, the cultural traditions of the Northeast region and for its economic consequences. Tourism, recreation (including skiing) and the forest products industry are important economic contributors throughout the region. Many of the forests are already under stress from forest dieback associated with acid deposition and regionwide air pollution.

One analysis concludes that an increase of 3°C in the global average temperature would eliminate both the northern tundra and the Canadian boreal forest from the North American continent (Emmanuel *et al.*, 1984). The temperature in Canada would become suitable for the mixed hardwood forest that currently dominates New England, although soil conditions might not be appropriate. The temperature of New England would, in turn, become suitable for the southern hardwoods now found in Virginia. These conclusions are based upon temperature increases that, at northern latitudes, may be double the global average changes. Work carried out for the Canadian government also projected major ecoclimatic zone shifts (Rizzo, 1988). Another study attempts to use general circulation models to take into account temperature and precipitation patterns to examine the altered range of individual species (Smith and Tirpak, 1989). Although limitations exist on the ability of computer models to predict regional rainfall in a future changed climate, this study also predicted radically altered ranges for specific forest species.

What was ignored by these studies was the consequence of climate change occurring at anything like the rate that is now projected. Current estimates suggest temperature changes 10 times more rapid than any known previously (IPCC, 1990), and this could lead to the rapid collapse of forest ecosystems rather than their migration, resulting in billions of dead trees. Higher temperatures would accelerate the release of carbon dioxide and methane through enhanced decay (Woodwell, 1990). This collapse could be hastened by the migration into the region of more southerly insect pests and disease organisms (Smith and Tirpak, 1989). Warming would increase present losses of native species such as maple, beech, oak, hemlock and spruce to imported pests for reasons that may have nothing to do with climate change. Perhaps increased CO_2 concentrations might partially offset the temperature induced decline through fertilization of plant growth. Increased pests and temperatures could also adversely affect the regionally important agriculture industry.

Melting of the tundra will be accompanied by major changes in landforms and may also release large quantities of methane currently trapped in the frozen soil (MacDonald, 1989). On the other hand, anticipated increases in clouds and precipitation in the far north as a result of global warming could partially slow further temperature rises by producing additional reflective cloud cover and snow.

Water Resources

The Northeast region of North America currently receives plentiful rainfall, typically 40 inches (100 cm) per year. Some of the major cities, such as Boston and New York, have developed elaborate water supply systems that date back more than a century and that bring water from rural lakes and reservoirs many miles distant. While current computer models cannot predict future rainfall patterns with confidence, it seems highly probable that the quantity and distribution of rainfall will be altered by climate change. Regardless of the nature of the change, high costs could be incurred by water districts that rely upon elaborate collection and distribution systems.

Altered rainfall patterns will also have implications for the region's extensive hydroelectric power production capabilities, which include traditional river-based systems, pumped storage, and the rapidly growing capacity of Hydro-Quebec's massive James

Bay project. The year 1989 was the sixth year of below-average water runoff for this massive system and resulted in a 3 percent decrease in electricity sales relative to an already low 1988 level (Hydro-Quebec, 1990).

Other Effects

Warmer temperatures are likely to reduce heating requirements in winter, but increase the demand for summer air conditioning. Warmer summers are certain to increase regional air pollution, which is already a cause of serious concern. The region exceeds current air pollution standards, and is far from achieving the new, more stringent requirements of the 1990 Clean Air Act.

Regional Contribution to Global Warming

Canada and the United States are already principal actors in defining the global climate change issue and will be crucial participants in shaping the solutions. The United States is the world's leading contributor of greenhouse gases, currently contributing an estimated 17.1 percent of the total warming, whereas Canada, with one-tenth of the U.S. population, is tied for twelfth place globally with 1.6 percent of the total (Hammond *et al.* 1991). One analysis concludes that per capita contributions to global warming by West Germany and Japan are only 59 percent and 39 percent, respectively, of U.S. contributions. Much of this difference lies in lifestyle and industrial practices, with North Americans requiring twice as much energy as do Western Europeans or Japanese to produce a dollar of gross national product (World Resources, 1990).

One might expect the Northeast region to be a major contributor of greenhouse gases. It is a region of cold winters and hot summers that demands both space heating and cooling. Much of the region's thermal electric power plants, manufacturing industry, commercial space and housing stock is old. Because of its high urban concentrations, the Northeast region is a major producer of methane from landfills and sewage treatment facilities. To a lesser extent, coal mines in Eastern Canada, natural gas pipeline leaks and the rural dairy industry also add to the atmospheric methane burden. The region also possesses a high tech industrial base that is a significant user of CFCs. The large number of automobiles produce carbon dioxide and additional greenhouse gases such as unburned hydrocarbons, CFCs and ozone.

Actual regional CO_2 releases are in fact generally well below the respective national averages, as can be seen in Table 17.1. On a per capita basis, New England produces only 62 percent of the 20.4 metric tons of carbon dioxide released by the average American, whereas New York, the most energy-efficient state in the union, releases only half as much. In Canada, only Nova Scotia and New Brunswick, with their heavy reliance on coal, slightly exceed the national average of 18.3

metric tons per person, while Quebec, which makes substantial use of hydroelectric power releases only 53 percent as much carbon dioxide per person as the national average (Saeger and Piccot, 1990; Statistics Canada, 1990, pp. xiv-xv).

Methane, which is 25 times more effective than carbon dioxide as a heat trapping gas, is increasing in the atmosphere at a rate of 1 percent annually and is now two and one-half times its preindustrial concentration (Ramanathan *et al.*, 1985). Methane releases from New England and New York account for an estimated 3.6 and 4.7 percent, respectively, of U.S. totals (Table 17.2), and are smaller on a per capita basis than for the rest of the nation. While landfills account for 46 percent of methane production in the United

TABLE 17.1 *Per Capita State- and Provincial-Level CO_2 Emissions* (tons/year/person)

State-level emissions in U.S. (1985)

	Total	Residential/ commercial	Industrial combustion	Utility	Transpor- tation
United States	20.36	2.98	5.37	7.13	4.87
New England	12.92	3.97	1.46	3.21	4.28
CT	11.95	3.77	0.81	3.09	4.28
ME	17.15	4.22	6.37	1.36	5.21
MA	13.09	4.05	1.02	3.94	4.09
NH	13.90	3.74	1.41	4.32	4.44
RI	8.68	3.75	0.75	0.48	3.71
VT	13.22	4.57	0.81	2.94	4.91
NY	10.27	3.60	0.95	2.46	3.26

Provincial Level Emissions in Canada for 1987

	Total	Industrial sources	Fuel Comb. Stationary	Transpor- tation	Solid Waste	Miscella- neous
Canada	18.31	1.61	8.80	5.56	0.40	0.61
Atlantic provinces	18.21	1.50	10.40	5.55	0.18	0.58
Nfld	14.75	0.14	8.46	5.72	0.05	0.40
PEI	12.00	0.00	7.15	4.31	0.23	0.31
NS	19.76	1.09	12.23	5.95	0.13	0.36
NB	20.18	3.38	10.29	5.15	0.33	1.04
Quebec	9.63	1.05	3.61	4.14	0.34	0.50
Ontario	16.51	0.95	10.08	5.07	0.13	0.28

States, they are responsible for approximately three-quarters of methane in the region. There is no contribution from coal mines, and the per capita contribution from farm animals is only one-quarter of the national average. Other energy-related sources of methane, including natural gas pipeline leaks, are estimated to be comparable to the rest of the country (Saeger and Piccot, 1990).

We were unable to obtain comparable figures for the Canadian provinces. Nationally, it is estimated that Canadians release 0.40 metric tons of methane per capita (World Resources, 1990), more than three times the U.S. figure of 0.12 tons (Saeger and Piccot, 1990). While most categories of per capita methane releases are comparable to the United States, the proportionally larger Canadian natural gas pipeline system substantially raises the amounts per person (World Resources, 1990). There also exist two additional potential sources of methane from Canada that have no similar analog in the United States. The first is the possible large scale release of methane from the melting tundra. The second is the generation of methane by anaerobic bacterial action on the organic matter at the bottom of recently constructed large hydroelectricity reservoirs. Additional work needs to be done in order to determine the potential magnitude of these two sources.

Both CFCs and other industrial gases constitute the other major source of greenhouse gases for the region. Massachusetts, with its heavy concentration of computer and defense industries, contributes nearly half of New England's CFC total. New England's 27,000 tons per year represents only an estimated 6.2 percent of the national total, but it consumes about 18 percent more CFCs per capita than the national average because of its industrial base. New York releases an amount of CFCs that just exceeds all of New England's (Saeger and Piccot, 1990). Canadian national per capita use for the two most common CFCs (CFC-11 and CFC-12) equals the 0.8 kg per capita figure for the United States (World Resources, 1990), but totals for all CFCs are probably two and one-half times this figure.

Other gases released in the region also contribute to global warming, but by unknown amounts. Ozone, produced by the release of hydrocarbons and nitrogen oxides from automobiles and stationary combustion sources, is both a greenhouse gas and a threat to human health and the environment. Carbon monoxide from auto emissions, while not itself a greenhouse gas, interferes with the destruction of methane in the atmosphere, and hence contributes indirectly to global warming by raising methane concentrations. These and other air pollutants are produced in large amounts in the densely populated corridors of the United and Canada.

TABLE 17.2. *Release Rates for Methane in Northeast States, Total and Per Capita (millions of metric tons/year, 1985)*

	Total	Combust. sources	Nat. gas distrib.	Coal mines[a]	Farm animals	Municipal landfills	% of total	Per capita (kg)
United States	29.94	2.54	4.26	2.99	6.26	13.61	100.00	122
New England	1.06	0.11	0.09	0.00	0.04	0.81	3.55	84
CT	0.26	0.03	0.02	0.00	0.01	0.21	0.88	83
ME	0.10	0.02	0.00	0.00	0.01	0.07	0.33	86
MA	0.48	0.04	0.06	0.00	0.01	0.38	1.61	83
NH	0.08	0.01	0.00	0.00	0.00	0.06	0.26	78
RI	0.08	0.01	0.01	0.00	0.00	0.06	0.26	79
VT	0.06	0.01	0.00	0.00	0.02	0.03	0.21	118
NY	1.36	0.11	0.20	0.00	0.12	1.00	4.55	77

Source: Saeger and Piccot, 1990; U.S. Bureau of the Census.
a: Underground coal mines only; surface mines not included.

Regional Strategies for Responding to Global Climate Change

A successful strategy for responding to the threat of global climate change for Northeast North America will require a gradual but significant shift in economic activity and a change in regional governance. Success will require a degree of cooperation among the states and provinces that will permit the region to take advantage of its complementary strengths as it develops a sustainable economy. States and provinces actually have significantly more control than their national governments with respect to policies that have direct impact on greenhouse gas production, including the setting of building codes, transportation and land use planning, energy development, and the regulation of gas and electrical utilities. In addition to changing their own activities, states and provinces can influence public, municipal, institutional and corporate behavior, as well as investments through specific regulations and tax policies. Regionwide, compatible taxing policies can shape consumer purchasing patterns and industrial and commercial choices. Placing higher taxes on energy facilities and fuels that contribute significantly to greenhouse warming and other pollution could reduce their use and help to encourage the introduction of alternative technologies and processes that are more environmentally benign.

To be effective, many of these activities will require an enhanced system of regionally consultative governance that will assist in coordinating and implementing public- and private-sector development strategies. Transportation and energy development are obvious examples where coordination is essential, but as will be indicated below, other greenhouse gas reduction strategies will also benefit from coordinated action. Decision making would remain with the states and provinces, but a permanent and expanded Northeast Regional Council for developing and administering regionally based sustainable development projects that address global climate change would be created from one or more of the existing institutions. Within this new framework, states and provinces would have negotiators meeting more or less continuously, somewhat like the loosely confederated nations of the European Community, to develop mutually supported climate change strategies. These agreements would then be subject to approval and implementation by each member government. It is clear that such a system would also allow for flexibility, by allowing a state or province to take individual actions and by deciding which actions required the approval and participation of all or part of the membership.

The analysis is premised on a reduction in greenhouse gas emissions by transforming the regional economy, first to provide the technologies and services required to meet regional needs. Eventually, these developments could be exported to other parts of North America and elsewhere in the world. The New England economy has become overly dependent upon the defense industry, which is now being scaled back as the result of the end of the cold war. This has resulted in large scale unemployment and economic recession. The high tech industrial base and the extensive university research community that remains can be combined with similar intellectual resources in Canada and the natural resource base of the region to develop a new environmental industrial economy. This new regional economy will depend upon a combination of alternative and more efficient energy supply and end use options, better local and regional transportation systems, revised land-use planning and more effective utilization of solid waste as a resource. In the following analysis, an attempt has been made to develop, wherever possible, regional and local solutions that have multiple benefits for economic development and other environmental and resource problems, in addition to addressing climate change concerns.

Energy Efficiency

There is substantial potential for energy savings in the heating and cooling of buildings and hot water. Some significant gains were achieved as a result of the oil shocks of the 1970s. In fact, average annual U.S. carbon dioxide emissions declined during the period 1973 to 1989, but have shown a sharp upturn more recently. Unfortunately, despite the development of new efficient technologies, many new commercial buildings and housing units that do not incorporate the most effective options have been constructed during the past decade, and large numbers of older buildings continue to leak large quantities of energy. Establishing regionwide building-efficiency standards, coupled with an aggressive utility and energy service company-based retrofit program, could dramatically transform the building sector.

One of the current barriers to action has been the lack of reliable information on performance, cost and quality control of the alleged energy saving potential of particular products. To address this problem, it is proposed that the states and provinces create and finance an independent, regional institute that carries out research and testing of technology and products appropriate to the development of a sustainable economy for the Northeast region. Assessing energy-efficient technologies and strategies, and developing model standards that could be adopted by the states and provinces, would be high on the institute's agenda. There are several examples of public and private institutions that perform some evaluation of products, such as TUV in Germany, Consumer's Union in the United States, the Energy Research and Development Administration of New York State and the now-abandoned building assessment program at the former U.S. National Bureau of Standards. This institute would be in a unique position to provide the necessary focus for the development of higher and more uniform standards that could stimulate a regional industry responsive to regional needs. No individual state or province could afford to set up such a facility on its own, and neither of the national governments nor the university or private sector seems capable of developing such a service.

Improving energy efficiency in all sectors is clearly the first priority in addressing the greenhouse effect, other air pollution problems, nuclear waste and economic security issues for the region. New England utilities, driven by challenges from the Conservation Law Foundation, have taken a more aggressive role in developing demand side management programs than has most of the rest of North America. Utilities are paying for cost-effective energy saving design and technology in all new commercial buildings and for some retrofitting in the commercial and domestic sectors. A critical factor in the success of such programs is the ability of the utility to obtain a rate of return on its investments that is at least as large as it would receive from constructing new supply options. The utility has also played the role of banker, meaning that the customer has not been required to come up with the necessary money. Maine's bidding procedure has strongly favored end-use efficiency over supply-side options, and New York State's public utility commission is among the first to include explicit carbon dioxide emission and other environmental factors in deciding among options. A study recently completed by Pace University attempts to quantify the greenhouse and other environmental costs of electricity production from different sources (Ottinger *et al.*, 1990).

A complete energy policy that simultaneously addresses greenhouse and economic development issues must also address the supply side of the equation. As shown in Figure 17.1, the concentration of the Northeast population along two parallel bands in southern New England and southern Canada creates both a problem and several opportunities for solution. The region is excessively dependent on petroleum for transportation, space heating and electrical power production. Because of the high population density, air-quality concerns and declining public support for nuclear power, it has become increasingly difficult to construct new central power stations. Yet the very density of the existing distribution of power stations creates an opportunity to reduce greenhouse gas emissions and address other concerns as well.

District Heating

District heating has a history of more than a century in this part of the United States, but unfortunately many of these systems are old and have become quite outdated. Advances in the technology of district heating in Europe, plus new developments in district cooling, have extended the range of such systems to 30 km and lowered the cost of installation. It is possible to utilize up to half of the waste heat from existing power stations for district systems. For new co-generation power stations, the costs of the district system can be partially offset by the reduced need for cooling towers. Because of the high population density, eastern cities are well-suited to district heating and cooling. A number of such systems already exist, and an expansion of the co-generation potential from the existing power stations that lie within reach of existing district distribution networks could dramatically reduce CO_2 emissions.

New York State is the only jurisdiction in the region that is actively supporting the district heating and cooling option. There are now successful new and retrofitted projects in Jamestown and other New York cities. As part of its urban development program, Hartford, Connecticut, has also introduced a highly successful district heating and cooling system that serves 35 percent of the downtown area. A number of ancillary benefits have been found for these programs in addition to substantial reductions in CO_2 emissions. Hartford found that providing district heating was a great draw for developers and their clients to locate within the central city rather than moving to the suburbs. District cooling systems are also attractive to building owners because they eliminate the need for large space-consuming cooling systems, thereby lowering the construction and operating costs of buildings. In addition, cooling systems operate on the basis of absorption chillers that do not use CFCs. Also, by satisfying the demand for cooling during peak periods, they offset some of the need for generating capacity, thereby reducing capital expenditures, as well as fuel use. This has the added benefit of substantially reducing other air pollutants. Stockholm has seen sulfur dioxide levels plummet from 55,000 tons annually in 1965 to as little as 3,000 tons in 1990 as a result of burning lower sulfur fuels and expanding their district heating system (Jacobsson and Westergaard).

Currently, most of the district heating and cooling technology is being developed and marketed by European firms. If the Northeast region were to make such systems a high priority in its energy planning, a sufficiently large market would be created to attract manufacturers to the area.

Fuels Policy

The hydroelectric generating potential of northern Quebec and Labrador and the significant coal reserves in Nova Scotia and New Brunswick could satisfy the demand for electrical power in highly populated, but resource short, southern Canada, New England and New York. There are also potential oil and gas reserves in the rich fisheries of Georges Bank. In addition, New England is heavily dependent on nuclear power, and eastern Canada and New England are the largest users of biomass fuels in North America. While none of these options is without controversy, they provide a rich mosaic of possible supply-side options to complement the enormous potential for demand side reductions. Table 17.3 shows the current mix of generating sources for each state and province.

An examination of fuels policy is an important component of any CO_2 reduction strategy. Within this region, coal resources exist only in Nova Scotia and New Brunswick, and these provincial governments are attempting to develop a coal-based export market for electricity to New England. Coal produces nearly twice the CO_2 as does natural gas and half again as much as oil for each unit of heat released during combustion. Utilizing this regional resource requires that only the most efficient generating facilities be built and that some method be found to mitigate the large CO_2 releases. One independent producer, Applied Energy Services, burns coal in clean, small, efficient stations to produce electricity and cogenerate heat. To fully offset its CO_2 emissions, the company is paying to plant CO_2 absorbing trees in Central America and elsewhere in the tropics. As for this specific linkage of forestry within the Northeast region and in the tropics to fossil fuel burning of all kinds, were it adopted as a matter of policy by the Northeast governments, it could significantly lower CO_2 emissions, help pay for needed forestry projects, and, by increasing the cost of these fuels, encourage conservation.

Switching to natural gas is an attractive transition strategy for reducing carbon dioxide. Not only does it release significantly less CO_2 than does coal or petroleum, but it is also being found in increasing abundance in the western United States and Canada. As of 1987, 4 percent of the natural gas utilized in the northeastern United States was of Canadian origin,

TABLE 17.3 *Patterns of Electrical Power Production by State and Province (1987)*

	Net generation (GWh)	Coal	Petroleum	Gas	Nuclear	Hydro	Other[a]
United States	2,572,127	1,463,781	118,493	272,621	455,270	249,695	12,267
New England	87,770	16,674	32,831	4,744	29,256	4,109	156
CT	33,172	1,940	9,648	707	20,540	337	NA
ME	8,346	NA	2,600	NA	4,043	1,703	NA
MA	34,741	11,687	18,069	3,602	1,136	247	NA
NH	6,186	3,197	2,092	1	NA	896	NA
RI	836	0	402	434	NA	0	NA
VT	4,634	NA	17	0	3,536	925	156
NY	117,630	19,384	31,389	16,385	22,926	27,546	0
Canada	481,864	78,630	11,809	5,304	72,888	313,233	—
Atlantic							
Nfld	40,090	0	2,317	0	0	37,773	—
PEI	56	0	56	0	0	0	—
NS	7,749	5,306	1,669	0	0	774	—
NB	12,601	1,227	4,043	0	5,112	2,219	—
Quebec	166,852	0	182	0	4,660	162,110	
Ontario	131,582	31,700	2,075	87	63,116	34,604	—

Source: U.S Energy Information Administration, *Electric Power Annual*; Canadian Electrical Association, *Annual Report*.
NA: not applicable.
a: Includes geothermal, wood, wind, waste and solar.

accounting for 9 percent of the 28 billion cubic meters of exported gas (Canada National Energy Board, 1988). Beginning in early 1992 the Iroquois Pipeline will bring additional Canadian gas to New York and New England. Landfill gas (roughly half methane) is a potentially significant source of low carbon fuel for the region and is close to densely populated urban centers. New England Electric has recently installed a small 12 MW landfill gas-fired electrical generating station in Rhode Island. This strategy has the advantage of not only preventing the release of a potent greenhouse gas, but also of utilizing highly efficient co-generation technology that releases relatively little CO_2 into the atmosphere.

Two sensitive issues in the energy equation that have important implications for greenhouse gases are nuclear energy and hydroelectricity. Both of these options have the advantage in the greenhouse context of

not producing any CO_2 except during construction, but each is highly controversial for different reasons. The nuclear option has been stalled in the United States by a lack of public and investor support. Canadians feel that their Candu reactor design is both safer and more reliable than the U.S. alternative. While there has been considerable development of nuclear capability in Ontario, at present there is only one complex in Quebec and one in New Brunswick for the eastern part of the country. Opposition to nuclear power appears to be growing in Canada as well, which means that the further introduction of this technology is likely to be slow.

The most rapidly growing regional electric generating source has been hydroelectricity from Quebec. Total current installed capacity at Hydro-Quebec's James Bay complex is 10,282 MW. Total capacity for this utility is 25,126 MW, of which all but 1,758 MW is hydroelectric. An additional 5,426 MW of capacity is available to the utility from Churchill Falls in neighboring Newfoundland. Because Hydro-Quebec can provide electricity at one-third the cost of that purchased in New York City, there has been great demand to create an export market. Hydro-Quebec has plans to develop an additional 18,810 MW of capacity to meet an expanding market in the U.S. and Canada (Hydro-Quebec Financial Profile, 1989).

Canada is the leading producer of hydroelectricity in the world, but the James Bay project and its proposed expansion have set off a storm of controversy. Initially, the remoteness of the project shielded it from criticism, but more recently the ecological impact on the local area and the consequences for the native Cree and Inuit people have sparked concern. The issue has been catapulted into a regional debate over low electricity rates in the northeastern United States and Canada and the environmental costs of that electricity for native peoples. Vermont, in its recent energy-planning process, expanded its environmental review to include the damage that might occur at the James Bay site if its utilities were to contract for more Quebec hydropower. Should methane emissions from the reservoirs prove to be significant, the issue of further development of Quebec hydropower could become a global one.

There is no easy way to evaluate differing environmental and social trade-offs. The Pace study mentioned earlier attempts to develop a consistent methodology that monetizes environmental externalities so that one can compare, on a common basis, the costs of disposing of nuclear waste with the costs associated with producing carbon dioxide or sulfur dioxide from burning coal. However, even that study could find no good method for evaluating the external costs of hydro projects, because each is so site specific (Ottinger *et al.,* 1990). It appears highly probable that the further development of hydroelectricity in Quebec and Newfoundland will be slowed by controversy, just as has occurred with nuclear power. Ultimately, each decision will be based on political considerations, which makes it even more important for the region's leaders to meet and agree on an energy strategy that aggressively improves the efficiency of production and use of energy.

Transportation

Developing a regional transportation plan is likely to be a more achievable task for the Northeast Regional Council. The high population density makes the area well-suited to mass transportation within local regions and along the populated corridors. Because of the pattern of suburban development, people tend to use their own vehicles for commuting and shopping. Surely suburban sprawl is not sustainable, and subsidized sprawl is not defensible, yet this is just what we have done to lock in an energy-intensive living and working arrangement that will be difficult to change, as Governor Kunin noted. Although the development of sustainable land-use patterns is among the most politically intractable of issues, subregional land use planning is essential to encourage patterns more like those in Europe that can support mass transit systems. Supplying a full range of concentrated services, like district heating and cooling, can help to encourage more efficient development patterns. Private van pooling by employers and other innovations need to be encouraged to strengthen the substantial public transit systems that already exist in the cities of the region.

Other strategies for reducing CO_2 and other vehicle-related greenhouse gases in local transportation is to replace petroleum-fueled vehicles with electric- and hydrogen-powered cars, trucks and buses. A scheme for solar generated hydrogen-vehicle fuel from water has been proposed that could become economically competitive with other synthetic fuels by the end of the century (Ogden and Williams, 1989). A prototype of such a system is being developed in Germany, and the Daimler Benz company has announced a hydrogen powered urban bus that not only produces no carbon dioxide, but is also a far cleaner alternative in terms of conventional air pollutants. One might also envision fleets of electric commuter vehicles that would have the advantage of producing no greenhouse gases during operation. The ability of either the hydrogen and electric vehicle schemes to address the greenhouse problem, of course, depends upon the CO_2-

free generation of electricity. A possible scheme for electric commuter vehicles based upon the region's successful photovoltaic industry has been described (Moomaw, 1991), and such a scheme has already been implemented in Zurich. Options for the region include hydro, solar, nuclear, wind, biomass and tidal generated electricity. The Northeast Regional Council could play a crucial role in developing a coordinated plan for providing additional carbon-free electrical power for the transportation sector.

It goes without saying that intercity transportation of freight and passengers would be of primary concern to the council. The high density of the U.S. corridor that runs from Boston through New York to Washington has long been seen as a prime candidate for a European or Japanese type high speed rail system. Development of this corridor might rest with the American members of the council jointly with other states to the south, just as development of intercity transportation along the Canadian corridor might rest with provincial leaders. If the goal of a more integrated regional economy is to be realized, however, the Northeast Regional Council will need to be involved in order to coordinate rail, highway and air links. If done properly, the excessive contribution of carbon dioxide by the transportation sector could be significantly curtailed.

Solid Waste as a Resource and Fuel

While Northeast North America is short of many raw materials, it imports massive amounts of metals, paper and glass that find their way into the solid waste stream. In fact, finding disposal sites has become a major problem for local officials. Turning this waste into raw material for regional industries requires major cooperation among government, industry and the public. Regional governmental cooperation can play a critical role in developing materials recovery facilities, such as the highly successful one near Springfield, Massachusetts, that will ensure an adequate raw materials supply for potentially interested industries.

Helping to develop a regionwide market for products manufactured by recycled products can be encouraged through direct governmental purchases and appropriate regulatory and tax policies. The recent agreements in Connecticut and Massachusetts for newspapers to use 50 percent recycled newsprint is one example of government mediated efforts. Because of the large existing pulp and paper industry in the Northeast region, recycled paper efforts need to be carefully coordinated to avoid major economic dislocation and to effect a smooth transition.

Unusable organic waste can still be incinerated in state-of-the-art, waste to electricity co-generation power plants, eliminating future methane production from anaerobic decay. Locating these facilities at existing landfills and sewage treatment plants will not only reduce siting problems, but, as mentioned earlier, will allow the utilization of methane released from those sites.

Finally, governments within the region need to set consistent standards that will encourage industrial and other economic activity that addresses global climate change and other sustainable development issues. For example, several states within the region have worked on a common policy to ban heavy metals in inks used in packaging so as to eliminate them from incinerator ash. Four New England states and New York have adopted the same bottle deposit law to reduce litter and encourage recycling. Developing standard beverage containers might further facilitate reuse and the return of a local bottling industry.

Conclusions

Ultimately it will be the task of the Northeast Regional Council to develop the standards and incentives that will encourage the development of a sustainable regional economy that effectively addresses the problems raised by global climate change. Several examples have been given of existing and potential industries that could flourish and meet both economic and environmental needs. There are in fact many other possibilities that would also help the transition out of dependence on the defense industry. One regional company has developed efficient fuel cells that operate on natural gas or hydrogen that is capable of on site generation of electricity and heat. High efficiency gas turbines developed for military aircraft by two companies with manufacturing facilities in the region are capable of converting more than half of the heat released from the burning of natural gas into electricity. Combining these turbines into co-generation packages would produce reductions in carbon dioxide by 70 percent compared to conventionally produced energy services.

The Northeast region of North America is of course not only imbedded in the North American continent, but is tied through economic activity and the greenhouse effect to the rest of the world. It has become a cliche to say that global problems require global solutions, yet such a perspective begs the question of how those solutions might be implemented. Development of a regional system of consultative governance has the potential to provide adequate local input to decision making, while maintaining sufficient distance from the turbulence of local politics. By examining Northeast North America specifically, it has been possible to determine both particular and general factors that can lead to regional solutions to global climate change. The general lesson from this study is that each region must define itself in geographical and political terms, and develop a sustainable economic development plan if it is to address issues like global climate change. It is also clear that what constitutes a workable definition of the region must remain sufficiently flexible to meet the alternative demands placed upon it as one attempts to develop specific solutions to climate change.

Actual progress in addressing the climate change issue has been slim to date. The U.S. administration has been openly hostile to taking any sort of action, although there has been considerable interest and some legislation introduced in the Congress. Most of the response has been at the state and local level. New York State has set a goal of a 20 percent reduction in CO_2 by the year 2000 and is attempting to develop implementing policies. Other states such as Vermont and Connecticut have passed legislation to promote energy efficiency and alternatives and to phase out or ban uses of CFCs. In Canada meetings have been held between the provinces and the federal government to determine whether it is possible to meet the Toronto goal of a 20 percent reduction in carbon dioxide by the year 2005. The report of those discussions concludes that such a reduction "is not practical in the time frame. It is believed that [such a reduction] could cause significant economic dislocation and would require significant changes in lifestyle" (Federal-Provincial-Territorial Conference of Ministers of Energy, 1990, p. 17). Of course, that is the point. It is not possible simultaneously to maintain our current lifestyle patterns and prevent changes in the atmosphere and climate. The proposals in this chapter lead more in the direction taken by Germany and other European nations who plan to exceed the 20 percent target and become more powerful economic powers in the process. With foresight and cooperation, the same opportunity is available to the Northeast region of North America. Perhaps a new "regionality" may be one answer to Governor Kunin's dilemma.

References

Canada National Energy Board, 1988. *Canadian Energy Supply and Demand, 1987-2005.* Minister of Supply and Services, Ottowa, Canada.

Canadian Electrical Association. *Annual Report 1987.* Montreal.

Emmanuel, W. R., H.H. Shugart and M.P. Stevenson, 1984. Climatic change and the broad-scale distribution of terrestrial ecosystem complexes. *Climatic Change,* 7:29-43.

Environment Canada, 1990. *National Inventory of Sources and Emissions of Carbon Dioxide, 1990.* Report EPS 5/AP/2, Ottowa, Ontario.

Federal-Provincial-Territorial Conference of Ministers of Energy, 1990. *Report of the Federal-Provincial-Territorial Task Force on Energy and the Environment April 2, 1990,* Document 830-358/021, Kananaskis, Alberta.

Gold, Allan R., 1990. Quebec Indians ponder true cost of electricity. *New York Times,* October 12, 1990, p. A10.

Hammond, A.J., E. Rodenberg and W.R. Moomaw, 1991. Calculating accountability for climate change. *Environment,* 33:11-15.

Hydro-Quebec, 1990. Electricity Demand in Quebec—Proposed Hydro-Qué bec Development Plan, 1990-1992—Horizon 1999. Montreal, Hydro-Quebec, 1990.

Hydro-Quebec Financial Profile 1989. Montreal, Hydro-Quebec, 1990.

Intergovernmental Panel on Climate Change (IPCC), 1990. Overview and conclusions, Climate change: A key global issue. Draft (July 1990). IPCC Special Committee on the Participation of Developing Countries, "Policymakers Summary" and "IPCC First Assessment Report Overview," August 31, 1990, prepared with the World Meteorological Organization and the United Nations Environment Programme.

Jacobsson, L., and B. Westergaard. Energy planning in Stockholm and its effect on the environment. Internal document of Stockholm Energi, Stockholm.

MacDonald, G. J., 1989. Climate impacts of methane clathrates. *Coping with Climate Change*. J. C. Topping, Jr. (Ed.). Climate Institute, Washington D.C., pp. 94-101.

Moomaw, W. R., 1991. Photovoltaics and materials science: Helping to meet the imperatives of clean air and climate change. *Journal of Crystal Growth*, **109**:1-11.

Ogden, J., and R. Williams, 1989. *Beyond Fossil Fuels*. World Resources Institute, Washington D.C.

Ottinger, R. L., D.R. Wooley, N.A. Robinson, *et al.*, 1990. Environmental costs of electricity. Report prepared for the New York State Energy Research and Development Authority and U.S. Department of Energy, Pace University Center for Environmental Legal Studies, New York.

Ramanathan, V., R.J. Cicerone, H.B. Singh, *et al.*, 1985. Trace gas trends and their potential role in climate change. *Journal of Geophysical Research*, **90(D3)**:5547-66.

Rizzo, B., 1988. The sensitivity of Canada's ecosystems to climatic change. *Canada Committee on Ecological Land Classification Newsletter*. No. 17. Canadian Wildlife Service, Environment Canada, Ottowa, Ontario.

Saeger, M., and S. Piccot, 1990. *National and State Level Emissions Estimates of Radiatively Important Trace Gases From Anthropogenic Sources*. USEPA, Washington D.C.

Smith, J.B., and D. Tirpak, 1989. *The Potential Effects of Climate Change on the United States*. A Report to Congress, U.S. Environmental Protection Agency, Washington D.C.

Statistics Canada, 1990. *Canadian Almanac/ Sourcebook*.

U.S. Bureau of Census. *A Statistical Abstract Supplement—1986*. U.S. Department of Commerce, Washington D.C.

U.S. Energy Information Administration. *Electric Power Annual 1988*. U.S. Department of Energy, Washington D.C.

U.S. Energy Information Administration. *State Energy Data Report 1960-1987*. U.S. Department of Energy, Washington, D.C.

U.S. Forest Service 1990. *Northern Forest Lands Study*. U.S. Department of Agriculture with the Governors' Task Force on Northern Forest Lands, Rutland, Vermont.

Woodwell, G., 1990. Personal communication with the author.

World Commission for the Environment and Development-WCED, 1987. *Our Common Future*, Oxford University Press (Brundtland Report), Oxford and New York.

World Resources, 1990-91. A report by the World Resources Institute in collaboration with the United Nations Environmental Programme and the United Nations Development Programme, Oxford University Press, New York.

18. CLIMATE CHANGE—NEED FOR NATIONAL AWARENESS AND COMMITMENT AMONG SMALL REGIONAL COMMUNITIES: CASE OF THE SOUTHWEST INDIAN OCEAN ISLANDS

Vijaya Lakshmi Saha

No country will be insulated from the new planetary threat of global warming, a phenomenon that has only recently been propelled to the forefront of the international scene. The primary impacts are going to vary from country to country and from region to region, ranging from beneficial to harmful, and creating winners and losers at the international level.

There will also be secondary ripple effects arising from these major primary impacts that will affect the subsystems of the economy. Despite the fact that there is enough concern to justify precautionary actions to be incorporated into current planning efforts, economic as well as political factors would appear, however, to mitigate against a concerted global effort.

Economics of Survival

Many nations are aware of impending global changes, but they are totally unconcerned by them. Others deliberately ignore them because it suits them to do so. For instance, environmental degradation is greatest among the fragile economies of the world because of their need to repay unending debts. However, there is a time-lag, stretching over one or more generations, between the depletion of environmental capital and the occurrence of irreversible effects. While the global ecosystem is now able to absorb these impacts, there will come a point when it will no longer be able to do so; degradation will almost certainly follow in an irreversible way.

On the other hand, adaptive measures suited to the new planetary threat would also progressively open up vast commercial prospects for those nations in the vanguard of the appropriate technology and, for this reason, some nations could be proceeding very cautiously with the process of information transfer.

Politics of Survival

Within the strongest economies politicians increasingly recognize that an open commitment to the "green" concept is essential for winning votes. The strong economies also command an articulate voice at international forums and play a leading role on the Intergovernmental Panel on Climate Change (IPCC), where their scientists are actively involved in global change research. Active participation kindles interest and brings together policymakers and scientists. In the environmental field this process is normally absent among the small nations, where politicians are more preoccupied with issues of immediate survival such as hunger, debt, job creation and so forth. In addition to their limited research capacity, the interest of small nations has not been kindled at the international level, resulting in a lack of active involvement.

Island Predicament

Island issues have been generally brushed aside in the IPCC working papers, yet the impact of climate change and sea-level rise (SLR) on high islands with high population growth and development pressure may be as significant as atoll islands that, by virtue of their origin and formation, have more limited development prospects. We shall examine here the small islands of the Southwest Indian Ocean—Mauritius, Reunion, Seychelles and Comoros—to illustrate our concern and our proposals.

Vulnerability of the Southwest Indian Ocean Islands

This region, located off the East African coast, consists of a mixture of major and minor islands of volcanic, granitic and coral origin (See Figure 18.1). History on these islands began, ironically, with a practical lesson in environmental management. Having evolved in the undisturbed environment of the remote islands into unique species over millions of years, endemic animals like the dodo and the solitaire disappeared from the world within decades of the arrival of colonial man.

Plant species met with no better fate. There was ferocious denudation of the hardwoods when ebony was fetching a high price on the world market. This action was subsequently compounded by later colonizers who felled more trees to make room for plantation agriculture.

Subsequent population growth has placed enormous strains on the fragile economies of the islands and their carrying capacity. Today the population stands at 2 million, with Mauritius, one of the most densely peopled islands in the world, supporting half this total. Total population of the islands is predicted to increase to over three and a half million by the year 2025.

Culturally this region is a very distinctive part of the world, consisting of geographically diverse islands with equally varied linguistic, political, cultural and economic backgrounds. All of these countries have a narrow natural resource base and the absence of mineral resources, as well as a considerable distance to major markets, affect their potential for economic development and condition their choice of economic activities. Thus, the economies are based on agriculture, tourism and fishing, with only Mauritius having a genuine industrial base. The fragility of the economies is also compounded by their constant exposure to tropical cyclones, which constitute a recurring cause of concern with the approach of each new warm season.

Reviews that the author has made on the

FIGURE 18.1. Map showing the location of the Southwest Indian Ocean islands discussed in the text.

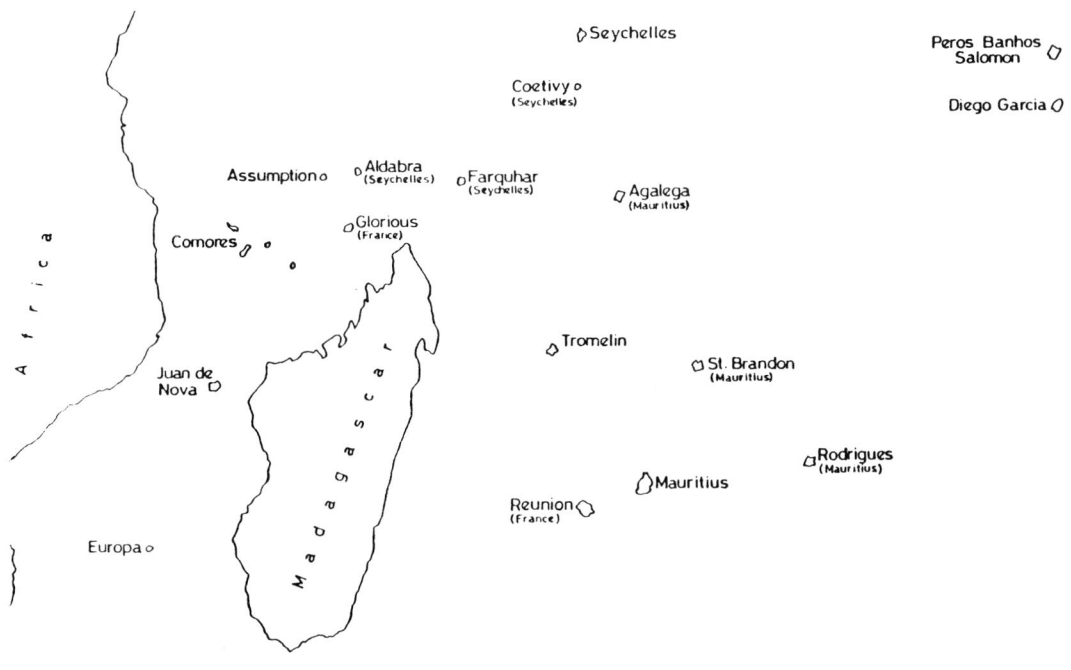

effects of global change on the area would suggest the need for concern and precautionary measures among the island communities. The combination of temperature rise and increasingly frequent extreme events, together with a SLR of about half a meter, is predicted to affect our island systems to a significant degree within the next century (ACTS, 1990; Bheenick, 1986; Nunn, 1988; Pernatta and Hughes, 1989; Perth Workshop 1990; SPREP, 1989). This would have an immediate impact upon our fragile terrestrial and marine ecosystems, and, therefore, cries out for an immediate, emergency response.

Terrestrial Ecosystems

The conservation of our outstanding rare and fragile ecosystems already requires very strict management policies and is quite costly. Temperature change, as well as its specific rate of change; alterations in the timing and intensity of precipitation; and indirect additive factors such as altered hydrological resources and soil chemistry may cause irreversible loss of species leading eventually to loss of biodiversity. Migration is not a viable option because of the overutilization of existing space, as well as the barriers formed by the surrounding water.

Besides their uniqueness, the natural terrestrial systems regulate essential ecological and life-support systems, like watershed protection, and provide resources for activities such as education and leisure. They are of psychological, economic and social value, not only to the island people, but also the world community. Their degradation would be an unquantifiable loss to present and future generations.

Inhabitants of these islands should be alerted to the impending risks, and the conservation and enhancement of all natural systems should, therefore, be a priority. Increasing their vulnerability through further exploitation will inevitably lead to a more rapid degradation of ecological systems.

Marine Ecosystems

There is already serious concern about the destruction of habitats, eutrophication and microbial contamination of marine ecosystems resulting from various current activities on our islands. Climate change will add further to the vulnerability of the marine ecosystems. For instance, fringing reefs and mangrove areas act as a wave breaker and form part of the islands' natural defense system. Long-term implications of global warming for marine ecosystems demand comprehensive coastal zone management, continuous monitoring and increased protection.

Secondary and Tertiary Effects of Climate Change

Some of the secondary effects of climate change could be even more complex than the primary ones. We shall elaborate here upon the likely impacts on the hydrology and agriculture of the islands. These impacts will, in turn, have far-reaching effects on human settlements and energy planning, as well as regional implications.

Hydrological Resources

A very critical indirect impact of climate change on small islands, high and low, could be on the hydrological resources. Our low coral islands are almost totally dependent on underground water, which is found in the freshwater lens below. The high islands are dependent on a mixture of surface and aquifer water, and there is already a low level of satisfaction with water availability on these islands: The likelihood of longer dry periods, changes in the intensity of precipitation and higher evapotranspiration would adversely affect both surface and aquifer resources and increase the need for more storage.

Furthermore, many of the aquifers are not on high ground, but on the coastal areas, where they are also vulnerable to decreased recharge or salt water intrusion, as on the low islands. However, on account of the denser population, the risk of contaminating the water and overdrafting the aquifers is greater.

Hydrological resources ultimately dictate most economic activities, including agriculture and energy planning. If the people on islands were more aware of the risks, they would see the need to review water policies, including pricing and recycling, and initiate hydrogeological training programs.

Agriculture

Our islands are essentially agricultural areas, with sugarcane as the lead crop in Mauritius and Reunion, maize in the Seychelles and subsistence crops in the Comoros. Significant impacts for agriculture could be created by increased levels of carbon dioxide, higher temperatures, changes in the seasonal distribution of rainfall, and an increased frequency of extreme events such as droughts or storms. In addition, indirect impacts will result from reduced land availability and hydrological resources. These factors would affect the regional distribution of crops on all of the islands. Long-term planning incorporating climate change predictions should, therefore, be extended to agriculture.

Human Settlements

The main criterion for economic survival on our low coral islands could be the state of the aquifer water. Degradation of this supply could occur rapidly after storm surges and make these islands completely uninhabitable.

On the high islands, the overall impact on socioeconomic activities would be less rapid, but more significant on account of higher investment levels. With the higher incidence of storms and other extreme events, the process of soil erosion and sedimentation would be accentuated in small islands with bare slopes like Comoros and Rodrigues. The infrastructure network along the coastal area would be at risk, resulting in increased rehabilitation costs after each major storm.

In response to sea level rise, the islands would need to adopt policies that might generally be categorized as retreat or adaptation, through hard, soft or a mixture of options (see Figure 18.2). At present coastal zone management needs to be adopted, and there should be an integrated strategy whereby new development is channelled away from vulnerable and low-lying areas. Proactive measures would ensure the efficiency as well as the longer life of current developments.

Energy Resources

Currently, our islands are net importers of fossil fuels and are expected to become in-

creasingly so in the future, despite the potential for development of renewable resources like hydro, wind, bagasse from sugarcane, solar and tidal energies. There is relatively little concern about pollution arising from greenhouse gas emissions.

Sensitivity to climate change, including the need to reduce reliance on fossil fuels and slow down the uneconomic use of those fuels, should be built into our energy planning. The search for alternative indigenous energy sources must be pursued with determination. Energy substitution can be justified, even without climate change, in the light of other factors like the drain on foreign reserves and sudden risks of disruption, as occurred during the Persian Gulf crisis.

Regional Concerns

Impacts can also be of a regional nature. All of the islands look toward the sea for part of their economy, but climate change may affect fish productivity in marine and coastal waters. The level of impact will depend on the attributes of the species and their regional specificity.

Another regional impact relates to economic boundaries. The islands of the area are just tiny specks in the Indian Ocean, but they have jurisdiction over a considerable economic area because of the location of their tiny islets. Should such islets disappear, the validity of these boundaries could be challenged.

Issues for the Southwest Indian Ocean Islands

The over-review just concluded indicates that all of the small islands, including the high ones, will be vulnerable to climate change and that responding to the problem will involve wide ranging changes in government policies (Figure 18.2). Climate change will affect virtually all areas of policy, and these interrelationships need to be managed right from the outset.

Paradoxically, smallness brings with it an added vulnerability for various reasons, such as coastal dominance, natural barriers for species migration, population pressure and elasticity of freshwater resources. But smallness has also acted as an excuse, among the

FIGURE 18.2. Policy options for adapting to global climate change.

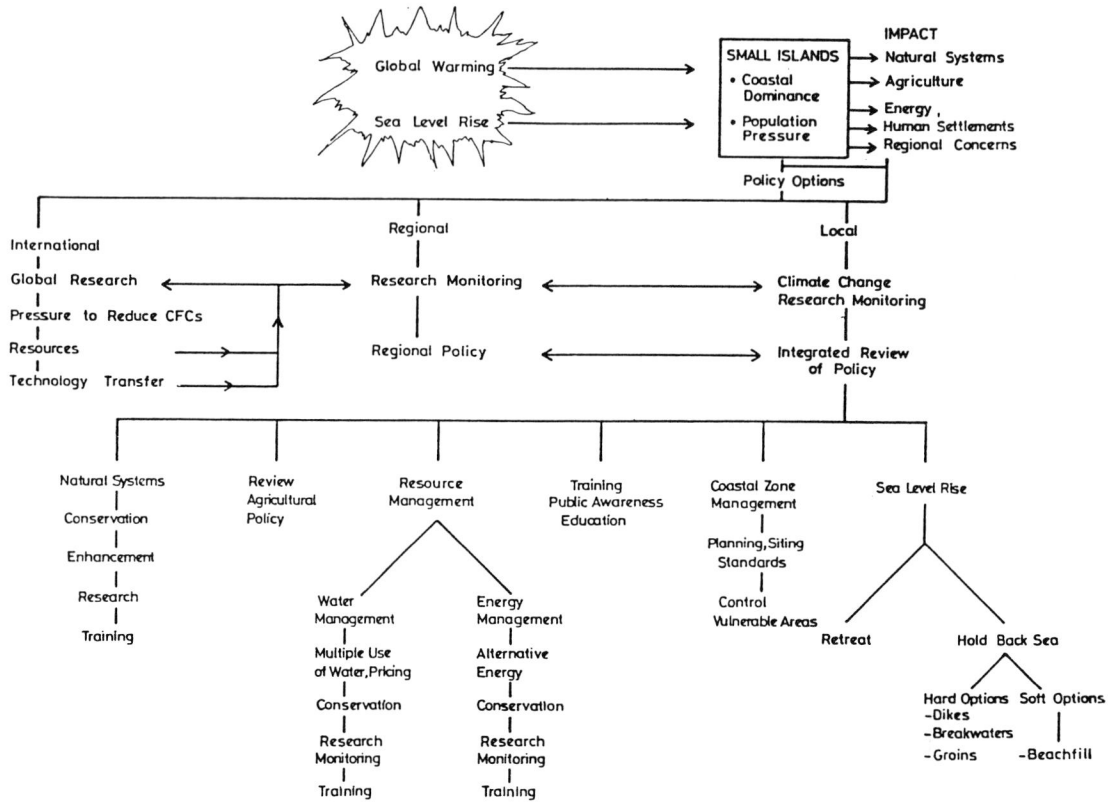

international scientific community, to play down likely impacts among small islands.

The saving grace for our region, and similar ones across the world, however, is that nature has bought us time. It may take several decades before the fundamental changes predicted for the area occur. This provides a certain lead time, by no means excessive, during which populations can intervene, adapt and manage to the best of their ability the resources that they themselves have endangered.

High islands, in the meantime, are acting in blissful ignorance of their long term prospects rather than exercising options on policy approaches right now. At the global level, high islands should, along with the atoll nations, pressure those large nations that are still dragging their feet over the Montreal Protocol and demand stricter control of emissions.

At the national level, they should have by now adopted a highly integrated approach to development. This would include reviewing virtually all areas of policy, ranging from basic sectors like energy, water and agriculture, to activities like planning and siting, as well as training and research. With regard to the latter, the need for a regional level of operation would immediately become obvious on account of the constrained territorial limits.

The final choice of policy options, at the global, regional and national levels (Figure 18.2), will be a political one. It will have to be built on accurate information, and not only on scientific information filtered from the international scientific community, but also on local research. Such information will only be obtained if there is a national commitment to the issue of climate change.

The Need to Develop a Comprehensive Awareness

Small nations, especially those with " island culture " are focused on internal concerns and, under normal circumstances, are unlikely to be moved by global "commonalities," especially " non-tangible " ones with a long time frame. Nations may well be conscious of climate change, but this would not necessarily propel them into action, unless they are sufficiently convinced that it is in their sovereign interest to act. Such a process would require an acute level of public awareness that would create a demand for political action, as well as a scientific effort that could address the critical issues in the debate.

The American Experience

James Titus reviewed the development of American awareness of climate change at the Miami Workshop in November 1989 (Titus, 1990). No one would dispute the fact that the Americans have had the largest exposure to the topic. Titus recalls that this first came about as early as 1979, but through failure of information transfer, it was dismissed as mere speculation.

The scientific community did not start examining potential responses to accelerated sea-level rise until the summer of 1982, and even by 1986 the primary goal of alerting people to the subject of SLR had failed. It was not until the emerging hole in the ozone layer over the Antarctic was discovered, that the U.S. Congress decided to hold hearings on global warming and that the American public became focused on the greenhouse effect and its potential impact on sea level. However, once the country was geared into action, the momentum for vastly increased efforts began to develop.

Based on the American experience, Titus has a few lessons to offer countries beginning to prepare for climate change. He emphasizes that it takes time to develop expertise and that it is important to designate at an early stage an individual to work full time on the issue. In America, it took some eight years to come to an appropriate first synthesis. For a start, informing the public is sufficiently important to warrant 10 percent to 20 percent of the total budget and 25 percent of

the project manager's time: "Anyone who views his time as too valuable to completely satisfy all inquiries is doomed to failure..." (Titus, 1990).

Finally, studies should begin with a socially relevant hypothesis before being funded. In the American case, the hypothesis was that a particular change in current activities was warranted, even if one allowed for the possibility that sea level might not rise. It is necessary to focus efforts on identifying actions that need to be taken today and to make sure that the knowledge becomes commonplace. Only through such a process can a national commitment be realized.

Increasing International Awareness

The time has now come for this awareness to be replicated around the globe. An aware public is essential for addressing climate change and for allowing decision-makers to focus their efforts. Awareness can be cultivated through well-defined education and information programs operating at all levels and in all regions, and supported by appropriate institutional frameworks.

Two compelling reasons exist why it is important to increase awareness of the implications of climate change. We have already elaborated on the first, namely that major development decisions are necessary and that it would be wiser and cheaper to take precautionary measures than rely on reactive and perhaps "hard" and expensive solutions at a later date.

The second reason is that negotiations toward a Global Convention on Climate Change are set to begin shortly, and the ultimate success of these negotiations will depend very much on the level of importance bestowed upon them by each nation of the world. Climate change cannot be managed by the industrialized world alone, and all nations will have to cooperate in this highly complex enterprise. The first stage of the IPCC work, relating to scientific research, impact and response strategies, was presented to the Second Climate Conference in November 1990 and from there it went to the UN General Assembly.

The next stage will consist of setting up the Global Convention on Climate Change. The convention is likely to follow the format of the Vienna Convention for the Protection of the Ozone Layer in laying down general principles and obligations. Convention goals should be framed in such a way as to gain the agreement of the largest number of countries, with protocols to deal with specific obligations. The convention should introduce technical concepts like net emissions budgets and tradeable permits which, to say the least, would require a high level of understanding from all nations concerned, including an informed awareness of such issues at international, regional and local levels.

Enforcement of these obligations will be the next major issue and, in view of practical difficulties likely to be encountered with compliance, a new kind of environmental cooperation, with built-in incentives, will be required from the entire community of nations. For the Global Convention to be successful, it is essential that there be as many signatories as possible to the entire framework, and no "free riders" from either the developing world (forecast to have the largest emissions in the future) or the developed world.

Currently, informed awareness of climate change is restricted to a limited number of nations, and it is only among these nations that the national competence to address all of the issues of concern has developed. Within such a context, it is unlikely that there will be a satisfactory consensus at the Global Convention. The protracted processes involved in arriving at the Montreal Protocol, even in the face of known scientific evidence about ozone depletion, indicate how much more difficult a Global Convention on Climate Change would be.

The deciding factors in arriving at agreements will be dictated by national and regional considerations and not by global concerns. Since impacts of predicted warming will be unevenly distributed across the planet, nations will require a specific and sound knowledge base before embarking on any agreement process and will not unconditionally agree to constraints that would be imposed upon their already thorny path of development. This will be particularly true for the small nations, who have been left behind in the ongoing debate.

All nations should, therefore, be made to participate fully and knowingly in this planetary decision making, including front end in-

volvement in the actual design of the forthcoming convention. It is the author's view that this aspect of comprehensive awareness and involvement needs to be given maximum priority, especially with respect to all those regions that have been left behind in the race for more information.

Within the lead period of some six decades for profound climate change, as much as one decade, if not more, could be taken up in initial awareness raising, institution building, organizing of scientific networks, training and progress in designing global agreements. Political considerations and disputes with respect to funding will only build further inertia into the system. The leaders in this process should not be the small endangered nations that have contributed little to the greenhouse effect, but those countries with the largest emissions and both the resources and capacity to act.

The Institutional Framework for Eliminating Regional Gaps in Awareness

The IPCC (1990) has made proposals aimed at implementing multilevel strategies of public education and information, with the core aims of promoting awareness and developing appropriate responses. However, this process has to be grafted onto an appropriate institutional structure. Where there are institutional weaknesses and no existing commitment, as within our region, it is also necessary to structure an appropriate institutional framework.

International Programs

Currently, the information process on climate change in the Southwest Indian Ocean islands is carried out through an institutional framework, which is replicated in many parts of the globe, consisting of direct linkages between the United Nations Environment Programme (UNEP) and the various environment ministries (which are generally new and not well structured) and between the World Meteorological Organization (WMO) and national meteorological departments. However, internal cross-linkages between the local organizations are minimal. In cyclone-prone areas the primary functions of meteorological offices are perceived as cyclone tracking, not long-term climate change. There is also no intermediate link between the governments and the head office of UNEP, nor is there a UNEP regional office.

UNEP is carrying out commendable work relating to climate change under its Regional Seas Programmes. Within our region, it has identified, with the approval of the various governments, a group of individual specialists. In June 1989 these individuals were organized into the East African Task Team to identify, in the first instance, the likely impacts of climate change on the East African mainland and the islands of the Southwest Indian Ocean. The group, of which the author is now the task team coordinator, has performed its initial work with the dedicated collaboration of UNEP staff. A number of reports have been prepared as a result of this effort (UNEP 1989a-d).

However, with hindsight, several issues can be raised:

- response from the countries was poor and, among the islands, only Mauritius delegated a representative;
- with respect to climate change, island issues are very different from continental issues and island concerns need to have a more specific forum;
- the work proceeds very slowly, because the joint tasks between UNEP and the various governments are confronted not only with the bureaucracies of the various member countries, but also that of UNEP, which has in its own way become a big bureaucracy, confronted with ever increasing responsibilities and an overworked staff;
- at the level of the country, work filtered through UNEP and other international organizations is diffuse and uncoordinated because the concept of climate change as a multifaceted problem is not clearly understood by management, which is used to compartmentalized thinking.

On top of that, the work does not get the political priority that is required on account of its perceived time frame vis a vis other government work. Compared to other survival issues, the environment in general does not command a high priority. It would ap-pear that, while UNEP tries to stimulate the maximum effort it can from its international stance, a corresponding level of effort is not evident at the national level. As a result, the international effort dies a natural death. Climate change will capture and retain the attention of policymakers and the public alike, only if the proposed framework promoting an education and information policy were to be supported by an additional institutional linkage with more local involvement.

Regional Commissions

Within the Southwest Indian Ocean, as in many other areas around the globe, a regional political organization already exists, known here as the Indian Ocean Commission (IOC). It became fully operational last year and oversees many regional projects. The IOC is a political, economic and cultural organization that includes all of the islands referred to above together with Madagascar. The Secretariat is at Quatre Bornes, Mauritius, and the Presidency is rotated among top political leaders (Commission de l'Ocean Indien, 1990).

The IOC is very close to the political pulse of the region and representation is at the highest level. Summit meetings of the IOC take place once a year, but can be called more often if required. The internal institutional structure of the IOC is shown as Figure 18.3. A specific climate change committee could be added to the standing Technical Committees, with the main objectives of coordinating long-term policy, developing international and regional linkages, initiating and catalyzing public awareness among the islands of the region, and formulating a coordinated research and monitoring program. Such a program would rapidly increase scientific knowledge on local trends, develop impact studies and lead to appropriate policy responses. A regional project leader, the services of which could be funded by global resources, could be appointed by the IOC to assist in the overall process.

There are innumerable advantages to a regional rather than a national emphasis within the proposed global network relating to climate change. Creation of a strong and central regional node for climate change

FIGURE 18.3. Organizational structure of the Indian Ocean Commission (IOC).

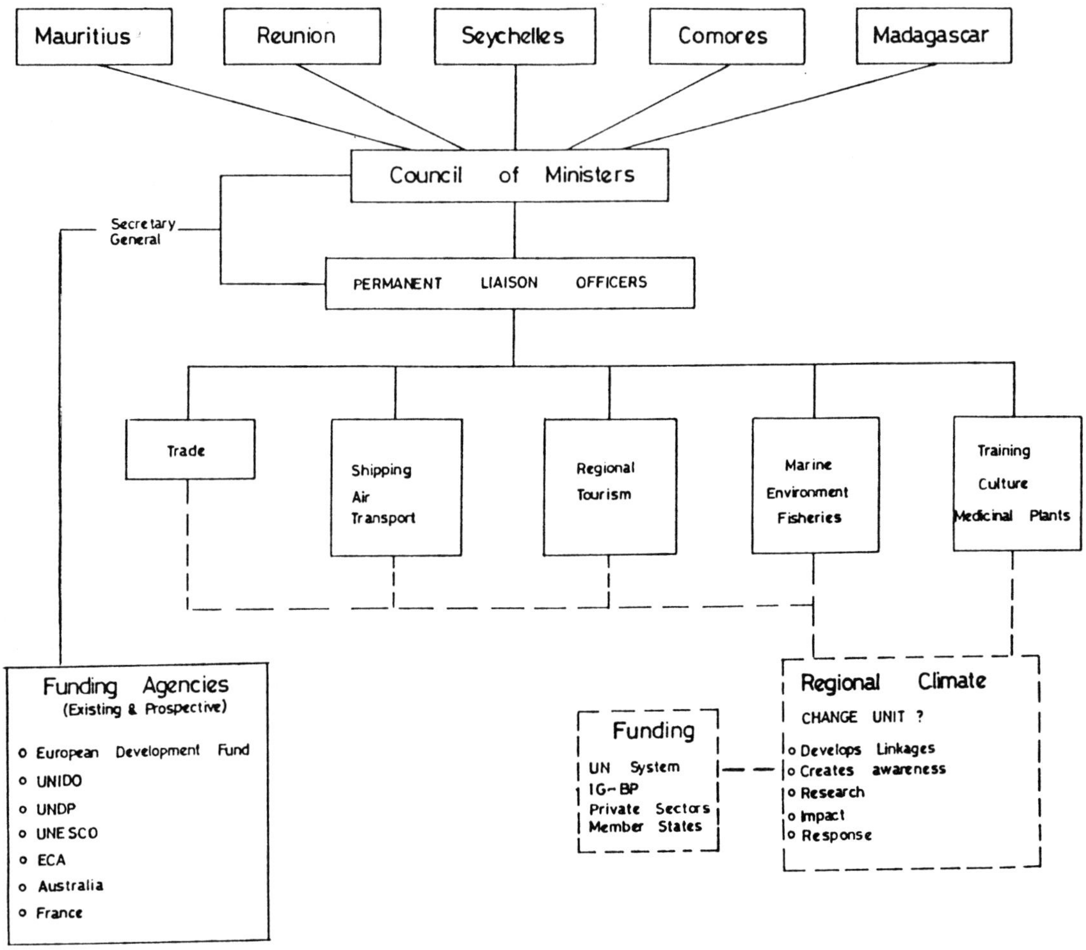

would lead to a more efficient use of the resources of the UN system, which would evolve both on the regional level (for operational purposes) and on the National level (for ratification). Within a multilinkage system, there are less risks of issues getting lost, as is currently happening.

Furthermore, for small islands, the regional boundary rather than the national one would be a more appropriate unit of research. This would apply inter alia to oceanographic data collection, climate and atmospheric research, tidal gauging, fisheries and agriculture. The regional level of work ensures that climate and global Earth systems would be more readily available for interpretation at in

a regional, as well as national, level.

An added advantage of this regional arrangement is that experts of the region would be able to pool resources on common areas. It is unlikely that the region would be able to furnish on its own all of the expertise necessary to address all aspects of regional change. However, the very institutionalization of the regional process could attract, say, one of the dozen regional research facilities that the International Geosphere-Biosphere Programme (IG-BP) is proposing to set up in the developing world. Such a facility would "stimulate global change research" (IG-BP, 1990).

At the island level, the IOC would

encourage countries to institutionalize multidisciplinary units that would coordinate all national research on key indicators and introduce programs of national awareness. These programs would be a key component of national projects throughout all sectors of the economy. Such climate units could be grafted initially onto existing National Disaster Committees, which already exist in cyclone prone areas and interact directly with a high-level government office, such as the office of Prime Minister (Figure 18.4). The need for local expertise would be strongly felt, and a recurrent amount of global resources would have to be deployed for training in the appropriate fields.

Developing a Comprehensive Program

An informed public is essential if decision makers are to obtain the support necessary to mobilize funds or to take bold decisions. Once the proposed institutional structure is in place, the strategy spelled out by the IPCC Response Strategy Group for the fostering of environmental sensitivity and awareness could be applied. Public information and public education activities would be formulated at the regional level, and subsequently developed by national climate units where the sociocultural diversity and specific requirements of particular islands could be incorporated. The target groups would ideally include everyone, from public- and private-sector policymakers to children's groups, the uneducated as well as the educated (Figure 18.5).

This systematic two-track process of developing appropriate institutions and increasing public awareness would provide the requisite support for government action. It would also bring about the attitudinal and behavioral changes necessary for ensuring that markets accurately take into account the potential consequences of, and opportunities created by, climate change.

The regional hierarchy and distribution of functions would stimulate the interest of the regional politicians, and would be backed at the technical level by organizations like UNEP, WMO and other affiliate organizations. Under such conditions, a healthy policy/science interaction would be created, while the mobilization of public opinion would stimulate the research process further.

Development of an adequate database would provide information on net regional and national emissions to date, local vulnerabilities and appropriate adaptation alternatives. This would provide the scientific base upon which meaningful decisions could be made with respect to long-term development planning in all sectors. The accumulation of high quality information and skills that would be a product of this continued process would augment the regional and national capacity to undertake meaningful negotiations at the Global Climate Convention.

Instead of coming up at the tail end of a fait accompli, small nations would confidently contribute at the front end so as to ensure the effective implementation of the new global policies. They would also be in a better position to assert their own regional position with respect to emissions strategies and suitable implementation time frames vis-a-vis their own development programs. However, because the priority target areas are also the least able to pay, it is essential that the strong nations provide adequate funding. This would be an important indicator of their commitment to ecological security.

Conclusions

1. There is limited awareness of the likely impacts of climate change among small island communities. Scientific capacity is limited and has to shoulder heavy responsibilities of an immediate nature. Policymakers are more preoccupied with problems of immediate survival than with issues having a long lead time. But these nations, least responsible for and least aware of climate change, may eventually be among the worst hit and among the least able to pay.

2. The international community, whether at the scientific or political level, has not done much to bridge the information gap. At the global level there is a regional discrepancy in the information available on climate change that defies comprehension.

3. The limited country awareness of the commonality of the climate change issue is also matched at the institutional level by a narrow compartmentalization of the problem. There is no perception

FIGURE 18.4. Functional distribution of responsibilities among different governmental entities.

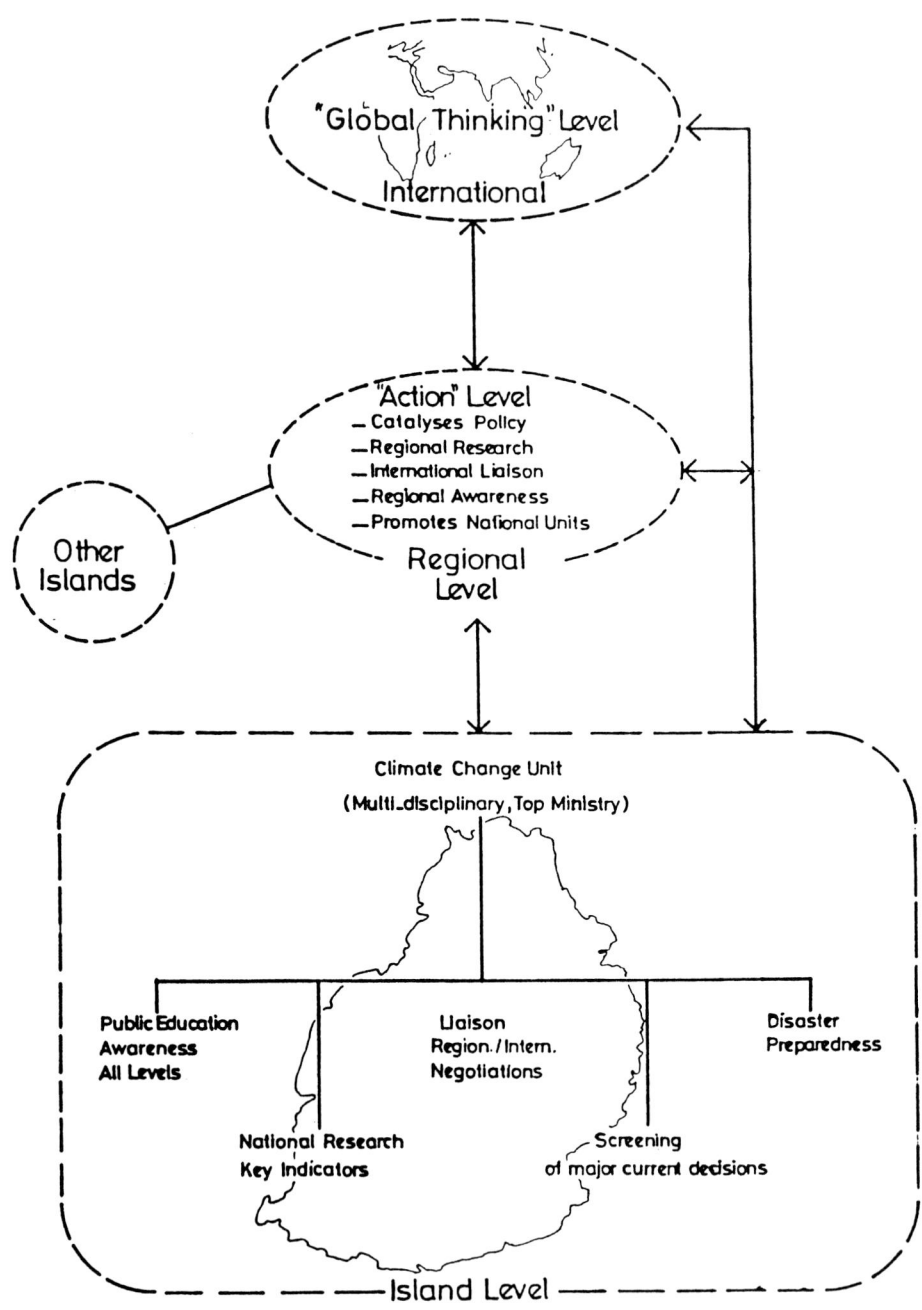

FIGURE 18.5.   Mulitlevel education and information tracks for disseminating information on climate change and its impacts.

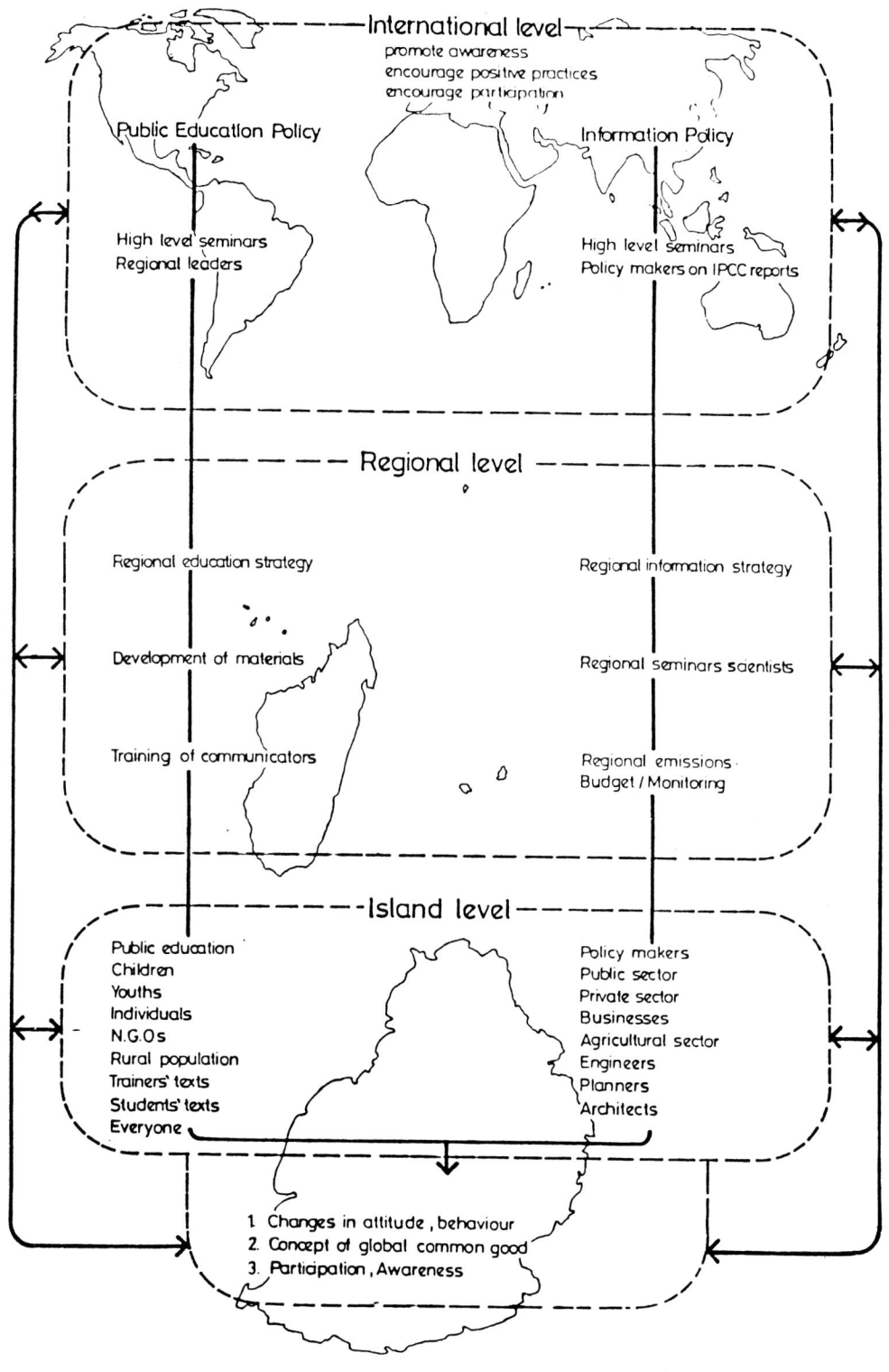

that climate change is cross-cutting and that it will affect almost every activity in the country.

4. Current scenarios, however, confirm that global warming and sea-level rise may have a significant impact over the entire area. Each region should be concerned and should be developing reliable information on probable local trends. Regions should be incorporating specific parameters, as of now, into each component of their development program that has a long-term physical effect.

5. The time has also come for global negotiations on a climate convention to start. However, each region has to negotiate from a detailed information base. The deciding factors in arriving at agreements will be dictated by national and regional considerations, and not by global concerns.

6. A comprehensive multilevel education and information process is required immediately. But there are also institutional weaknesses. It is necessary to ensure that an appropriate multilevel institutional structure, likely to stimulate both political and public awareness, as well as research, be the backbone of this process.

7. For the Southwest Indian Ocean islands, the author has suggested that a regional climate change function be slotted in at the level of the Indian Ocean Commission (IOC), in between the global and national levels. A three-tier hierarchy would stimulate awareness at all levels and make more efficient use of the resources of the UN and related organizations.

8. Accumulation of high-quality information and skills through a regional process would enable the small nations to incorporate climate change issues into decision making processes, starting immediately. This would augment the regional and national capacity to undertake meaningful negotiations at the convention and in the formation of protocols that will emerge over the next few years. It would also bring a more effective political solution to the problem of funding, as well as technology transfer.

9. The absence of these measures will lead to the widening of the information and technology gap between North and South and create further stress. It will also lead to the formulation of an international convention that will operate to the disadvantage of uninformed nations, thereby limiting international consensus.

10. Early formalization of the appropriate awareness mechanism—its timing, its fortitude, funding—is very important. In the wake of the tremendous task ahead, the author is convinced that this development would greatly augment the chances of a global strategy for climate change.

References

ACTS, 1990. The Nairobi Declaration on Climatic Change—International Conference on Global Warming and Climatic Change, African Perspectives, May 1990.

Bheenick, R., 1986. *Sustainable Development and Environmental Management of Small Islands—Man and Biosphere Series Vol. 5.* W. Beller, P. d'Ayala and P. Hein (Ed.), Parthenon Publishing Group.

Commission de l'Ocean Indien, 1990. Factsheets, April 1990.

IG-BP - Global Change Report No. 12, 1990—The Initial Core Projects.

IPCC Draft Reports April 1990.

Nunn, P., 1988. *Potential Impacts of Projected Sealevel Rise on Pacific Island States.* ASPEI, Split, Yugoslavia.

Pernetta, J.C., and P.J. Hughes, 1989. *Studies and Reviews of Greenhouse Climatic Change Impacts on the Pacific Islands.* Association of South Pacific Environmental Institute, Marshalls.

Perth Workshop, 1990. *Adaptive Responses to Climate Change.* Coastal Zone Management Workshop, February 1990.

SPREP, 1989. Report of the SPC/UNEP/ASPEI intergovernmental meeting on climate change and sealevel rise in the South Pacific. South Pacific Commission, Noumea, September 1989.

Titus, J., Ed., 1990. Miami Workshop. *Changing Climate and the Coast, Vols.*

1 & 2.

UNEP, 1989a. Regional Seas—Action Plan for the Carribean Environment Programme. *UNEP Regional Reports*, **109.**

———, 1989b. Regional Seas, G. Sestini, L. Jeftic and J.D. Milliman (Eds.). *UNEP Regional Report*, **103.**

———, 1989c. Implications of Climate Change in the South Pacific Region. J.C. Pernetta and P.J. Hughes (Eds.). *UNEP Report* , **128.**

———, 1989d. Regional Seas. J.C. Pernetta and G. Sestini (Eds.). *UNEP Report*, **104.**

19. POLICY RESPONSES TO CLIMATE CHANGE IN SOUTHEAST ASIA

Ferenc L. Toth

The culmination of five years of intensive and widespread research efforts on climate change, the Villach Conference (WMO, 1986) called for studies to look at regional impacts of climate change in more detail and to analyze possible policy responses based on regional impact assessments. One response by the UNEP was the project on "Socio-economic impacts and policy responses resulting from climate change: A study in Southeast Asia" (hereafter called the "Southeast Asia study" or simply "this study") involving three National Study Groups (NSGs) assembled under the leadership of a designated government agency in Indonesia, Malaysia, and Thailand; a core group of consultants headed by Prof. Martin Parry of the University of Birmingham (U.K.); and UNEP's Regional Office for Asia and the Pacific (ROAP) in a coordinating role. The project was concluded at the end of 1990 and the results published by UNEP in 1991.

This chapter draws on results of the project, but cannot be considered even an incomplete summary of the findings. In fact, the analysis presented here is one level of generalization higher, as the author seeks to present what are believed more general conclusions that can be derived from the project in the overall context of the greenhouse policy debate. Still, the focus is regional and the support material was generated in the Southeast Asia study.

Organization of the Chapter

Our analysis begins with a short overview of the most relevant economic and environmental concerns in the region so as to set the stage for the discussion of policy issues in the second half of the chapter. The section titled "Sensitive Areas" explains the social and economic importance of the areas selected for the impact assessments. It is followed by a short description of how regional scenarios of climate change were developed for the study. The next three sections provide an overview of the assessment techniques that were used to prepare the first- and higher-order impact assessments and to generate the policy responses. These sections also present a few examples of the potential impacts and policy responses. The final section is a succinct summary of some of the "general lessons" concluded from the study.

Socioeconomic Profile of the Study Region

The three countries participating in the study (Indonesia, Malaysia, and Thailand) belong to one of the most dynamic regions of the world. While the world in general could only achieve an average annual economic growth rate of 3 percent in the 1980s, the Asia-Pacific region grew 7 percent, more than three times as fast as the 2 percent average of the developing world as a whole. (Selected basic indicators are summarized in Table 19.1.)

This impressive economic performance was not without a price. Many countries in the region, including those participating in this study, achieved a considerable share of their economic growth by supplying an increasing quantity of raw materials to industries in the developed world and, thus, depleting their natural resources. The other major source of growth—industrial development—has also left its mark on the environment in these countries, as is evident from the air and water pollution, and the exhaustion of both renewable and nonrenewable natural resources. The result is that the region has become a significant contributor to global environmental change, including climate change.

TABLE 19.1. *General Indicators of the Three Countries in the UNEP Study*

	Indonesia	Malaysia	Thailand
Area, 1,000 km^2	1,905	330	514
Population, million (mid 1987)	171.4	16.5	53.6
Population density (per/km^2) (mid 1987)	90	50	104
Average annual growth rate, % (1980-87)	2.1	2.7	2.0
Projected avg. annual growth rate, % (1987-2000)	1.7	2.2	1.5
GNP per capita (U.S.$,1987)	450	1810	850
GDP growth rate (1980-87)	3.6	4.5	5.6

Source: World Bank (1989).

According to recent estimates, the CO_2 emissions from industrial sources in both South Asia and Southeast Asia amounted to 166 million metric tons, that is 3 percent of the world total in 1985 (WRI, 1988). Table 19.2 presents CO_2 emission data from industrial and commercial energy sources in the three study countries. The actual figures are much higher due to high ratios of non commercial energy use in the region. Greenhouse gas emissions from paddy fields and deforestation are also significant contributions to the global emission figures.

While the dynamics of growth are similar in the three countries, they represent markedly different stages of economic development. According to the World Bank classification, Indonesia (GNP per capita in 1987, $450) belongs to the group of low-income economies, while Thailand ($850) and Malaysia ($1,810) are in the lower-middle-income group, in the bottom half and close to the top, respectively. Despite a variety of problems in both the domestic and the world

economy, each country has shown an impressive performance through the 1980s. Moreover, each country is fighting a different set of social, economic, and environmental problems with varying degrees of success. The differences in current levels of development and the diversity of other current problems create very different social and political climates that have to be taken into account when addressing issues characterized by lengthy time scales and significant uncertainty.

Sensitive Areas

Numerous studies have been conducted worldwide over the past few years to assess the impacts of climate change on various ecological, natural resource systems, and social systems (Kates *et al.*, 1985; Parry *et al.*, 1988). The geographical scales of these studies range from small regions of a few thousand hectares of agricultural land, important

TABLE 19.2. *Carbon Dioxide Emissions From Fossil Fuel Consumption and Cement Production*

	Indonesia		Malaysia		Thailand	
	Total	Per capita	Total	Per capita	Total	Per capita
1970	8.072	0.067	3.876	0.357	4.190	0.115
1980	20.810	0.137	6.699	0.487	10.921	0.235
1986	28.127	0.165	9.205	0.578	13.522	0.259

Source: Marland *et al.* (1989).
Note: Total in million metric tons of carbon, per capita in metric tons.

river basins, or short stretches of coast (see the case studies in Glantz, 1988) through larger regions up to global-scale studies (Prentice *et al.*, 1989). Based on a preliminary assessment of possible climatic vulnerabilities in Southeast Asia, this study focused on three major impact areas: agriculture, river basins, and coastal areas.

Virtually any form of agriculture is sensitive to longer-term changes and shorter term fluctuations in climate attributes. The most important climate component and the tolerable magnitude of change varies from crop to crop and also depends on the genetic characteristics of the varieties. Agriculture has traditionally been an important sector of the Malaysian, Indonesian, and Thai economies. The most important policy-related concerns over the impacts of climate change on agricultural development can be presented as three major issues:

1. National food self-sufficiency is a strategic political issue considered to be part of the overall national security. In Malaysia the 1988 level of self-sufficiency in rice production was 72 percent, exceeding the Fifth Malaysia Plan expectation of 55 percent to 60 percent (EPU, 1989, p. 137). Currently this is considered to be the desired minimum level of domestic rice production that should be maintained over the longer term. Indonesia has reached self-sufficiency in rice relatively recently; therefore, the current five-year plan points out: "Efforts to increase food crop production, covering rice and non-rice, are to be carried out to consolidate the country's food self-sufficiency achievements" (BAPPENAS, 1989, p. 44).
2. Contribution of agriculture to GNP and exports is an important economic concern in each of the three countries. The contribution of agriculture to GNP in 1986-1988 averaged 22 percent in Malaysia, 26 percent in Indonesia, and 16 percent in Thailand.
3. Employment and earnings provided by agriculture continue to be of strong social concern in these countries. Although the contribution of agriculture to GNP is only about 22 percent in Malaysia, the sector provides employment for one third of the labor force.

These ratios are similar in the other two countries. As the process of industrialization is expected to continue in the region over the coming decades, the share of agriculture in GNP and employment will gradually decline. However, the decline in agriculture related employment has been much slower than the decrease in the sector's contribution to GNP and exports. Therefore, the long-term prospects in agriculture as a source of income for a significant fraction of the population requires special attention.

The NSGs selected a set of crops to analyze the impacts of climate change on agriculture. Rice is the only crop that was selected by each group. It is the major staple food in the region; hence, it is a strategic crop in achieving or preserving national food self-sufficiency targets over the long term. At the pre-interviews in preparation for the policy analysis component of the study, senior officials of the Economic Planning Unit (EPU) and the Department of Agriculture in Malaysia, and of equivalent government agencies in the other two countries, expressed strong concerns over the potential impacts of climate change on paddy production and on the economic conditions of the affected social groups.

Climate-induced changes in precipitation regimes and evapotranspiration rates are expected to create stresses in the management of water resources. Early studies on possible impacts of climate change have identified the importance of water resources and management as a key impact and response area (Revelle and Waggoner, 1983). Many parts of the study region are characterized by changes in dry and wet seasons. Therefore, a thorough assessment of the impacts of floods, droughts, water quantity and water availability constituted an important part of the project.

Coastal areas in the participating countries are usually important. The most fertile agricultural lands are located in the coastal regions, together with aquaculture and coastal fisheries. Coastal areas support high densities of population and have a long history as centers of infrastructure, settlements, defense, transportation, and other considerations. The rapid increase in beachfront tourism has added further to their economic importance. The length of coastline as a single indicator shows the potential threats from sea-level rise in these countries: Indonesia, 54,716 km; Malaysia, 4,675 km; Thailand, 3,219 km (WRI, 1988, p. 327).

Scenarios of Climate Change

The starting point for studying impacts of climate change is one or more forecasts of the most important climate attributes, typically as the result of a scenario in which the CO_2 concentration doubles. The most advanced climate simulation models, the three-dimensional General Circulation Models (GCMs), are generally agreed to perform relatively well at predicting changes in global mean values of climate parameters at annual or seasonal levels. Detailed (weekly or daily) predictions of climate change at the regional level are much more difficult, and their reliability is admittedly much lower. In addition, forecasting changes in the local climate in Southeast Asia is complicated by a series of human activities (deforestation, changes in land use) that interact with the global climate system and also significantly affect the local climate.

Given all these problems related to forecasting regional climate under double CO_2 conditions, the only sensible strategy in a policy oriented research would be to use a series of scenarios of climate change, assess their impacts, formulate possible policy responses, then compare and analyze them with a view to which strategies are expected to work under a range of plausible scenarios. Detailed scenarios of climate change for the region were prepared by Amos Eddy (Panturat and Eddy, 1989; Eddy *et al.*, 1989).

The input data sets used in this project included three monthly GCM output (GISS, GFDL, OSU); the United Nations Development Programme (UNDP) ASEAN monthly meteorological archives of the three countries (UNDP, 1982); and the daily values of maximum and minimum temperatures, as well as 24-hour precipitation data, from the meteorological station network in each of the three countries. The daily climate scenarios for a 25-year period were produced for the individual project sites as follows (Eddy *et al.*, 1989). The monthly GCM model output data were used in the form of ratios, and they were interpolated to the station location from

the four surrounding GCM grid point values using bilinear interpolation. The resulting 12 monthly long-term mean ratios for temperature and for precipitation were interpolated to produce 366 daily ratios. In the final step, the 25-year historical record of daily data was modified using the 366 daily ratios, the GISS daily, GFDL daily, and OSU daily climate scenarios under double CO_2 conditions. Additional scenarios have also been developed to reflect the special climatic characteristics of the region (the El Nino Southern Oscillation) or special concerns proposed for analysis (increasing variability of precipitation as a result of climate change).

These scenarios were supposed to be used as inputs to plant process models for selected crops in each country. One clear deficiency of the national studies was that only one scenario was actually used to simulate crop yield responses. This fact has imposed severe limitations on the assessment of higher-order impacts as well as on the policy implications (see section titled "Policy Responses").

Most analyses carried out by the NSGs were based on the GISS GCM results. The Malay and Thai experts have actually developed daily scenarios for the double CO_2 conditions based on the procedures described above for use with the crop process simulation models. Monthly scenarios were used for crop studies in Indonesia and the river basin studies in Malaysia and Indonesia.

First-Order Impacts

Detailed results of the biophysical impact assessments conducted in the three nations are documented in the project report published by the UNEP. This section presents a few selected examples of first order impacts of climate change to support the discussions of socioeconomic impacts and policy responses.

Rice production

The CERES RICE model, Version 2.00 (Godwin and Singh, 1989), was used by expert teams in Malaysia and Thailand to study the impacts of climate change on rice production (the Indonesian team used different approaches, as outlined below). The model provides a daily time-step simulation of crop growth and development based on four classes of input data: climate, soil, plant genetics and cultural practices. Daily climate data for double CO_2 conditions were generated from the GISS model output according to the procedure described in the previous section. Data on soils, plant genetics, and cultural practices were obtained directly from regional and local agricultural agencies.

In Malaysia, the area selected for the study was the Muda region, the country's rice bowl, on the coastal plains of peninsular Malaysia, in the states of Kedah and Perlis. The area has a long tradition of rice cultivation and a large-scale irrigation project has been implemented in several phases since 1965 to provide irrigation, drainage, and other facilities for double-cropping of rice. The annual production is about 700,000 tons of rice from two harvests a year. That is more than 60 percent of the national total.

Detailed results of the modelling effort are presented in a report prepared by members of the NSG (Salim *et al.*, 1989). The rice model was calibrated for the region by 16 model runs using historical climate data and different combinations of main or off-season crops, transplanted or direct seeding methods, irrigation or rainfed options, and fertilized or unfertilized cultivations. The next eight runs simulated the impacts of double CO_2 conditions on the crop using a GISS-based climate data set for the combinations listed above. The maturity period was shorter by up to 10 days for the main season crop. As a result, the average reduction in grain yield was estimated to decline by 12 percent to 22 percent for both transplanted and direct seeded rice. The overall average reduction in grain yields was about 19 percent, while the variance of crop yields under double CO_2 conditions was higher for the main season crop than for the off-season crop. The Malay analysts reported problems related to handling the impacts of plant water availability in the RICE model, but they estimated that irrigation demand would be 15 percent to 20 percent higher under the GISS scenario.

In Thailand, two studies were conducted to estimate the impacts of climate change on rice production. The first study was carried out by members of the core team of consultants, together with local collaborators, and focused on upland rice in northern Thailand (Eddy *et al.*, 1989, Panturat and Eddy,

1989).[1] The second study was conducted by the Thai NSG and selected a typical lowland rice region in central Thailand (Thai NSG, 1990).

The study area for upland rice was the province of Chiang Mai, northern Thailand. The analytical procedure was the same as the one described above for the Muda region, but the model was the upland version of the CERES RICE model (Version 1.10, Ritchie 1988). The model was calibrated for the base climate series of 25 years of daily data for two soil types, and combinations of fertilized versus nonfertilized, irrigated versus nonirrigated crops. The results show that yields would decline under all management variations by 4 percent to 14 percent, but in most cases this would be offset by the direct fertilization effect of higher atmospheric CO_2 availability to the plants (Eddy *et al.*, 1989, p. 61). The exact impacts of higher CO_2 availability are, however, subject of a major debate, so the modelling results should be treated with care.

Results of the same group from their experiments with the lowland rice model (CERES RICE V 2.00) show a 15 percent to 20 percent decrease in lowland rice yields for both rainfed and irrigated crops, while the irrigation demand is 15 percent to 20 percent higher under the GISS double CO_2 climate. Their results also confirm that "the practice of transplanting from seed beds produced a higher average yield than does the direct seeding practice for both rainfed and irrigated strategies under both climate scenarios" (Eddy *et al.*, 1989, p. 64).

The above results seem to contradict those produced by the Thai NSG in their study of impacts of climate change on rice yields in Ayutthaya province, central Thailand (Thai NSG, 1990). This group also used the lowland version of the CERES RICE model and a daily scenario of climate change based on the GISS model. The analyses included two soils series, five rice varieties, transplanting versus direct seeding, and fertilized versus nonfertilized practices. The group's conclusion is that "if CO_2 was to be doubled in the next 30 to 40 years, rice culture in Ayutthaya Province, in general, would benefit from such a change, except for some

---

[1]See Chapter 11 by S. Panturat and A. Eddy.

varieties. The average 25-year yields increase up to 8 percent. The benefits, in most cases, would be very marginal, however. The average per year change is insignificant" (Thai NSG, 1990, pp. 11-12). This conclusion holds for both the main season and off-season crops. The latter is grown only in the lowland irrigated areas and full irrigation is assumed.

Sharp contradictions are common in the climate change and climate impact literature, but the Thai lowland case is a unique example. We have at least one plausible explanation, however. Authors of the NSG report readily admit flaws in their analyses: The absolute values of yield from past climate and the GISS scenarios are not consistent with general observation because the modelled yields of transplanted rice are much higher than the observed yields, whereas they are much lower for direct seeded rice than the observed values. "Such a high fluctuation and inconsistency is mostly due to the model itself. Normally, any model to be used in a particular area should be validated by field data and observation. This has not been done for the Thai case" (Thai NSG, 1990, p. 10).

Because of the difficulties in creating the climate scenarios at the level of detail required by the CERES model and the problems of obtaining other necessary data, the Indonesian group used an empirical/statistical approach and an ecophysiological process model to assess impacts of climate change on rice yields. Based on results from an earlier study in Indonesia, the report concludes that rice yields in the January-June season will decrease by an average of 2.5 percent, but this will be offset by the average increase of 5.4 percent for the July-December crop (Blantran de Rozari et al., 1990, p. 33). The results suggest that the direct impact of climate change on rice yields will be fairly limited. As we demonstrate later, there are major threats to agricultural production in Indonesia coming from other sources.

Impact assessments of climate change on yields of other crops and plantations were conducted in each country. These studies relied on a variety of assessment techniques from simple empirical-statistical approaches to complex process models. The crops included maize (Malaysia and Indonesia), soybean (Indonesia), and rubber and palm oil (Malaysia). These efforts, however, did not

provide the level of detail and sophistication presented in the rice studies above.

River Basins

Possible impacts of climate change on the water resource regimes are of major concern in Southeast Asia, especially in regions with a marked difference between wet and dry seasons. One general conclusion from the GCM results is that under double CO_2 conditions the wet seasons would bring more precipitation, while the dry season would aggravate existing water stresses through several months. This will bring additional challenges for the government agencies and institutions responsible for water management.

The Malaysian NSG selected the Kelantan River Basin in the northeastern part of peninsular Malaysia to analyze the impacts of climate change on water resources. The annual rainfall in the Kelantan region is high (2,200 to 3,000 mm), up to 50 percent falling in the monsoon months of October to December. The result is severe floods in the wet season. "In a major flood such as that experienced in 1967, 300,000 hectares (or some 20 percent of the total state area) were inundated, 540,000 residents were affected, of which 125,000 had to be evacuated, 30 lives were lost, and flood damage to public property was estimated to be over M$30 million" (Lim and Salmah, 1989, p. 4). The shortage of water in the dry season affects agriculture, hydropower generation, and industrial and residential water supplies.

The Malaysian experts used the Storage Function Model to assess the impact of climate change on flooding, and the Thornthwaite and Mather Water Balance Model to calculate the changing balance of inflows and outflows. Both models were properly calibrated for present climate using historical data, then the climate parameters were modified based on the GISS GCM results.

The first-order impact assessment found an increase in both flood peaks and duration in the Kelantan River Basin. While the increase in flood duration is not significant, the increase in peak discharges would be approximately 9 percent, resulting in a larger overbank spill and more widespread flooding. In terms of flood recurrence, this means that a

30-year return period flood under double CO_2 conditions would have similar impacts as a 50-year return period flood under present climate. The water resource study, on the other hand, concluded that water deficits in the dry season would increase by 30 percent to 35 percent. This would seriously exacerbate the present water supply conditions because the amount of water available, especially for irrigation, is already at a critical level.

The Indonesian NSG also used the Thornthwaite and Mather model to assess impacts of climate change on water resources in three river basins: Upper Citarum in West Java, Upper Brantas in East Java, and the Upper Saddan River Basin in Southwest Sulawesi. The focus of their study was the impact of increasing monthly average rainfall (by 7 percent to 33 percent in the Citarum, 5 percent to 50 percent in the Brantas, and 8 percent to 59 percent in the Saddan Basin) and increasing temperature (1 percent throughout the regions) on seasonal soil water deficits and surpluses in these areas. The results show that soil water deficits will decrease in all three basins (and completely disappear in the Saddan Basin), with dramatic increases in soil water surplus in all regions. Unfortunately, the NSG did not investigate the impacts of the increasing runoff on flooding.

Coastal Areas

The impacts of climate-change-induced sea-level rise (SLR) in coastal areas of Southeast Asia would involve complex processes associated with shoreline retreat, permanent or temporary inundation, coastal erosion, increased flooding, and increased saline intrusion inland. Forecasts of the rate and magnitude of SLR are subject to considerable debate in the scientific community, but the potential threats were deemed serious enough to call for an assessment of possible impacts in this study.

The Malaysian NSG selected the area covered by the West Johor Agricultural Development Project in the southern part of peninsular Malaysia to study the impacts of SLR (Zamali and Lee, 1989). The area includes some 150,000 ha of agricultural land behind an almost continuous stretch of mangrove-fringed muddy shoreline. The group identified increasing coastal erosion as the most severe impact of rising sea level that will further aggravate the already ongoing processes of coastal erosion. Increased tidal flooding resulting from higher sea level may inundate about 16 percent (20,000 ha) of agricultural land in the project area and approximately 1,000 km^2 of fertile agricultural land along the west coast of peninsular Malaysia. Associated increases in backwater flooding are expected to prolong flood duration in the upland areas, resulting in lower crop yields. Theoretically, the mangrove forests could migrate landward, driven by a gradual increase in sea level. But the hinterland areas are already developed, leaving the mangroves threatened by extensive drowning. Increasing saline intrusion will threaten water abstraction facilities. Depending on the topographical, geological, hydrological, and other properties at specific locations, these four processes (erosion, flooding, mangrove loss, and saline intrusion) will produce a variety of complex dynamic processes with far-reaching impacts on other ecological systems and socioeconomic activities in the region.

The Indonesian NSG selected the lower Citarum Basin (Bekasi, Krawang and Subang districts in West Java) as their study site for SLR (Blantran de Rozari et al., 1990). Unlike the Malaysian NSG, this group attempted only to identify the area in the low-lying coastal plain that would be inundated by elevated sea. The coastal area in the three districts is characterized by a 2.5-km to 4-km wide belt of brackish-water fish ponds, while the area behind that belt is mainly wet rice fields. The current practice of aquaculture in the region is to construct fish or prawn ponds 0.5 m above the level of lowest tide. This means that as a result of a 0.6-m SLR, all the ponds will be inundated at normal tides, and the paddy fields and other agricultural areas behind it will also be seriously affected.

Higher Order Impacts

The previous section presented a few examples that were generated by the NSGs and the core team to assess first-order impacts of climate change on selected, valued, environmental components and resource systems. The direct impacts on biophysical processes (temperature, precipitation, runoff, flood return periods, saline intrusion in coastal areas and many others) are of little interest to policymakers who are concerned with shorter- or longer-term development of the economic sectors or specific regions for which they are responsible. Even the more sensitive issues like decreasing yields or inundated coastal areas are of limited policy relevance, unless we demonstrate the effects of first-order impacts on the most directly affected economic activities and social groups. Thus, analyses of economic and social consequences of first-order impacts are essential for preparing a complete assessment of risks and threats caused by climate change, as well as for providing the necessary inputs to the policy response component of the study.

Despite the large number of studies conducted to assess regional impacts of climate change, no generally accepted methodological blueprint exists for preparing the assessments of economic and social impacts. The NSGs followed the author's proposal to use a bottom-up approach. Based on the biophysical impacts generated in the first phase of the project and an inventory of economic activities in the study region, subsets of heavily affected and/or economically important activities were selected. The next step was analysis of the biophysical impacts on economic performance, identification of the most vulnerable social groups affected by these changes, and identification of potential economic and social conflicts arising from the changing natural resource and economic conditions.

Rice Production

The economic impact assessment started with an analysis of current distribution of farms by size and tenure group; the economic performance (e.g., net cash returns) of farms; the technological options available and economically feasible to different types of farms; and the most characteristic trends, longer-term changes in these attributes. The next step was the assessment of economic performance of various types and sizes of farms under the changing input/output conditions due to double CO_2 climate. Which farms would survive under double CO_2 climate? What are their options to adapt to the changing conditions? Which current trends in land ownership and tenure point toward conditions better adapted to double CO_2 climate? How would the biophysical and the resulting microeconomic changes affect the national food self-sufficiency and food security objectives?

The Malaysian study (Salim *et al.*, 1989) identified two major first-order impacts in the Muda region. First, the average reduction in grain yield was estimated to be between 12 percent and 22 percent for both transplanted and direct-seeded rice. The overall average reduction in grain yield is expected to be around 19 percent (decreasing output). Second, average irrigation demand is estimated to increase by 15 percent, with transplanted rice showing a smaller increase in irrigation demand than direct seeded rice (increasing input).

The impact of 12 percent to 22 percent reduction in yield will seriously affect the total net income of paddy farmers. With a lower total production, the amount of rice they can sell will decline and this will decrease the amount of cash subsidy they get from the government when they sell their produce. The situation will be exacerbated by the increasing demand for irrigation. It will not only increase production costs, but can limit the double-cropping of rice to a much smaller area due to the constraints in the amount of water available for irrigation in the dry season. At the national scale the result is that the present target of 60 percent rice self-sufficiency set by the Malaysian government will not only be threatened by decreasing yields, but also by decreasing area of double-cropping in the most fertile region of the country.

Earlier in this chapter it was pointed out that the overall economic importance of paddy is relatively small in terms of its contribution to agriculture value added, but that it is a major staple food, and therefore a strategic crop with respect to national food self-sufficiency. Yet, there seem to be problems with the paddy production. It has shown significant fluctuations over the past five years and declined by almost 10 percent between 1985 and 1990 (from 1,826 thousand tons to 1,665 thousand tons). The decline in the cultivated area was less dramatic, only 3.4 percent in the same period (from 661,400 to 638,700 ha), but the implications are likely to be more dramatic than suggested by these figures.

One long-term trend in land ownership in general, and paddy production in particular, is the joint ownership, subdivision, and fragmentation of peasant landholdings (Sundaram, 1988). Surveys conducted in the study region in 1976 and 1980 show that the average farm size decreased from 5.6 relongs to 4.22 relongs (Soon, 1983). This is far below the officially determined optimum two-cultivator peasant family farm size. The trend of land fragmentation is likely to continue because of the practice of Islamic and *adat* inheritance systems. The most typical solution to increased farm size is tenancy. While these arrangements clearly produce solutions to the farm size problem for some farmers, the number of people earning their living off the paddy is decreasing. Parallel to this process is the trend of land accumulated by a relatively small, well-to-do group. Concentration of land ownership and the spread of tenancy clearly point toward a capitalist type of agricultural management with larger land areas operated as a single unit employing wage laborers.

Climate change will affect these processes in several ways. First, the projected increase in input requirements and decrease in output will have the worst impact on the small farmers whose agricultural income has been insufficient, even under present conditions. In addition, deteriorating economics of paddy farming is likely to pull better-off farmers, operating somewhat larger farms, back under the poverty line. Paddy farmers constitute the largest group in terms of poverty incidence in Malaysia. In 1987 50.2 percent of paddy farmers were under the poverty line in peninsular Malaysia, 79.4 percent in Sabah, and 56.2 percent in Sarawak (EPU, 1989, pp. 52-53). Thus, impacts of climate change are likely to speed up the already ongoing processes of land fragmentation to below the economically feasible size, with a resulting land abandonment and land concentration.

The process involves both challenges and opportunities for the development of macro-scale social and economic policies in general, and agricultural policies in particular. To begin with the latter, both land concentration and increasing farm size, together with the shift from landlord-tenant to landlord-wage laborer relationship, are likely to increase the efficiency of paddy production. This has been an important objective in the previous Malaysia plan and is expected to be even more important for the Second Outline Perspective Plan (SOPP) as the nation moves from an extensive to an intensive development path. On the negative side, still related to the agricultural policy, what are the options for those who become landless? With increasing farm sizes, mechanization will accelerate, displacing wage laborers. The demand for wage laborers will increase far slower than the rate of increase in (land-)free agricultural labor. In addition, to operate the technology requires skilled workers, not very typical of the social strata becoming landless.

In the context of broader social and economic policy issues, we might consider another possibility for the landless: moving to urban areas, where the fast rate of growth in the industrial and service sectors will need a new labor force. One problem here is whether the rate of savings, capital formation, and investments will be able to keep up with the speed of transition in the rural areas. The other is how fast these people can be trained so that they can qualify to fill those new openings. The government can control the speed of these processes on both the "push" and the "pull" side. Providing increased support to the agricultural population to live off their agricultural income would slow down land abandonment and migration. Enhancing capital formation and investments and providing training and education programs for the newcomers to urban areas would make this transition smoother.

Kelantan River Basin in Malaysia

This study represents one rare example of climate impact assessments where possible impacts of climate change were combined with future trends in socioeconomic development projected for the same time horizon as the full impacts of double CO_2. The starting point was the present socioeconomic profile of the region, to which the NSG developed forecasts of changes in land use up to 2010, and water demand by sectors up to 2030 were added.

At present, the economy in the State of Kelantan in Malaysia is dominated by agriculture (70 percent of the population, 50 percent of the labor), accounting for 30 percent of the state's GDP. The most important crops are presented in Table 19.3.

TABLE 19.3. *Agriculture in the Kelantan Basin of Malaysia*

Crops	Area (ha)	Production (tons)
Paddy	70,000	200,000
Tobacco	10,000	8,000
Rubber	130,000	45,000
Oil palm	60,000	84,000

Source: Sea and Salmah, 1990, pp. 1-2

Rice production in the state accounts for 13.5 percent of the total national production figures, while tobacco production accounts for more than 80 percent of the total. In addition, there are expanding manufacturing and already substantial commerce and service sectors in the region.

The most severe damages caused by floods in 1967, 1983 and 1986 were estimated and then combined with the projected future expansions in the impact areas (affected population, agriculture, houses and buildings, infrastructure) by the year 2030, together with a 9 percent increase in flood peaks due to CO_2-induced climate change. The results show that there will be a 5 percent to 8 percent increase in the probable inundation area and a 10 percent to 12 percent increase in total flood damage.

Similar projections were made to identify the future demand for water by various sectors in the region (Table 19.4), by combining the increasing water demand figures with the estimated 30 percent to 35 percent water deficit in the Kelantan Basin resulting from climate change. The current priority order in fulfilling water demand is domestic, industrial, river maintenance, and irrigation. What this suggests is that the first three sectors are not likely to be affected by water shortages in the dry season, but that there will be major consequences for agriculture. "It has been estimated that the total irrigable area for paddy in the year 2030 would be 50,000 ha, requiring an annual peak demand of 84.6 m^3/sec of water. The deficit would, therefore, result in the abandonment of 65 to 70 percent of the off-season crop. The consequent loss of between 32,500 to 35,000 ha of paddy crop would affect the livelihood of tens of thousands of people and losses of millions of ringgit" (Sea and Salmah, 1990, p. 7).

Policy Responses

The main objective of the policy analysis phase of the project was to develop and evaluate strategic policy responses with a view to how societies in Southeast Asia might respond to the potential impacts of climate change identified in the previous phases so as to protect their environmental and natural resource base, their economic vitality and their

TABLE 19.4. *Projected Gross Water Demand for Kelantan River Basin* (m^3/sec)

Year	Domestic and industrial	Irrigation	River maintenance	Total
1985	0.5	35.0	70.0	105.5
1990	2.1	72.7	70.0	144.8
2000	4.3	84.6	70.0	158.9
2010	6.5	84.6	70.0	161.1
2030	9.2	84.6	70.0	163.8

Source: Sea and Salmah, 1990, p. 24

prosperity. The goal was to identify possible short-term adaptive moves and longer-term strategic responses, together with the potential gains and losses involved in those alternative policy responses. The central problem for the policy component of the study can be reformulated as generating national and local responses to impacts of a global change. The problem involves:

- *various spatial scales*: from individual farms to river basins to watersheds to regions and provinces to the national level;
- *various temporal scales*: changes in farming practices, selection of crops grown (1 to 2 years); crop varieties (breeding), changes in land use patterns, land conversion (few years); plantations (e.g., rubber), national agricultural policies (1 to 2 decades); canals, reservoirs, dams; coastal structures (several decades);
- *various economic sectors*: agriculture, fisheries, water management, tourism, energy;
- *various jurisdictional levels*: county, state/province, nation;
- *uncertainties and surprises*: even the best GCMs we are currently using are admittedly imperfect tools to predict climate change; but even if they were perfect, their input data and scenarios are highly uncertain (global energy use, deforestation, the role of wetlands and terrestrial vegetation); the crop models used by the NSGs are state-of-the-art but rely on imperfect data and outputs from the GCMs; a considerable amount of uncertainty exists regarding the relationships between climate change and sea-level rise;
- *management/policy orientation*: global climate change seems to be inevitable; therefore, an appropriate response is inevitable; appropriate response means developing policies that are robust with respect to potential surprises; because some actions involve long lead times, it is better to prepare now and act soon.

The Approach

Over the past 30 years various attempts have been made to develop approaches related to synthesizing scientific information from different disciplines relevant to a practical man-

agement problem and communicating it to the policymakers in an appropriate form. Two major approaches that have been widely used can be considered as extremes: computer models and expert committees (blue ribbon panels). Both approaches have their own merits and shortcomings, but there is still considerable room for improvements. To overcome some of the shortcomings and to complement existing methods, the policy exercise approach was developed at IIASA (Toth, 1986, 1989) and it was used in the Southeast Asia study to generate and analyze policy responses.

A policy exercise is a flexible structured process designed as an interface between academics and policymakers. Its function is to synthesize and assess knowledge accumulated in several relevant fields of science for policy purposes in light of complex management problems. It is carried out in one or more periods of joint work involving scientists, policymakers, and support staff. A period consists of three phases (preparations, workshop, evaluation) and can be repeated several times. At the heart of the process are scenario writing of "future histories" and scenario analysis via the interactive formulation and testing of alternative policies that respond to challenges in the scenarios. These scenario-based activities take place in an institutional setting, reflecting the institutional features of the problem at hand. There are two basic types of participants: policymakers as members of one or more policy team(s), and experts serving on the control team. Their activities at the policy exercise workshop are coordinated and moderated by a facilitator. If there is more than one policy team their relationships can be cooperative or competitive based on the nature of the problem with which they are dealing.

The policy exercises (PEs) in Malaysia and Indonesia were carried out in three phases. The preparations phase took almost two years and included data collection, modelling, completing the first order impact assessments, analyzing the socioeconomic impacts, several meetings of the NSGs, and conducting preinterviews with many to-be policy participants so as to ensure proper targeting of the exercise.

The workshops of the PEs were intensive, two-day meetings in Genting Highlands (Malaysia) and in Jakarta (Indonesia).

Because of the special characteristics of the project and the unusual circumstances of preparations, the original PE protocol was only loosely followed. In Indonesia, the input material for the workshop was sent out to participants prior to the workshop; in Malaysia it was not. Still, both workshops started with a session of introductory summaries presented by members of the NSGs and extended discussions of the reports per se. Because we had only one scenario of climate change and impact assessments with which to work, we made consecutive iterations on the same scenario by modifying the focus and key questions for the discussions and by regrouping participants accordingly.

The final phase of the policy exercise involves a careful analysis of the workshop results, preparing various forms of documentation from the exercise, and an overall evaluation. Products of a policy exercise include:

- a summary report dubbed as a cabinet briefing document (CBD), consisting of a short and succinct presentation of the problem, analysis, and major results;
- detailed policy assessments especially relevant to technological initiatives required to mitigate potentially adverse local and regional impacts of climate change, institutional changes (e.g., land ownership, water rights, legal considerations, government agencies) required to cope with the problems identified by the analysis; and
- research and monitoring necessary to acquire improved knowledge and to identify the most vulnerable economic sectors and geographical areas.

Context

In preparing for the policy exercises, the author conducted a series of preinterviews with the senior policymakers responsible for strategic policy formulation at ministries, and with directorate generals whose area or sector of responsibility is expected to be most heavily affected by the impacts of climate change. One important conclusion from these interviews was that it would be inappropriate to invite these individuals to consider and develop serious policy responses to highly uncertain events expected to occur in the dis-

tant future. Policymakers in these countries discount long-term risks and, therefore, there is a pressing need to improve our techniques and approaches so as to put long-term, uncertain, but high-risk issues into a policy-relevant context.

One possible strategy to overcome these difficulties will be an improved understanding of the transient processes involved in climate change. These improvements are expected to come from the new generation of high resolution GCMs that are also expected to decrease the level of uncertainties related to current predictions of regional climate change. More reliable impact assessments are more likely to capture the attention of policymakers, and, thus, their level of involvement and contribution to the analyses of climate policy issues can be considerably enhanced.

For the purposes of this project, the author proposed a different approach to make the impact assessments more relevant to senior policymakers. This proved to be highly successful even though the level of implementation varied across the three participating countries. The strategy was to link impacts of climate change to four major sets of issues in current policy-making.

1. *Link impacts of climate change to current problems and strategies to solve them.* The objective was to identify long-term, large-scale, and complex social and economic problems (equivalent in scales to those of climate change) and evaluate whether the proposed solutions and strategies remain valid under plausible impacts of climate change and explore how they could be enhanced according to the newly emerging threats and opportunities triggered by climate change. The objective of the exercise was clearly not to fine-tune these strategies to assumed double CO_2 conditions, but rather to evaluate whether they were robust with respect to different patterns of climate change and variability.

The previous section presented two examples of this approach. The area of abandoned paddy fields in the Muda region, Malaysia, was already increasing in the past because the reduced farm sizes did not permit economically feasible farming. Appropriate government strategies will be required to solve the problems of land fragmentation and land market rigidity resulting from traditional inheritance practices. But will the proposed strategies provide a long-term solution if we consider the gradually emerging impacts of climate change? Floods in the Kelantan River Basin, Malaysia, had repeatedly caused major damage to agricultural areas, property, and infrastructure in the past. Large-scale flood control and water management schemes are now in the planning phase to solve or mitigate these problems. Would these massive investments provide a long-term solutions if dam sizes, gates, and other engineering works are based on precipitation and run-off data derived from historical climate records?

2. *Link impacts of climate change to ongoing or planned long-term government programs to clarify whether the objectives remain valid if climate is changing and whether the strategies are robust to those changes.* What are the perceived modifications to these programs to deal with the new threats and opportunities emerging from impacts of climate change?

The Indonesian government has been carrying out an ambitious long-term program to reduce the population pressure in the most densely populated regions of Java, Bali, and Lombok. The transmigration program moves millions of people each year to newly opened areas in Sumatra, Sulawesi, Kalimantan, and other islands where they are provided with 2 ha of land, a house, farming equipment, and other assets to start a new life. The current five-year plan, Repelita V, is projecting the resettlement of 550,000 families between 1989 and 1994. This involves 1.1 million ha of land development. The question again arises: Are the resettled families expected to conduct economically feasible farming operations under double CO_2 climate, water, and sea level conditions?

3. *Link impacts of climate change to the long-term objectives for overall socioeconomic development to identify components and areas that might be threatened by those impacts.*

Poverty eradication is an important long-term objective in all three countries. The incidence of poverty is most severe among the rural population, and the resource base they rely on will be subject to impacts of climate change.

Increasing agricultural efficiency is one overall objective set by the government for the Malaysian agriculture and paddy production. Another important objective is to eradicate poverty, with a special emphasis on rural poverty. The New Economic Policy brought major achievements in poverty eradication, not without costs though. The Revised Fifth Plan allocated over M$13.6 billion to poverty eradication and 55.7 percent of this budget went to programs directly related to agriculture. Moreover, M$2,179 million was spent on new land development, and almost M$1.1 billion on the Integrated Agricultural Development Project. Other major programs included drainage and irrigation (M$478 million), replanting (M$643 million), and rehabilitation (M$873 million).

Impacts of climate change are likely to affect most of these programs over the long term. The major question is whether the program objectives remain valid under the changing climatic situation. Will the newly developed land be suitable for agricultural production, and will the production be economically efficient in these areas as key climatic attributes are changing over the coming decades? Will the newly constructed drainage and irrigation schemes be appropriate to handle markedly different precipitation patterns? Are the newly established and rehabilitated plantations appropriate for the new climate or will they have to be replaced well before their economically feasible lifetimes? Our present knowledge on these issues is rather vague, but these questions brought the long-term risks into the time horizon of strategic planners in ministries and departments of agriculture, irrigation and drainage, forestry, primary industries, and others in all three countries.

4. *Identify potential new economic or social problems arising as a result of climate change.* Most climate impact studies focus on these issues: What are the newly emerging problems resulting from climate-induced changes in the biophysical system?

Issues in these four areas are closely interrelated and in some cases they overlap. Nevertheless, the approach seemed to provide a useful framework to focus policymakers' attention on the relationships among their strategic objectives, current endeavors and the slowly evolving but long-term threats of climate change. Policy participants were asked to formulate responses on behalf of their own organization in five major categories: economic, technological, institutional, research and monitoring. Participants provided their responses according to a structured form, including specification of the problem and the category of response; a short description of the proposed policy; what government agencies are involved and who makes the final decision; a broad outline of the implementation; the source and amount of resources to implement it; and the potential side-effects of the proposed policy.

Examples of Results from Malaysia and Indonesia

Participants at the Malaysian policy exercise (PE) proposed a series of adaptive measures to mitigate adverse impacts of climate change on rice production. The initial focus of the discussions was, not surprisingly, the possible "technical fixes" stimulated by the undisputable success of breeding new, high-yield varieties over the past two decades. Participants proposed that objectives for future research and breeding efforts should also include increased resilience of the new varieties to climatic stresses. Among the many unknowns, the potential impacts of changing climate on pests and diseases were highlighted. Research on these relationships should be started very soon in order to get the relevant data to the breeders who are currently working with historical climate data and present climate-pest-crop relationships. More reliable information on the nature and magnitude of the expected climatic stresses affecting the crop is also needed.

The second set of examples of possible *technological* responses is related to the water stress and irrigation problem. The current level of irrigation efficiency is estimated to be around 55 percent to 60 percent, and it could be significantly improved by new irrigation techniques such as water recycling. This would involve establishing a large number of mobile pumps at strategic points in the irrigation scheme and transferring unused water back to the storage facilities for redistribution across the channels. Further analysis is required to determine the potential improvements in irrigation efficiency, the costs of different implementation strategies, and the potentially deleterious sideeffects of the new irrigation practices.

A series of possible *economic* responses was also discussed. The most straightforward solution to compensate the paddy farmers for losses in their income would be to increase the net cash subsidy paid to them. However, the negative sideeffects of this arrangement should be further explored. These include an increasing drain on the government budget and the secondary impacts on the efficiency of rice production and higher food prices. An indirect form of subsidy to offset yield losses would be an increased supply of free or low-priced fertilizers to the farmers. In addition to the economic sideeffects above, however, this would create additional environmental stresses: soil contamination, nutrient leaching, and eutrophication of the irrigation canals.

A larger scale *strategic* response to the already existing problem of reduced farm sizes, economic and technological inefficiency of small farms, and the emerging impacts of climate stresses would be to open up new agricultural areas and resettle the population in other regions. When this alternative was discussed, policymakers were concerned about the costs of new land development and the social and political stresses associated with relocating large number of people. Based on previous experience, they also raised the possibility of changes in local climate as a result of land-use changes: "We opened up jungle areas for annual crops and generated a 1°C annual mean temperature increase in the region. If we had known that, we would have opened it up for agroforestry," said a senior official from the Department of Agriculture, Malaysia.

Land fragmentation, economically inefficient sizes of landholdings, and resulting high incidence of poverty characterize the rubber sector as well. The recently introduced new clones increase yields by up to 50 percent and this will offset potential losses due to expected climate stresses. The proposed institutional response strategy to increase the efficiency in the rubber sector (and other plantations, and, perhaps even paddy) would observe the Islamic inheritance traditions. Smallholders would retain ownership of their land but they would be encouraged to let large plantation companies manage their land in return for rent and to look for employment elsewhere.

Discussions on the water-management

problems in the Kelantan River Basin focused on technical solutions to the current as well as increased future risks of flooding and water shortage. It was concluded that urban river stretches of more than 1,000 people/km^2 should be protected from the 50-year-return-period floods, while the remaining areas should be protected from the 20-year floods. A large number of alternatives (combinations of river improvements and dam constructions) were evaluated to identify the best solutions for flood mitigation, hydropower generation, water supply, and river maintenance. Policymakers from the Department of Irrigation and Drainage and other government agencies felt confident about their models and identified the scenarios of climate change as the most important sources of uncertainty in the sector. Nonetheless, some results will have to be reevaluated in the light of SLR (backwater effects, flooded areas and saline intrusion in the estuary of the Kelantan River), changes in land use and demand for irrigation water induced by climate change.

Some of these concerns emerged from the "looking outward" session of the PEs when participants were asked to identify key linkages to other sectors: what they would need to know about climate induced changes in the other sectors to improve their own strategic responses. Examples of the most important cross-sectorial impacts included water supply for irrigation and soil erosion (between agriculture and river basins); drainage problems (between agriculture and coastal management); and problems affecting estuaries (between SLR and river management). Participants were rearranged into new groups to analyze these cross-sectorial impacts, and in many cases they found that indirect impacts of climate change via other sectors would cause more serious problems to them than the direct impacts on their own sector.

Just to mention two examples of these cross-sectorial impacts: In Indonesia, increasing erosion from upland agriculture due to increasing rainfall under the double CO_2 scenario will create more severe problems for flood mitigation and river-basin management in the form of increased sedimentation in streams and reservoirs than would the otherwise significant increase in the amount of precipitation. The solution would involve substantially improved erosion control pro-

grams in the upland areas not only to protect soil fertility but also to prevent devastating impacts on flood control, on the life span of the already existing dams, and on the hydropower plants. In Malaysia, drainage problems caused by only a few inches of SLR in the coastal areas will severely affect coco, oil palm, and other plantations in the most fertile agricultural areas. The resulting losses in yields and the shorter lifetime of plantations are expected to cause higher-magnitude losses than the temperature and precipitation changes. Research into new technologies of drainage and breeding salt resistance crops should be considered, together with relocating the most sensitive crops and developing new land use zoning in the affected areas.

Many other response strategies were discussed at the PE workshop, and they were all related to present strategies or longer-term objectives set for the agricultural sector, water management, and coastal regions in Malaysia. Despite all the uncertainties readily admitted and fully revealed in the impact assessment reports, policymakers seemed to take the potential threats seriously. The PE is not intended to produce official decisions or new strategies with full-scale analytical underpinnings. Rather, it is a preparatory activity for effective participation in official decision processes. In this respect, the Malaysian exercise achieved its objectives by creating awareness among senior policymakers in key government agencies by identifying the most important risks and by sorting out the usable knowledge and usable ignorance related to the adaptive policy responses to climate change.

General Lessons

It would be premature to draw far-reaching conclusions on how developing countries balance the social costs of adaptation and prevention in the context of climate policy. Yet, one is forced to organize the experience accumulated in this study into an oversimplified framework that might prove profoundly wrong as more studies of a similar nature are conducted in different parts of the developing world. Based on the indicators presented earlier in this chapter, Malaysia is the most developed among the three participating countries. The taste of increasing welfare and the promising outlook for fast development in the future seemed to create the highest risk perception and awareness of the possible threats that might undermine past achievements or jeopardize future development prospects. The expert team provided a thorough analysis of the potential impacts, and policymakers who participated in the PE had the necessary long-term vision to consider the remote, but high stake, threats seriously.

In Thailand the prevailing growth enthusiasm seems to outweigh longer-term risks associated with some patterns of economic development both at the local and global scales. The impact assessments did not reach a level enabling them to serve as inputs to PEs, and the NSG did not seem to be very concerned with involving policymakers.

Indonesia is the least developed among the three countries and struggles with the most severe environmental problems in the region. Despite, or perhaps because of that, they developed some preliminary impact assessments and presented the results at a PE to members of the National Committee concerned with impacts of climate change.

The varying degrees of success in transferring the assessment techniques (climate scenarios, crop and hydrological models, PEs) fit well into this framework. The Malaysian NSG successfully calibrated and used the CERES RICE model. The Thai group was not able to calibrate it, and the Indonesians did not even attempt to use it. The hydrological model proposed by the core team was not used at all by the NSGs. Instead, the Indonesian and the Malaysian teams used the much simpler Thornthwaite and Mather Water Balance Model. The lessons for future climate impact assessments in developing countries is that complex models with extensive input data requirements are not appropriate when local expertise and historical data are in short supply.

The awareness and perception of climate associated risks have been changing both in the scientific and policymaking communities over the two-year life span of the project. Probably the most dramatic development occurred in Thailand, where public and policy concern even over apparent and immediate environmental hazards was limited, at best. In the traditional speech on the eve his birthday, Thailand's king "showed special concern about the so-called Greenhouse Effect, cited

figures for the amount of pollution being discharged into the atmosphere by the burning of wood and fossil fuels" (*Bangkok Post*, December 5, 1989). In his New Year speech, the Thai Prime Minister echoed these concerns, and when the preinterviews were conducted in January 1990, senior policymakers of several government agencies expressed their interest in the issue and the results of the study. The question why the Thai NSG was not able to capitalize on this momentum and involve more policy people in the exercise remains unresolved. Beyond this striking example, however, just a casual overview of the English language papers in the three countries shows an increasing coverage of environmental problems and also of global environmental issues like climate change.

The three nations seem to have markedly different institutional mechanisms to manage complex policy issues like those associated with climate impacts. Driven by the values of performance and efficiency, Malaysian government agencies readily setup interministerial or other interagency committees whenever the problem at hand extends beyond their own area of jurisdiction. The success of the Malaysian part of this project is partly attributable to the excellent collaboration among a wide variety of government agencies and academic institutions. Institutional rigidities appear to be a problem in Indonesia. Based on the premises that information means power and institutions are reluctant to share power, limited flow of information among government agencies cause problems in managing complex issues. This is partly due to the present institutional setting in which ministries that wish to launch major development programs need to convince BAPPENAS to secure the necessary funds. Unlike the ministries, BAPPENAS does not have hierarchical networks going down to the district, subdistrict, and village levels. As a result, the ministries reveal only that part of the information necessary to make their proposals more attractive. Moreover, environment (Ministry of Population and Environment) and development (BAPPENAS) issues in policymaking are separated, even in their international contexts. This will have to change as new strategies of sustainable development get more emphasis in future development efforts.

The current practices of climate impact

assessment are not very helpful in attracting the attention of policymakers. The use of landmark CO_2 conditions as the starting point for the assessment work limits the analyses of full-scale impacts and practical implications. The evolving impacts of climate change would not be the only dynamic process affecting the natural resource systems, and it is easy to envision serious statistical studies conducted at the agricultural universities in Bogor (Indonesia) or Kuala Lumpur (Malaysia) 10 to 20 years from now analyzing the causes of changes in yields attributable to changes in soils, erosion, irrigation, and different climatic parameters. Dynamic patterns of climate change would be superimposed on other variables affecting the resource base (degradation, depletion), on redevelopment and rehabilitation efforts (drainage, land reclamation), and on factors in the social and economic systems (values, technologies, cultural practices). We know very little about these complex interactions, but we can certainly expect surprises due to the feedback relationships and unknown thresholds. Therefore, we need to put more emphasis on the dynamic nature of both climate impact studies and socioeconomic development, in order to understand their interactions and to formulate better strategies to manage their complex linkages.

Recent research efforts on impacts of climate change follow two major directions: one group of studies is concerned with sectorial implications (agriculture, water); the other group is looking at regional consequences. There are attempts to link results from sectorial studies conducted in different countries (impacts of changing agroclimatic conditions on agriculture) at the international level to assess global implications (on the world food system and international trade of agricultural products). Impact assessments conducted in Southeast Asia have identified cross-sectorial linkages of indirect impacts of climate change (water availability for irrigation, soil erosion on rivers and reservoirs) that might produce more severe consequences than the direct impacts on the sector. The resulting policy responses to the full-scale impacts might be profoundly different from those based on just the direct impacts. There is a need to link results from the two types of research because broad-scale studies often omit the rich diversity of potential adaptive

measures at the local scale.

In retrospect, both moral and scientific concerns are related to the policy component of this project. The NSGs were not able to carry out full-scale analyses for at least three scenarios of climate change as originally planned. Despite all warnings and disclaimers, going into PEs with impact assessments based on just one scenario created the impression that the scenario is a forecast. This led to credibility problems at policy workshops both in Malaysia and Indonesia. Although it was repeatedly emphasized that the proffered material was merely a "not impossible" scenario, policymakers treated it as the prediction of impacts of climate change. The result was that instead of exploring and analyzing policy responses to plausible scenarios of impacts, we had long discussions on climate modelling, reliability of forecasts, the relationship between global and local climate change, and other issues irrelevant to the debate on response strategies. Familiarity of some participants with drafts of the IPCC report was not very helpful either. The report scaled down the magnitude of changes in most climate attributes, making the initial scenarios used in the study (and based on our 1988 knowledge) appear to be at the high end. This also diverted some of the debates at the exercises from the policy implications to credibility issues.

The policy exercises identified a series of adaptive measures to mitigate adverse impacts of climate change. Although participants were repeatedly encouraged to give serious consideration to economic and institutional responses beyond the most apparent technological fixes, the most specific and possibly most valuable outcomes were the guidance and recommendations from the policymakers to the analysts regarding future research to provide the kind of information they will need to formulate more specific responses. This confirmed earlier observations: "Until we know more about the magnitudes and rates of change of temperature and precipitation in specific regions . . . it will be hard to make a convincing case for more drastic measures of abatement or adaptation" (Revelle, 1989, p. 173). The second type of specific result that came out of the PEs outlined monitoring needs to identify the most vulnerable regions, sectors, and social groups. The proposed economic and institutional response strategies will require further work, but the policy exercises were successful in increasing the awareness of senior policymakers who represented the most relevant government agencies.

Given our present knowledge and the magnitude of uncertainties related to climate change, perhaps the most promising response strategy on the prevention side is the so-called tie-in strategy developed and presented in detail by Kellogg and Schware (1981) and more recently discussed by Schneider (1989). Pursuing this strategy would reduce greenhouse gas emissions from various anthropogenic sources and simultaneously would provide generally agreed social, economic, and environmental benefits, even if current predictions of climate change turned out to be highly overestimated. By linking the potential regional impacts to current problems, ongoing programs, and long-term objectives, the PEs in Malaysia and Indonesia have identified a series of tie-in strategies on the adaptation side. Tie-in adaptations would involve:

- a series of protective and rehabilitative measures in resource management;
- modified price, subsidy, export, and import strategies in economic policy;
- changes in the legal and government systems to enhance the institutional mechanisms; and
- numerous R&D options in agricultural, coastal-protection and water-management technologies.

The resulting improvements in reduced soil erosion, drainage, flood mitigation, and many other areas would prove greatly beneficial, even if local impacts of climate change were far less dramatic than indicated by some of the impact assessments generated in this study.

Acknowledgements

This chapter draws on results of a research project initiated and financially supported by UNEP with the participation of Indonesia, Malaysia, and Thailand. The opinions, observations, and conclusions presented here are those of the author and do not necessarily reflect the views of UNEP or of any governmental or nongovernmental agencies of the participating countries. Any errors or misinterpretations, though inadvertent, are the sole responsibility of the author.

References

Bangkok Post, 1989. Save Thailand from becoming a desert. *Bangkok Post* (December 5).

BAPPENAS (National Development Planning Agency), 1989. REPELITA V. Indonesia's Fifth Five-Year Development Plan 1989/90-1993/94. Percetakan Negara Republik Indonesia, Jakarta.

Blantran de Rozari, M., N. Koesoebiono, D. Sinukaban, *et al.*, 1990. Socioeconomic impacts of climate change (manuscript).

EPU (Economic Planning Unit), 1989. *Midterm review of the Fifth Malaysia Plan 1986-1990*. National Printing Department, Kuala Lumpur.

Eddy, A., S. Panturat, S. Salim, *et al.*, 1989. Climate change impacts on food and water resources in S. E. Asia (manuscript).

Glantz, M.H., (Ed.), 1988. *Societal Responses to Regional Climatic Change. Forecasting by Analogy*. Westview Press, Boulder, Colo.

Godwin, D. and U. Singh, 1989. *CERES-RICE V2.00*. International Fertilizer Development Center, Muscle Shoals, Alabama.

Kates, R.W., J. Ausubel, and M. Berberian (Eds.), 1985. *Climate Impact Assessment: Studies of the Interaction of Climate and Society*. John Wiley, New York.

Kellogg, W.W., and R. Schware, 1981. *Climate Change and Society, Consequences of Increasing Atmospheric Carbon Dioxide*. Westview Press, Boulder, Colo.

Lim, T.K., and Md. S. Salmah, 1989. Impacts of climatic change on the water resources in the Kelantan River Basin (manuscript).

Marland, G., T.A. Boden, R.C. Griffin, *et al.*, 1989. *Estimates of CO_2 Emissions from Fossil Fuel Burning and Cement Manufacturing, Based on the United*

Nations Energy Statistics and the U.S. Bureau of Mines Cement Manufacturing Data. Oak Ridge National Laboratory, Oak Ridge, Tenn.

Panturat, S., and A. Eddy, 1989. Some impacts on rice yield of changes in the variance of regional precipitation (manuscript).

Parry, M.L., T.R. Carter, and N.T. Konijn (Eds.), 1988. *The Impact of Climate Variations on Agriculture, Vols. 1 and 2.* Kluwar Academic Publishers, Dordrecht, The Netherlands.

Prentice, I.C., R.S. Webb, T. Mikhail, *et al.*, 1989. Developing a global vegetation dynamics model: Results of an IIASA summer workshop. Publication No. RR-89-7. International Institute for Applied Systems Analysis, Laxenburg, Austria.

Revelle, R.R., 1989. Thoughts on abatement and adaptation. In: *Greenhouse Warming: Abatement and Adaptation.* N.J. Rosenberg, W.E. Easterling, P.R. Crosson, and J. Darmstadter (Eds.) Resources for the Future, Washington, D.C., pp.167-74.

Revelle, R.R., and Waggoner, P.E., 1983. Effects of a carbon dioxide-induced climatic change on water supplies in the western United States. In: *Carbon Dioxide Assessment Committee. Changing Climate.* National Academy Press, Washington, D.C., pp. 419-32.

Ritchie, J., 1988. User documentation of the CERES-RICE Simulation Model. Version 1.20. Upland Condition, 8/14/87 (draft).

Salim, S., E.F. Yong and P.K. Yeow, 1989. Effects of climatological changes on rice production in the Muda area (manuscript).

Sea, C.H., and Md. S. Salmah, 1990. Socioeconomic impact assessment on the water resources in the Kelantan River Basin resulting from climate change (manuscript).

Schneider, S.H., 1989. The greenhouse effect: Science and policy. *Science*, **243**:771-81.

Soon, W.H., 1983. Muda II evaluation survey: An impact evaluation study of the Muda II Irrigation Project. MADA Publication, February 1983.

Sundaram, J.K., 1988. *A Question of Class.* Monthly Review Press, New York.

Thai NSG (National Study Group), 1990. Effects of climate change on rice production in Ayuthaya province, Thailand (manuscript).

Toth, F.L., 1986. Practicing the future: Implementing the policy exercise concept. Publication No. WP-86-23. International Institute for Applied Systems Analysis, Laxenburg, Austria.

Toth, F.L., 1989. Policy exercises. Publication No. RR-89-2. International Institute for Applied Systems Analysis, Laxenburg, Austria.

UNDP, 1982. *The ASEAN Compendium of Climate Statistics.* Colorcom Grafik Sistem Sdn., Kuala Lumpur, Malaysia.

WMO (World Meteorological Organization), 1986. Report of the international conference on the assessment of the role of carbon dioxide and other greenhouse gases in climate variations and associated impacts. WMO, Geneva.

World Bank, 1989. *World Development Report 1989.* Oxford University Press, New York.

WRI (World Resources Institute), 1988. *World Resources 1988-89.* Basic Books, New York.

Zamali, M., and S.Ch. Lee, 1989. Impacts of sea level rise on the West Johor Agricultural Development Project Phase I (manuscript).

Appendix

Report of the 1991 Woodlands Conference

THE REGIONS AND GLOBAL WARMING: IMPACTS AND RESPONSE STRATEGIES

At the close of their discussion, the participants in the 1991 Woodlands Conference, held March 3-6, 1991, at the Houston Advanced Research Center's Center for Growth Studies, reviewed as a group the following statement. This statement represents a general agreement, although individual opinions varied on specific issues.

Thinking Regionally About Global Climate Change

The composition of the earth's atmosphere is being altered by human activities. During the past 150 years, industrial societies have added large quantities of carbon dioxide through the burning of fossil fuels and, more recently, have released industrial gases such as chlorofluorocarbons. Other gases, such as methane and nitrous oxide, are also increasing as the result of expanded agriculture throughout the world. As the developing countries have enlarged their industrial base in recent years to improve their standards of living and converted more land from forest to agriculture to meet the needs of growing populations, their contribution of greenhouse gases has grown rapidly. These gases have the capacity to trap radiant heat from the earth, raise the average global temperature and significantly alter the climate. The consequences for human societies and natural ecosystems could be severe.

To date, much of the scientific research and policy attention in this area has focused on the global picture. International negotiations and agreements are important, but by themselves are not sufficient to address global warming. Efforts to reduce greenhouse gas emissions, measures to mitigate the consequences of global warming and adaptation strategies will be most effective when implemented through coordinated regional actions. In many countries, subnational jurisdictions such as states and provinces or community organizations can already take effective actions without direction from their national government or waiting for international agreements.

In fact, some national governments have posed substantial barriers to the development of effective policies that address global warming. Traditionally, nation-states are the decision making units that respond to international problems either by themselves or through multinational organizations of their own creation. However, nation-states are often reluctant to take necessary actions if that means giving up a measure of sovereignty, and international bodies lack the authority to do so. While it is nations that can ultimately develop treaties and pass laws, we have concluded that both subnational and international regions need to play an expanded role in order to address the climate change problem effectively. National barriers may be further lowered through the expanded participation of nongovernmental, scientific, corporate, and other organizations.

Ultimately, all climate change effects are experienced locally and much effective response depends upon local action. Yet individual localities cannot solve a problem of global proportions by acting alone. The political challenge will be to equip existing regional institutions with the capacity to act on climate change and to create new structures capable of effective action.

Regions are in some sense self-defining. In addressing climate change, regions need to be appropriate to the particular set of issues of concern. Similar or contiguous geography such as that of East Africa or Europe may be appropriate if the concern is altered regional weather patterns. Regions that are similarly sensitive in their response to climate change, such as tropical forests or low-lying coastal areas, can also constitute an appropriate region. Areas that are vulnerable because of natural or economic adaptation constraints may constitute a different type of regional grouping. An additional factor that can overshadow geographically determined regions is the disparate consequences of climate change for developed and developing or industrial and agricultural regions. Wealthy industrial regions may be better able to develop capital-intensive, adaptive infrastructure than regions with less discretionary resources where people's lives are more vulnerable to the vagaries of weather patterns. On the other hand, regions that rely on indigenous knowledge and local resources may be better placed to make incremental adaptations and be more willing to modify life-styles.

While many regions will suffer from climatic change, some regional benefits may accrue. For example, even though many areas may experience increased desertification, others may develop more grasslands. But although changed conditions may provide some new opportunities, the transitions are likely to be difficult, and, in general, to cause regional problems.

Anticipating Regional Consequences of Climate Change

Several tools are available for anticipating the regional consequences of global climate change. Perhaps the simplest approach is to utilize modern or historical correlations among regional temperature, precipitation variations, and vegetation growth to predict the effects of future climate change on agricultural production and natural vegetation. Projecting other changes, such as possible future soil moisture, stream flow and availability of water resources, can be done with the use of specialized hydrological models. Future sea level rise can also be predicted on the basis of anticipated temperature increases. Complex, global climate models have also been developed that can simulate present, past, and future global climates as they respond to differing concentrations of greenhouse gases. While these computer models were not developed with regional climates in mind and their spatial resolution is still poor, they have proven useful in generating possible scenarios in sufficiently large geographic regions. Nevertheless, the linkage between global and regional climate change remains extremely uncertain as does the timing of future change.

Regional Effects of Climate Change

Climate change may alter temperature, precipitation, and sea level. As a result, numerous regional consequences have been identified for agriculture, forestry, natural ecosystems, water and coastal resources, and energy use. The conference participants discussed each of these, but emphasized agriculture, forestry, and coastal resources.

Taken individually, the consequences of climate change for natural and managed systems may be quite different, but they overlap. Changes in agricultural productivity brought about by climate change may place additional stress on natural ecosystems because production may be expanded over larger land areas to compensate for the loss of productivity. Therefore, forest area may be lost to land use changes as well as to the effects of climate change.

In a similar way, climate-induced changes in natural ecosystems may affect agricultural productivity. First, the loss of biodiversity from climate change could mean the loss of primitive cultivars and germplasm necessary for growing improved plant varieties. Second, natural ecosystems adjacent to agricultural systems can assist in biological pest control. The loss of these ecosystems can mean devastation to agricultural systems. These factors suggest that changes in forests and agricultural systems must be linked when assessing the impacts of climate change.

Most of the world's agricultural and forest regions are sensitive to climate change, but the sensitivity varies with regional and local factors. Altered temperature and precipitation patterns are likely to shift the distribution of weeds, insect pests and other pathogens, thus creating new resource management challenges and reducing biological productivity.

Implications for Agriculture

The changing climate may lead to extreme weather events that change the length of growing seasons, alter the range between maximum and minimum temperatures, and create rainfall variabilities. The most sensitive agricultural regions are those in which warmer, drier conditions affect drought-sensitive soils, where agriculture is rainfed or is reliant on limited irrigation, or where there is strong competition for water supplies. With global warming, farmers will compete with foresters, conservationists, industries, river transport, hydropower, and urban areas for an ever decreasing supply of fresh water.

Agricultural regions in semiarid climates may be especially sensitive to even slight drying, and coastal agriculture may be adversely affected by sea level rise that brings saline intrusion.

It is important to categorize sensitive regions in terms of economies and social factors in addition to their environmental response. In these terms, subsistence agriculture and human settlements with limited options will be most sensitive if global warming brings agricultural decline. For example, those regions where farmers lack money, good land, irrigation, drought-tolerant seeds, alternative employment options, or reasonable prices for inputs and products will be particularly sensitive. Importing regions may be sensitive to production declines in exporting regions.

Implications for Forestry

Global climate models project that global warming will lead to the greatest changes in temperature at mid-to-high latitudes. Forests in these regions are likely to be severely and adversely affected.

Highly specialized species—those with very specific habitat and food requirements—are likely to be quite vulnerable to rapid climate change. If global warming occurs, species are likely to become extinct at an accelerating rate.

Tree species are likely to be especially sensitive in the warmer and drier parts of their ranges, particularly in soils that are thin, coarse, and in upland areas. Where soil water is depleted by an increase in evaporative demand, trees will be under stress and rapid changes can be expected. Ecotones—transitions between ecosystems—are likely to be especially sensitive.

Climatic alterations may be particularly strong in the tropics. But our ability to assess the possible effects of global warming on tropical forests is much poorer than our ability to make these assessments for temperate, more northern forests. Drought or lengthy dry seasons are likely to drastically alter tropical regions. A recent simulation of climate change indicated that the Amazon rain forest is likely to suffer significant decline as a result of the lengthening of its dry seasons. Historically, natural ecosystems have responded to long-term climatic change by migration of individual species at differential rates. This implies dispersal and establishment on the leading edge of the forest and decline on the retreating edge that is induced by competitive displacement or drought. But current, human-induced climatic changes appear likely to occur over a time period too short for trees and other plants to migrate. Plant migration is also limited by human-made barriers to migration. Today, natural habitats are often islands among agricultural and urban landscapes that impede the natural dispersal of vegetation, thus blocking migration.

Increased temperatures, sustained over a period of years, could produce a widespread, catastrophic, drought-induced decline in the world's temperate and boreal forests. Widespread drought would likely result in increased incidences of disease and large catastrophic fires in the world's forests. The subsequent release of methane and carbon dioxide from the decaying vegetation into the atmosphere would add significantly to the atmosphere's burden of greenhouse gases.

As a forested region dries out, other regional resources will be affected. Forests using up moisture in the soil will reduce volume in nearby streams, thus killing fish and causing water quality to decline. Upland soil erosion and nutrient loss would increase sediments and nutrient loadings downstream and suspended particulates from forest fires would increase air pollution.

Declines in tropical rain forests have ecological implications for other regions. The number of migrating birds that summer in temperate and boreal regions and winter in the tropics could decline precipitously, thereby affecting predation on insects and transport of seeds, as well as pollination in both parts of their range.

The deforestation of tropical rain forests can directly alter regional climates because the loss of vegetation cover can significantly reduce regional rainfall. For example, 50 percent of the rainfall in the Amazon region has been reported to come from the evapotranspiration of the forest itself.

Sea Level Rise

More than 50 percent of the world's population, 15 of the world's 20 largest cities, and a large share of the world's resources are concentrated on the coastal zone. And human migration toward coastlines is accelerating. Coastal fringe production of food will become increasingly important as we reach the limits of terrestrial food production. Seven major biophysical zones are threatened by the adverse effects of rising sea levels. These include deltas, small island nations, cities on coastal plains, barrier islands, estuaries, coral reefs, and coastal wetlands. Examples of these include the Ganges-Brahmaputra delta in Bangladesh and India; the Chesapeake Bay, which affects five U.S. states; and the 15 large metropolitan areas in coastal zones (among them Taipei, Tokyo, and Shanghai).

Coastal erosion associated with sea level rise threatens human settlements and infrastructure, including highways, freshwater supplies, power plants, and sewage treatment facilities. The rich estuarine breeding grounds that support much of the world's fisheries as well as regional mariculture can also be lost through sea level rise. Damage from storms to both coastal facilities and natural ecosystems will be made more severe by higher sea levels.

Geographic, ecologic and socioeconomic factors combine to create unique coastal regions. They need regionally tailored strategies but are frequently controlled by several different political entities. At the same time that regional planning is needed, however, local people should be involved in the process of mitigating the negative effects of sea level rise in their region.

By focusing on the dual need for regional planning and local implementation, a cooperative strategy can be developed for sharing studies, resources, and efforts to mitigate the impacts of rising sea levels. Careful coordination would avoid the duplication of studies and facilitate the dissemination of ideas for similar problems elsewhere throughout the region.

Regional strategies could be even more effective with stronger international dissemination of information that would assist other regions in developing strategies. Nations should set priorities for different coastal systems within their jurisdiction and allocate funds according to those priorities.

Integrated Regional Impact Assessments

Environmental policymakers often are constrained by resource-based analyses that do not adequately portray the problems that need to be addressed nor the opportunities that might result from change. To make informed decisions, policymakers need improved analytical tools that incorporate societal views and ecological principles. Such an approach will allow us to better project the effects of human activities and to develop more realistic management policies that do not view biological resources as if they could be harvested indefinitely without decline.

Integrated regional assessments should be interdisciplinary in scope and utilize the combined strength of local expertise and outside experts. Assessments should also address the potential for abrupt change resulting from the effects of multiple, smoothly changing stresses in dynamic environmental and social systems.

Integrated regional assessments need to identify and assess a range of policy options. A reference methodology for integrated regional assessment of the consequences of climate change should be developed with these characteristics: (a) be region-specific; (b) include trends in regional population, technology, economics, environment and relevant institutional conditions, and identify the region's social, technological, and economic vulnerabilities and strengths; (c) include a baseline scenario of regional changes in the absence of climate change; (d) utilize more than one scenario of climate change, with range rather than point estimates; (e) use traditional/local knowledge where appropriate; (f) include both quantitative and qualitative methods, where appropriate; (g) focus on vulnerability to climate change as well as on the impacts of such change; (h) include public participation in both early detection of global climate change at the local and regional levels as well as in the design and implementation of programs to respond to these changes; (i) results should be disseminated to decision makers in government and industry, to researchers and to the general public.

Policy Recommendations

We recognize that global agreements to address climate change are highly desirable and need to be negotiated and implemented by nation states. At the same time, regions must play a central role in reducing risks associated with global warming. They should do so by reducing the release of greenhouse gases through the development of mitigation and adaptation strategies. Regionally based strategies can take advantage of the community of interests that exist as the result of common geography, likely consequences of climate change, interconnected economies or similar history. Many regions have problems that seem more urgent than global warming and the best policies are those that help solve contemporary problems like deforestation, soil erosion, air pollution, and poverty while reducing vulnerability to global warming. Global warming responses should be linked to existing programs for energy conservation, soil and water conservation, sustainable agriculture, and sustainable urban development. The recommendations described below focus on the discussions that took place at the conference, and address agriculture and forestry, sea level rise, population and nutrition, energy and research needs.

Agriculture and Forestry

Maintaining and increasing the diversity of ecosystems, crop varieties, technologies, and farming systems are keys to reducing vulnerability to global warming.

1. Policies should emphasize the protection of biodiversity through species and ecosystem conservation programs. Special conservation programs for habitats in which the microclimate is cooler than the regional macroclimate are needed.
2. Crop genetic diversity and respect for traditional agricultural knowledge and technology can be enhanced by developing flexible agricultural systems and options for new food supplies.
3. Each region should monitor its forests, streams and soils for early drought warning signs. Should drought occur, policy measures to reduce water use should be implemented.
4. Changes in forests and agricultural systems must be linked when assessing the impacts of climate change. These linkages can be studied by examining the impacts of climate change on agroforestry. Agroforestry can utilize a traditional system that combines aspects of agriculture with forest management procedures.
5. To overcome the barriers to natural ecosystem migration, policies should be aimed at introducing new species into these systems that can expand buffer zones in preserved or managed areas. Transplanting or assisting in the migration of species might be useful in commercial fishing and agriculture, and to save some endangered species.

Rising Sea Level

1. Improved regional storm prediction is needed, especially in regions that are particularly vulnerable to typhoons, hurricanes, tropical storms, and other severe storms. One example is the reported increase in the number of tropical storms to hit the China coastline over the past century.
2. Cooperative research should be strengthened on the many areas that share both biophysical characteristics and vulnerability to the adverse effects of rising sea levels. This should include closer cooperation among international agencies and regional assessment entities.
3. Strategies to limit future development in vulnerable coastal environments must be given a high priority. Consideration should be given to locating new, major infrastructure inland from vulnerable areas. This could be enhanced by removing government incentives like insurance subsidies that allow development and

redevelopment in vulnerable areas. Accelerated sea level rise will aggravate problems associated with existing development.

4. Integrated coastal management zones are essential. These should include not only coastal fringe areas, but also those areas inland that affect the coastal fringe. Land use strategies should protect natural ecosystems and avoid, where possible, nonsustainable, rigid solutions. They should include measures to minimize pollution and storm risks. Once designed, coastal land use management strategies should move forward immediately.

Population, Nutrition and Health

Both excessive population growth and increases in per capita consumption levels contribute to the growth in greenhouse gases and increase the vulnerability of regions and societies to global warming. In some regions, population growth, immigration, and the conditions of land and resource distribution mean that pressures exerted by local populations are approaching or exceeding the carrying capacity of land and water. Where global warming is likely to further degrade a strained resource base, excessive population growth could bring regional crises and migration to other regions. Some of these relocations could be to ecologically vulnerable regions, such as the Amazonian interior, the Indonesian coastal regions, and other coasts, forests, and marginal agricultural and grazing lands. To enhance sustainable development it is important that policies to reduce rapid population growth be encouraged and integrated with other programs that reduce vulnerability and excessive per capita consumption of energy and resources.

The temperature and precipitation changes brought about by global warming also have the potential to alter the range of disease carrying insects from one region to another, which would have profound health effects for native populations. Monitoring programs should be designed to detect shifts in incidence of disease or those with important parasite vectors like malaria.

Other human health effects of global warming could include heat stress, immune response depression, and inadequate nutrition. Nutritional problems might stem, in part, from long-standing dietary preferences and their sensitivity to change as agricultural and livestock growing patterns are forced to change. A need exists for vegetation monitoring and early warning systems to detect malnutrition and famine as well as the gathering of relevant mortality and morbidity statistics.

Energy

Society's accelerating use of fossil fuel energy is the principal contributor to greenhouse gases. In industrialized regions, the extraordinarily high level of energy consumption is not sustainable and creates additional environmental problems in addition to contributing to climate change. Significant energy efficiency gains are possible, and major opportunities exist to develop both new energy-efficient supply and end use technologies. The development of renewable and other alternative energy supply technologies also needs to attract greater support from governments and industry. Industrialized, economically affluent nations have a special responsibility to carry out the necessary research and development.

Developing countries are caught in a bind as they attempt to industrialize or provide electricity to their people. Often they feel obligated to purchase the least expensive technology, and ignore the environmental, health and other costs associated with it. In the energy sector, this tends to relegate developing countries to the use of old, inefficient polluting technologies. In fact, some of the least efficient energy use is found in the industrial sectors of developing countries. Not only does this disproportionately increase the contribution of these nations to global climate change and other pollution, it also requires more fuel imports, which absorb scarce hard currency, and locks their economies into obsolescence.

Developing economies have a unique opportunity to leapfrog over several stages of industrialization and build their energy infrastructure properly the first time through the use of efficient, state of the art technology. In fact, developing countries have an advantage over mature, industrialized nations which can only introduce new technology slowly through expansion, abandon existing power plants and infrastructure, or replace it gradually as it ages. Wealthier, more technologically advanced regions should assist poorer and less developed regions in developing efficient, modern, low-polluting energy, transportation and industrial technologies. This will help to raise the economic level of developing countries in a sustainable manner that keeps greenhouse gas emissions from developing regions in check.

Research and Education

A shocking lack of basic data on natural ecosystems exists at all levels. Programs to obtain statistically valid, temporally consistent data monitoring for forests and other natural ecosystems are essential in developing realistic policies for managing these resources. To better understand complex ecological systems and their interactions with climate, we need an international network of regional-global environmental research centers. We also need improved or new regional climate monitoring by networks dedicated to global and regional climate change studies and their impacts. Scientists should be encouraged to develop finer resolution and mesoscale models, shed light on weather patterns, storm frequencies, and drought persistence at the regional level. There is also a need to move beyond the static doubled CO_2 model to recognize the dynamic nature of the buildup and the likelihood that we will exceed a doubling of CO_2.

We emphasize the importance of integrated studies that address interactions and competition among sectors of human activities like agriculture, forestry, industry, and urban areas. Studies should be site specific and focus on people. Water resources are a good integrating focus for many studies. Educational programs that address local, regional and global environmental problems and their solutions must be developed. Adequate education and training must be provided to assure that sustainable agricultural and forestry practices are implemented in all regions and that energy and other technologies are effectively utilized.

We also emphasize the need to support and establish international transfers of technology and information from technologically and financially equipped nations to those that are in need. Technology and information should be adapted to local conditions in developing regions, and financial support should be provided through international agencies, or when more appropriate, bilaterally. Data should be distributed to and from developing regions without cost, and should be easily accessible.

Regional, interdisciplinary impact assessments of the short- and long-term adverse effects of climate change should be carried out in areas likely to be adversely affected by climate change. The impact assessments should take into account the biophysical, human, socioeconomic, and ecological characteristics of each region, including the region's ability to respond and adapt.

Studies of the impact of global warming must be integrated with non-climatic-induced changes that will affect future vulnerability to global warming, such as changes in population, economic conditions, technology, and environmental degradation.

We have emphasized that impact assessments should be undertaken by research teams from within regions. At the same time, teams can learn from each other. A network to facilitate information exchange and regular meetings should be created. The benefit for improved study methodologies, evaluation of findings, and interaction with decision makers can be substantial.

A more ambitious coordinating effort would be to create an international network of global change research centers at key sites around the world. These centers could be coordinated by a single agency that provided a high-quality, homogeneous database for global climate models as well as the infrastructure needed to investigate the regional causes and effects of climate change. The creation of global change centers was a principal recommendation of the Second World Climate Conference in 1990.

Conclusions

International agreements among nation-states

will be important in assuring the widespread participation of all parts of the globe in slowing climate change and mitigating and adapting to its consequences. The actual implementation of strategies and the adoption of technologies will, however, be carried out locally and regionally. Regional responses are essential if global warming and climate change are to be addressed effectively. Broad international agreements cannot possibly take into account the unique regional opportunities and problems that exist. Nor can they ensure local ownership of the solutions. Inappropriate or unacceptable solutions are less likely to be developed locally and regionally than if they are imposed from a great political distance. Regional approaches also have the potential to bring indigenous knowledge and insights to bear. Another advantage of regional responses to global problems is that subnational regions need not wait for national policy to be formed or for international agreements to be reached in order to begin taking action. In many countries, the jurisdictional authority to act on many aspects of climate change rests with regional or local authorities. In fact, some states, provinces and corporations have already acted unilaterally to limit CFCs, CO_2 and some air and water pollutants without waiting for national or international direction. Such actions are likely to drive national policies in the case of nations that are slow to agree to international accords.

The international debate that has taken place to date illustrates that indeed regional groupings such as island nations have already formed because of commonly held concerns with sea level rise. Other coalitions of common geography or concern can be expected to occur as well. There is, however, also a danger that "regionalism" can drive a wedge among groups that define their climate change interests differently from others. This can best be addressed through improved knowledge, sharing of technology and resources and the realization that we are all suffering the consequences of a degraded planet.

One of the largest potential issues that must be addressed is the tension felt in many developing countries between the need to raise standards of living through development, often synonymous with industrialization, and protecting the environment. In many parts of the world, other problems are seen to be far more acute than climate change, suggesting that tie-in strategies that simultaneously address climate change, poverty and economic development be developed whenever possible.

A second issue is that of funding the cost of addressing global problems by developing countries that often lack either the economic resources or technologies to respond to problems like climate change. Developed countries can play a critical role in assisting the development process by providing economic and modern, efficient technological assistance. Human-induced climate change is indeed a global issue. Greenhouse gases will continue to increase in the atmosphere regardless of where they are released; temperatures will increase on all continents and sea levels will rise on all shores (as they now appear to be doing). As one speaker declared, "Assisting the developing countries to develop sustainably and respond to global climate change is not philanthropy, but enlightened self-interest."

INDEX